New Characterization Techniques
for Thin Polymer Films

SPE MONOGRAPHS

New Characterization Techniques for Thin Polymer Films

Edited by

HO-MING TONG
Thomas J. Watson Research Center
IBM Research Division
Yorktown Heights, New York

LUU T. NGUYEN
Philips Research Laboratories
Signetics Corporation
Sunnyvale, California

A WILEY-INTERSCIENCE PUBLICATION

John Wiley & Sons, Inc.

NEW YORK / CHICHESTER / BRISBANE / TORONTO / SINGAPORE

Library of Congress Cataloging-in-Publication Data:

Tong, Ho-Ming Herbert, 1954–
    New characterization techniques for thin polymer films / Ho-Ming
Tong, Luu T. Nguyen.
      p. cm.—(SPE monographs, ISSN 0195-4288)
    "A Wiley-Interscience publication."
    Includes bibliographical references.
    ISBN 0-471-62346-6
    1. Plastic films—Testing.   2. Polymers—Testing.   I. Title.
II. Series.
QD381.7.T66   1990   89-22681
621.381'52 – dc20   CIP

Printed in the United States of America

10 9 8 7 6 5 4 3 2 1

TO MAI-LIN, SHIRLEY, CARL, AND IRENE

CONTRIBUTORS

N. J. Chou (Chapter 11), IBM Research Division, Thomas J. Watson Research Center, Yorktown Heights, New York 10598

B. Chowdhury (Chapter 10), M & T Chemicals, Inc., Rahway, New Jersey 07065

H. Coufal (Chapter 9), IBM Research Division, Almaden Research Center, San Jose, California 95120

D. R. Day (Chapter 1), Micromet Instruments, Inc., Cambridge, Massachusetts 02139

B. L. Doyle (Chapter 6), Sandia National Laboratories, Albuquerque, New Mexico 87185

P. F. Green (Chapter 6), Sandia National Laboratories, Albuquerque, New Mexico 87185

K. K. Kanazawa (Chapter 5), IBM Research Division, Almaden Research Center, San Jose, California 95120

M. J. Matthewson (Chapter 13), Fiber Optic Materials Research Center, Rutgers University, Piscataway, New Jersey 08855

L. T. Nguyen (Chapters 3 and 8), Philips Research Laboratories, Signetics Corp., Sunnyvale, California 94088

D. H. Reneker (Chapter 12), Polymer Division, National Bureau of Standards, Gaithersburg, Maryland 20899

J. E. Ritter (Chapter 13), Mechanical Engineering Department, University of Massachusetts, Amherst, Massachusetts 01003

K. L. Saenger (Chapters 2 and 4), IBM Research Division, Thomas J. Watson Research Center, Yorktown Heights, New York 10598

Barton A. Smith (Chapter 7), IBM Research Division, Almaden Research Center, San Jose, California 95120

H. M. Tong (Chapters 2 and 4), IBM Research Division, Thomas J. Watson Research Center, Yorktown Heights, New York 10598

SERIES PREFACE

The Society of Plastics Engineers is dedicated to the promotion of scientific and engineering knowledge of plastics and to the initiation and continuation of educational programs for the plastics industry. Publications, both books and periodicals, are major means of promoting this technical knowledge and of providing educational materials.

New books, such as this volume, have been sponsored by the SPE for many years. These books are commissioned by the Society's Technical Volumes Committee and, most importantly, the final manuscripts are reviewed by the Committee to ensure accuracy of technical content. Members of this Committee are selected for outstanding technical competence and include prominent engineers, scientists, and educators.

In addition, the Society publishes *Plastics Engineering, Polymer Engineering and Science (PE&S), Journal of Vinyl Technology, Polymer Composites*, proceedings of its Annual and Regional Technical Conferences (ANTEC, RETEC), and other selected publications. Additional information can be obtained from the Society of Plastics Engineers, 14 Fairfield Drive, Brookfield, Connecticut 06804.

ROBERT D. FORGER

Executive Director
Society of Plastics Engineers

Technical Volumes Committee

Robert E. Nunn, Chairman

Lee L. Blyler, Jr.

Thomas W. Haas

Louis T. Manzione

George E. Nelson

James L. Throne

Lewis B. Weisfeld

Bonnie J. Bachman

PREFACE

The trend of device miniaturization in the rapidly expanding microelectronics industry has stimulated an ever-increasing demand for polymer thin film materials, and a need for their characterization and property monitoring during material screening or processing. Fueled by this need is a frenzy of recent research activities in various industries (electronics, aerospace and automobile) and universities aimed toward the development of highly sensitive techniques capable of studying ultrathin polymer films which are beyond the reach of conventional methods.

Because of the multidisciplinary nature of microelectronics packaging, studies of thin polymer films are particularly well suited to pioneering cooperative programs involving researchers of different disciplines including polymer and material scientists, engineers, chemists, and physicists. It is due to this intensely collaborative environment that many new characterization techniques are shaping up and are gradually becoming industry standards. In addition to providing fundamental insight into the physics of thin polymer films, the information provided by these techniques can have a significant impact on the development of new processing strategies of these films.

The main impetus for compiling this book, which is possibly the first of its kind, is to provide researchers dealing with thin polymer films a working description of the instrumentation, principle of operation, areas of application, and data analysis methods for some of these newly developed techniques as well as sufficient references for future in-depth studies. Special efforts were made to select techniques which are of general applicability and typify current advances in the measurement of bulk and surface/interface properties associated with thin polymer films. Most techniques presented herein allow in situ measurements, making them ideally suitable for process monitoring in an industrial environment.

In this book, the thin film techniques are divided into two groups, depending on whether they are primarily suitable for bulk property measurements (Chapters 1-9) or surface/interface property measurements (Chapters 11-13). Each chapter is self-contained. It is hoped that sufficient information is provided to the reader so that he or she can decide on which technique to use, construct the setup, collect the data, and perform data analysis. To most effectively use this book, the reader is advised to go directly to the chapters of interest after browsing through the abstracts of the chapters below.

Bulk Properties

- Chapter 1: Microdielectrometry

    Microdielectric sensors are well suited for monitoring chemical changes occurring in thin polymer films ranging from simple drying to polymerization and cross-linking as well as process control involving these changes. Cures may be analyzed in terms of changing dipole relaxation times or ionic conductivity. If enough is known about the material under test, microdielectric sensors may be used to observe changes in glass transition temperature, crystallinity, moisture content, as well as most other chemical changes. These sensors may be custom mounted to facilitate certain film deposition techniques or they may be used from within a standard polyimide flexible package.

- Chapter 2: Bending-beam Technique

    In the bending-beam technique, film stress is measured from the deflection of the unclamped end of a beam coated on one side with the polymer film of interest. Deflection changes observed during curing, thermal cycling, and/or exposure to gaseous or liquid penetrant fluids can be used to obtain the film's fundamental properties (e.g., elastic stiffness, thermal expansion coefficient) and/or characteristics (e.g., fluid diffusion coefficient). The technique is simple, in situ, and yields information which can have a significant impact on the choice of new materials and processing strategies.

- Chapter 3: X-ray Diffraction Technique

    X-ray diffraction is a versatile method of determining stress levels in crystalline structures. It measures the strain in a given lattice under either thermal or mechanical loading. With the use of a suitable thermomechanical model, stress may then be derived from the strain. This method is enjoying wide acceptance as a reliable way of determining radii of curvature of silicon wafers coated with polymer containing films. Such information would ultimately be useful for the design of low-stress polymer coatings. Instrumentation for X-ray diffraction ranges from portable diffractometers to laboratory devices equipped with special position sensitive scintillation detectors.

- Chapter 4: Laser Interferometry

    Laser interferometry has proven itself to be a powerful in situ technique in the areas of polymer science and technology. In the applications considered here, changes in thickness and/or index of refraction of a transparent thin polymer film are detected from changes in film reflectivity as measured by variations in the intensity of reflected laser light. Polymer properties and processing characteristics accessible with laser interferometry are those relatable to changes in film thickness and/or refractive index, such as solvent uptake/swelling and/or dissolution, thermal expansion coefficient, thermal stability, and shrinkage rates during drying/curing.

- Chapter 5: Piezoelectric Resonators

    The major use of AT-cut quartz resonators has been for microgravimetry, often coupled with electrochemical control of the interface. Thus, the growth, solvation and dissolution of polymer films as well as ion exchange during oxidation/reduction of polymer films can be studied in situ. While recent theoretical advances hold promise for use of the resonator in polymer rheology, the emphasis here is on the precautions necessary to obtain frequency data relatively free of instrumental distortion, and on the considerations for the quantitative interpretation of the frequency data.

- Chapter 6: Ion Beam Analysis

    Elastic recoil detection and Rutherford backscattering spectrometry may be used to determine the composition versus depth profile of various elements in polymer films. A number of applications were chosen from the field of polymer/polymer diffusion since most of the work of ion beam analysis has been concentrated in this area. Other important applications include studies of polymer/solvent diffusion and ion implantation of polymers. These ion beam analysis techniques have facilitated major advances in industries such as the microelectronics and biomedical technology.

- Chapter 7: Fluorescence Redistribution After Pattern Bleaching Technique

    In the fluorescence redistribution after pattern bleaching technique, an intense burst of laser light is used to irreversibly photochemically destroy (bleach) fluorescent dye molecules in the illuminated regions of the polymer sample. Patterns can be created with dimensions comparable to the wavelength of light or longer. The concentration gradients are created in situ, without any mechanical manipulation of the samples. Diffusion is measured along one or more specified directions. Anisotropic diffusion can be thoroughly characterized.

- Chapter 8: Surface Sensors

    Sensors created by microfabrication techniques have gained wider acceptance and usage in various applications related to electronic packaging. To study the moisture diffusion characteristics in polymer films, either a "volume effect" sensor (e.g., metal oxide based transducers) or a "surface effect" sensor (e.g., interdigitated comb structures or triple track patterns) can be used. In contrast, piezoresistive gages have been the preferred means for characterizing the nature of the stresses encountered during encapsulation.

- Chapter 9: Photothermal Analysis

    Photothermal methods rely on the detection of thermal or acoustic waves generated by the absorption of optical radiation and subsequent radiationless deexcitation. Since optical, thermal, and acoustic properties of the sample are involved in the signal generation process, any one or all of these properties can be probed by photothermal methods. Laser-induced transient heat sources combined with the extreme sensitivity of acoustic and thermal detection schemes enable thermal analysis of extremely thin and temperature sensitive samples.

- Chapter 10: Thermally Simulated Discharge Current Technique

    The effect of an impressed voltage field on the dielectric nature and the thermal properties of polymers in producing charge domains in the polymer is utilized in this technique. Thermal relaxation of polarizations in polymer micro-domains originating from the charging phenomenon can be resolved with sufficient sensitivity for the technique to be useful for very thin samples. Applications of this technique may range from determination of structure/property relationships of polymers to the practical utility of their charge storage capacity.

Surface/Interface Properties

● Chapter 11: XPS/SIMS/AES

A basic understanding of X-ray photoelectron spectroscopy (XPS), secondary ion mass spectroscopy (SIMS), and Auger electron spectroscopy (AES) techniques for polymer surface-related studies is provided. Due to the similarity existing among the three techniques, (in terms of sample preparation, vacuum requirement, instrumentation, etc.), the presentation of material is carefully arranged to avoid repetition. The advantages and limitations of each technique are examined, and various practical approaches are suggested for overcoming these limitations. The general information provided by this chapter will enable the reader to make proper choice of the equipment and experimental approach.

● Chapter 12: Scanning Tunneling and Atomic Force Microscopes

Early results from the application of the scanning tunneling microscope and the atomic force microscope to polymer molecules are described. Prospects for the elucidation of the structure of polymer molecules on the atomic scale as well as at the level of side branches and crosslinks are discussed. Observations of related substances such as small molecules, liquid crystal molecules, molecular membranes, biological molecules such as DNA, and biological structures are summarized.

● Chapter 13: Indentation Technique

An indentation technique is described for measurement of thin film adhesion. The mechanisms of adhesion loss are examined and the phenomenon is viewed as an interfacial fracture between dissimilar solids. The indentation technique is unique in being able to measure the two fracture mechanics parameters, namely, interfacial strength and resistance to fracture or toughness, which are required to fully specify adhesion properties. The method is also simple to use, requires small amounts of specimen, and can be used on extremely thin, fragile films.

As is behind every edited book, this book would have been impossible without the cooperation and invaluable contributions of the authors. We enjoyed very much our working relationships with them. A number of societies and universities are also gratefully acknowledged for permission to reproduce their copyrighted materials.

We are very thankful to Wiley-Interscience for publishing this book and to its editorial staffs, Mr. Thomas P. Henninger, Mr. Edward Cantillon and Dr. Theodore P. Hoffman in particular, who were extremely helpful during the preparation of this book. Finally, special thanks are due to the reviewers for their valuable comments and to Mrs. Joan Dunkin for her professional and timely help in various aspects of text formatting.

Ho-Ming Tong   Luu T. Nguyen
Thomas J. Watson Research Center   Philips Research Laboratories
                                               October 10, 1989

CONTENTS

B. Chowdhury

N. J. Chou

D. H. Reneker

New Characterization Techniques for Thin Polymer Films

1
MICRODIELECTROMETRY

D. R. Day

Micromet Instruments, Inc.
University Park
Cambridge, Massachusetts

Microdielectrometry is a technique for making measurements of relative permittivity and loss factor, classically known as the dielectric properties, on a very small and local scale (1,2). Materials that are suited for microdielectric analysis generally fall in the semiconductor to insulator categories and are usually organic polymers. The dielectric properties are influenced by many material properties such as chemical constituents, viscosity, crosslink density, temperature, moisture content, and others (3-6). The measurement of dielectric properties enables a skilled user to investigate the state of a material or changes in any of the above properties with time.

The classical polymer dielectric measurement consists of two parallel plates, situated around the material under test (7). One plate is excited with a sinusoidal electrical potential and the resulting sinusoidal current passing from the second electrode to ground is monitored. The dielectric properties may then be easily calculated based on the amplitude and phase of the measured current, and the area and separation of the plates. In cases where extreme accuracy is required, a guard ring may be used to eliminate current from leaking around the edges of the sample and deposited metal electrodes may be used to ensure the best possible contact to the material under test. To obtain high signal levels with the classical parallel plate technique, thin samples with large areas are required and the dielectric properties are an average of the bulk. In addition, low-frequency measurements, which reveal much about the mechanical properties of the material, are often difficult to obtain since the signal current is proportional to the frequency of test.

The microdielectric measurement consists of two interdigitated electrodes, amplifying electronics, and a temperature sensor on an integrated circuit. This design permits a one-sided, very local measurement of dielectric properties, and is particularly well suited for thin films. The circuitry and method of measurement in the microdielectric sensor also circumvent the traditional lower-frequency limits, thus allowing dielectric analysis down to at least 0.001 Hz. Lower frequencies are particularly important for monitoring property changes near the end of cure reactions. In this chapter, the sensor design, calibration, and modes of operation are discussed.

Several examples of the dielectric response as a function of cure state, dipole relaxation time, crystallinity, moisture content, and other variables will be discussed.

1.1 CONSTRUCTION OF SETUP AND PERFORMANCE

The current version of the commercially available microdielectric sensor (available from Micromet Instruments, Inc., Cambridge, MA) is a microelectronic circuit which has the ability to make both dielectric and temperature measurements. The chip is approximately 0.5 x 0.25 x 0.025 cm. (0.2 x 0.1 x 0.01 in.) with eight bond pads for external electrical connections (Fig. 1.1). The circuit was conservatively designed and there is no particular reason that the same circuit could not be made significantly larger or smaller. The details of the on-chip circuitry will be discussed in Section 1.2. In order to use the chip alone (wire bonded) or in a protective package, the circuitry is segregated so that the sensing electrodes are at one end of the chip while the amplifying field effect transistors (FETs) and bond pads are at the other. After bonding the chip into a protective package, the amplifier and bonds are encapsulated, leaving only the sensing electrodes exposed (Fig. 1.2). The currently available package consists of a polyimide flexible ribbon cable with eight conductors attached to the chip at one end and a widened, stiffened connector on the other end (Fig. 1.3).

For most applications, the film under test (polyimide, for example) is deposited onto the exposed electrodes. Intimate contact to the sensor surface is necessary so measurements on preformed films may be difficult. Measurements are best carried out if the film is solvent or melt deposited onto the electrodes, or at least pressed on while in a rubbery state. The protective housing around the chip creates a small depression down to the sensing electrodes. Under some experimental conditions, such as spin coating, this depression and the sensor cable itself cause a deposition problem. In these cases the film can be deposited on the bare chip (no package) and the chip can be subsequently wire bonded. Such a process facilitates conventional film deposition techniques such as spin coating or chemical vapor deposition.

The sensor is connected to interface electronics (Fig. 1.4, described in Section 1.2.1) which function in conjunction with the on-chip dielectric and temperature circuitry. The distance between the sensor and the interface electronics depends somewhat on sensor design and frequency used, but current sensors have functioned at up to 10 m at 10,000 Hz. The interface electronics, which consist of a box 10 x 10 x 3.7 cm, are then connected to the main dielectric instrument by a cable of up to 10 m in length. The main dielectric instrument houses a digital frequency synthesizer and autocorrelator, a sophisticated microdielectric calibration function (Section 1.2), and temperature measurement capability. The microdielectrometer is then connected to a controlling computer which initiates measurements, and then receives, plots, and logs data. The commercially available microdielectric system may also include a programmable oven so that dielectric measurements can be recorded during controlled thermal ramps.

The microdielectric chip has been tested and found to function well from 0.001 to 100,000 Hz. The temperature limits extend from liquid nitrogen temperatures to 250°C. However, the sensor package has been observed to fail in some cases at temperatures below -100°C due to thermal mismatch between the silicon sensor and the organic potting material. The temperature sensor ceases to function above 250°C,

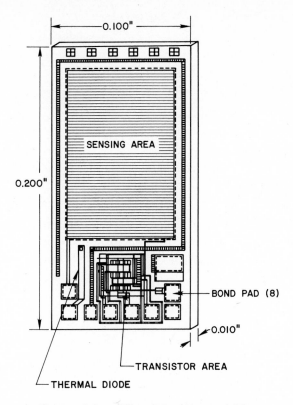

Figure 1.1:   Microdielectric sensor chip.

while dielectric measurements have been recorded up to the low 300°C range. Again, the package or bonding generally fails in the high 200°C range. The polyimide package and epoxy encapsulation are resistant to most chemically abusive environments. The microdielectric sensor has been used at isostatic pressures up to 34 kg/cm^2 (500 psi). However, it has very little resistance to bending stresses in the region surrounding the chip.

The commercially available microdielectric sensor measures permittivity (dielectric constant) from 1 to 50 with an accuracy of about 10%. The 10% variability is caused mainly by the fact that the commercially produced microdielectric sensor is calibrated (Section 1.2.2) in air due to time and cost issues. Factors such as humidity and surface contamination can cause slight inaccuracies of the base calibration value. Accuracy can be improved to about 3% if calibration (Section 1.2.2) is carried out in an organic solvent where contamination and moisture effects are removed. The chip is sensitive to changes in permittivity of about 0.3%. Loss factor can be measured from 0.01 to 1000 with similar accuracy. It should be noted that these limits are a function of the sensor, and that proper design could shift these ranges in either direction.

The design of the currently available sensor permits calibrated dielectric measurements on films down to 10 μm in thickness. Below this thickness, dielectric trends

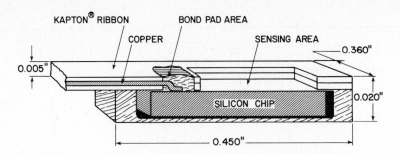

Figure 1.2: Cross section of chip mounted in polyimide housing.

are observed, however, the resulting dielectric values are no longer absolute. This limit is related to the spacing between the interdigitated sensing electrodes which is also about 10 μm. Using state of the art technology, there is no reason why this could not be reduced to 1 μm.

1.2 PRINCIPLES OF OPERATION

1.2.1 Sensor Circuitry

The sensor is based on an interdigitated comb electrode in combination with a depletion mode FET amplifier (1). One side of the comb electrode is excited with a sinusoidal potential. Current passes through the sample under test to the second electrode which is in turn connected to the gate of an FET. The potential that appears at the gate of the FET will depend on the geometry of the comb electrode, the dielectric properties of the material under test as well as the substrate (SiO_2), and finally, the gate capacitance of the FET. The circuit can be simply viewed as a capacitive divider split between the comb electrode and the gate capacitance (Fig. 1.5). In reality, the comb electrode is more complicated due to coupling of the electric field to the ground plane underneath the SiO_2 layer. If gate capacitance and the dielectric properties of the sensor substrate are known, then the dielectric properties may be determined from the measured amplitude and phase of the gate potential (relative to the excitation) using a finite difference calibration process (8) (see Section 1.2.2).

The performance of the sensor will be dependent on the dimensions of various critical parts of the sensor. The further apart the sensing electrodes, the further out into the material under test will the sensor sense. Roughly speaking, the sensor is sensitive to the dielectric properties of the material under test (deposited on the electrodes) to a penetration depth of about the same distance as the electrode separation. In other words, by increasing the electrode separation, dielectric property sensing depth can correspondingly be increased. However, due to the increased volume of material between the electrodes, the resulting current flow, and thus signal level, decreases. To counteract this, the area of the electrode array must be increased. In general, to increase the signal penetration depth by a certain factor, the electrode separation must be increased by that same factor and the area of the

Figure   1.3:   Microdielectric sensor in polyimide package.

electrode array must be increased by the square of that factor to keep the total equivalent current flow constant.   Other factors such as the oxide insulator thickness between the electrodes and the underlying ground plane have a more complex influence on sensor response which must be determined through finite difference modeling as described in Section 1.2.2.

It is impossible to measure the gate potential by conventional methods since the act of making the measurement would change the potential.   Instead, the gate potential can be monitored through the sensing FET source-drain current.   Off-chip electronics sense that current with a feedback circuit that consequently adjusts the gate potential of an on-chip reference FET.   Thus the circuitry matches the reference FET current to the sensing FET current.   Since the two FET geometries are identical and the source-drain currents are identical, then the two gate potentials must be identical at any given instant.   Through this dual FET design, the voltage of the sensing FET gate becomes well buffered by the feedback electronics and the effects of temperature and pressure on transistor characteristics are compensated out.   The potential of the reference gate may then be analyzed for amplitude and phase.

Figure 1.6 shows a typical amplitude (represented as gain in decibels) and phase response for a 5-min epoxy curing reaction on the microdielectric sensor.   Early in the reaction, the material is reasonably conductive so the gate capacitance at low frequencies is nearly fully charged.   This results in a gain approaching 0 dB.   As the material hardens, the conductive current is reduced leaving only the small capacitive current (resulting from the permittivity of the epoxy) and a much lower gain reading is observed.   Conversion of the raw gain (amplitude) and phase data into dielectric properties is described in Section 1.2.2.

In addition to the dielectric sensor, a temperature sensor is included on the chip. The temperature sensor is a simple diode that is forward biased.   Off-chip electronics drive the diode with a constant current.   Under these conditions, the voltage required to drive the constant current is nearly proportional to temperature, with a slope of about -2 mV/°C.   At temperatures above 150°C, leakage current through the polyimide package becomes comparable to the diode current.   The commercially available sensors incorporate a nonlinear calibration function which takes effect at higher temperatures to accommodate the package leakage.   Due to package to package variation, measured temperatures are less accurate above 150°C.

1.2.2 Calibration

Analyzing the response can be rather complex due to the nonstandard electrode geometry encountered in the microdielectric sensor.   Figure 1.7 shows a cross section of the chip and the orientation of the electric field in various regions (8).   Due to the

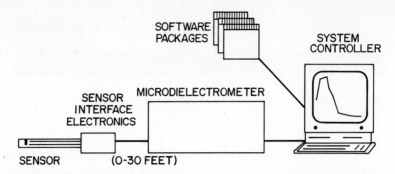

Figure 1.4: Schematic diagram of microdielectric measurement system.

arrangement of the SiO$_2$ substrate and the ground plane, the shape of the electric field will change as the dielectric properties of the material under test change. One way to account for these changes is to analyze the circuit with a finite difference model. The model incorporates the geometries used in the sensor as well as the dielectric properties of the sample under test and the SiO$_2$ insulator. Figure 1.8 shows the element model that was used for calculating the response of the microdielectric chip taking into account the distances and complex permittivities of the material under test and the SiO$_2$ substrate. Using standard iterative techniques of finite difference methods, the new potential for each element is calculated based on the following equation (8)

$$V_0 = \frac{\dfrac{V_1}{h_1(h_1+h_3)} + \dfrac{V_2}{h_2\left[\left(\dfrac{\varepsilon_{*2}}{\varepsilon_{*1}}\right)h_2 + h_4\right]} + \dfrac{V_3}{h_3(h_1+h_3)} + \dfrac{V_4}{h_4\left[\left(\dfrac{\varepsilon_{*2}}{\varepsilon_{*1}}\right)h_2 + h_4\right]}}{\dfrac{1}{h_1 h_3} + \left(\dfrac{1}{h_2 h_4}\right)\left[\dfrac{\left(\dfrac{\varepsilon_{*2}}{\varepsilon_{*1}}\right)h_4 + h_2}{h_4 + h_2\left(\dfrac{\varepsilon_{*2}}{\varepsilon_{*1}}\right)}\right]} \tag{1}$$

where V_x, ε_{*x}, and h_x are the potential at node x, the complex permittivity of medium x, and the distance from node zero to node x, respectively. Note that for nodes within the bulk of the material under test or the substrate, medium 1 and medium 2 in Fig. 1.8 are identical, thus simplifying Eq. (1).

Using these equations with the appropriate boundary conditions (8), a finite difference analysis is carried out for possible permittivities ranging from 0.5 to 1000 and loss factors from 0.001 to 10,000. The result of the analysis is a large table of permittivities and loss factors with corresponding gate potential amplitudes and phases. Such a table is represented graphically in Fig. 1.9 for gain/phase to permittivity/loss factor conversion. The microdielectrometer houses several of these compiled tables so that not only can response amplitudes and phases be converted into dielectric properties, but also small chip to chip geometric deviations can be taken into account. The chip is assigned a calibration value, the "CHIP- ID", by a

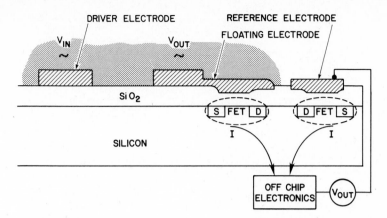

Figure   1.5:   Cross section of polymer coated microdielectric chip.

single calibration in air or some other "known" dielectric material. It has been found that chip to chip deviations are generally small so that omission of the CHIP-ID produces a maximum error of 20% in the reported dielectric values.

The temperature sensor is also calibrated through the CHIP-ID. During sensor fabrication, a measurement is taken at room temperature and part of the CHIP-ID is assigned to account for temperature offset. The CHIP-ID also has the ability to adjust the slope of the thermal response. Higher accuracy may be obtained through a two point temperature calibration and subsequent adjustment of the slope parameter in the CHIP-ID. Figure 1.10 shows the voltage response and calculated temperature from the sensor diode as a function of a measured thermocouple reference temperature.

1.2.3 Measurement

The microdielectric measurement is carried out by selecting a frequency, sampling the sensor response, and finally, converting the raw amplitude and phase data into dielectric values. The microdielectrometer has been designed for precise frequency generation from 0.001 to 100,000 Hz as well as accurate amplitude/phase analysis. A digital frequency synthesizer is used in combination with a digital to analog converter to generate a sine wave composed of small square steps (Fig. 1.11). This is passed through a filter so that a smooth undistorted sine wave is sent to the microdielectric sensor. A high-speed analog to digital converter samples the response sine wave at 3600 points. The individual samples are triggered by precise timing pulses from the frequency synthesizer. The 3600 voltage samples are then autocorrelated (single-frequency Fourier transform), resulting in the amplitude and phase of the response signal. The microdielectrometer then converts the amplitude/phase data into the dielectric permittivity and loss factor.

Depending on the sequence of a particular experiment, a second measurement may be taken, a temperature measurement may be initiated, or a different frequency may be selected. Various types of experimental sequences are reviewed.

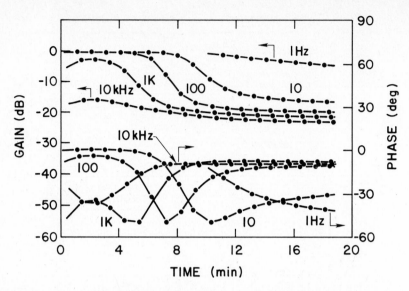

Figure 1.6: Gain (dB) and phase (degrees) signal on sensing FET gate during reaction of 5-min epoxy for 1, 10, 100, 1000, and 10,000 Hz.

1.2.4 Modes of Operation

Perhaps the most common use of the microdielectric sensor is the monitoring of dielectric response and temperature as a function of time. In diffusion experiments, a single frequency (usually 10 kHz) is selected and monitored during a period from minutes to days, depending on the diffusion coefficient and film thickness (as will be discussed in Section 1.3.8). The user may choose a cycle time for how often the dielectric properties and temperature will be recorded, and an overall experimental length. In the case of diffusion analysis, only one frequency is required to monitor the changes in permittivity. In the case of curing films, it is often necessary to monitor more than one frequency so that dipole relaxation times and ionic conductivity values may be extracted from the dielectric response over the course of the reaction. It is not uncommon for both the dipole relaxation time and ionic conductivity values to change by more than six orders of magnitude during a curing reaction. For this very reason, it is beneficial to monitor the dielectric response over an equivalent range of frequencies. The low frequencies accessed by the microdielectric sensor enable measurement of dipole relaxation times and ionic conductivities that are otherwise extremely difficult to obtain. During a typical experiment under this mode, frequencies at even decades are often monitored (0.001, 0.01, 0.1, 1, 10, 100, 1000, 10,000, 100,000).

Recently developed high-speed electronics (9) now enable measurement of the dielectric response at frequencies above 1 Hz every 10 ms. This can be useful for very-high-speed reaction studies where changes can be monitored which are simply not observable by any other analytical technique.

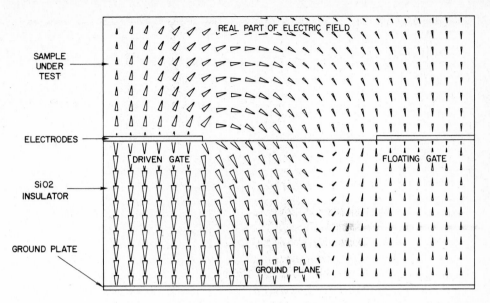

SAMPLE
UNDER
TEST

ELECTRODES

SiO2
INSULATOR

GROUND PLATE

Figure   1.7:   Computer simulation of electric field on the microdielectric sensor (from Ref. 8).

Both the standard and high-speed modes include temperature recording so that in postexperiment data analysis the dielectric response may be plotted as a function of time or temperature. In addition, other parameters can be automatically extracted from the dielectric response, such as conductivity, resistivity, or first derivative of conductivity and these may also be plotted as a function of time or temperature.

Another mode of operation, known as frequency scanning, is generally used on unchanging or slowly changing samples. In this mode many frequencies (8 decades at up to 10 frequencies per decade) are monitored in a single scan. This mode is useful for locating dipole loss peaks (for determining the relaxation time) and for determining ionic conductivity. Small changes that occur with long-term aging may also be studied by this method.

The chip housed in the polyimide sensor package is commercially available for about $30.00 and is generally considered to be disposable, especially when used with thermosets. However, thermoplastic films which are readily soluble in mild solvents may be removed, thus allowing the sensor to be reused.

1.3 APPLICATIONS IN THIN FILMS

1.3.1 Dielectric Response

The dielectric response consists of two measured properties, the relative permittivity and loss factor. The relative permittivity of a material (ε') is simply the absolute permittivity of the sample divided by the permittivity of free space (8.85×10^{-14} F/cm)

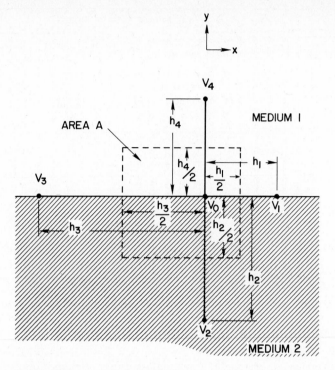

Figure 1.8: Basic element of finite difference model in terms of distance (h), and potential (V) of media 1 and 2 (from Ref. 8).

$$\varepsilon' = \frac{\text{absolute sample permittivity}}{\text{permittivity of free space}} \qquad (2)$$

Since air is a very-low-density material, it has a permittivity nearly equal to vacuum, and so it has a relative permittivity of 1. Most organic resins have relative permittivities ranging from 2.0 to 10, although there are some that lie outside this range. The relative loss factor (ε'') is the absolute loss factor divided by the permittivity of free space

$$\varepsilon'' = \frac{\text{absolute sample loss factor}}{\text{permittivity of free space}} \qquad (3)$$

Since air is not conductive, it has an ideal loss factor of zero. Polymers, if well below the T_g, generally have relative loss factors less than 0.1. When warmed up to the T_g and above, they can have relative loss factors as high as 1,000,000,000. Both the relative permittivity and relative loss factor (hereafter referred to only as permittivity

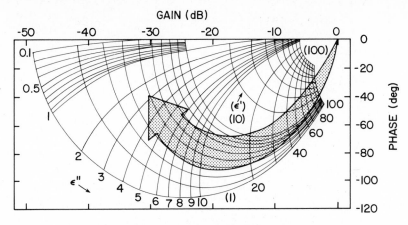

Figure 1.9:   Lines of constant permittivity/loss factor plotted against gain/phase for one electrode geometry as determined by the finite difference model.

and loss factor) may be influenced by several different phenomena occurring within a material under test.

During curing reactions changes occur in both the permittivity and loss factor. However, loss factor is usually used since it is influenced by ionic conductivity (Section 1.3.3) which can change by four or more orders of magnitude during a cure. On the other hand, permittivity is usually monitored during moisture diffusion in polymers. Although both the permittivity and loss factors are altered by the introduction of moisture, the change in permittivity (from the water dipoles) has been found to be proportional to the change in moisture concentration. Depending on the phenomenon being studied, either the permittivity or loss factor may prove to be the most useful value to analyze. Several examples are reviewed in this section. The molecular mechanisms that contribute to changes in both the permittivity and loss factor include induced dipoles, static dipoles, ionic conduction, electrode polarization, electronic conduction, and various combinations of these in an inhomogeneous material (Maxwell Wagner effect). The first four are common occurrences in standard polymers and are discussed below.

1.3.2 Induced and Static Dipoles

In the presence of an electric field, electron clouds may be slightly shifted, inducing a slight polarization which is aligned with the electric field. This acts to store energy and contributes to the capacitive nature of a material. The response times of the electronic shift are extremely fast so that at normal measurement frequencies (0.001 to 100,000 Hz) the effect is always present. These induced dipoles are responsible for nonpolar or symmetrically polar polymers having permittivities of 2 or greater. Static dipoles consist of inherently polar moieties (not induced by the electric field) within the polymer such as carbonyl or alcohol groups. The static dipoles, if mobile enough, may rotate in an electric field, thus also storing energy and contributing to

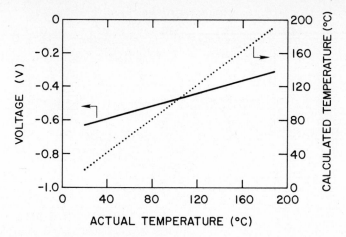

Figure 1.10: Sensor diode voltage and calculated temperature (°C) as a function of actual temperature (°C) during a thermal ramp.

the capacitive nature of the material. If a material, such as an epoxy resin with many polar groups, is warmed so that an immobile dipole becomes mobile, then an increase in permittivity is observed as the dipole starts to oscillate in an alternating electric field. This effect is referred to as a dipole transition and has a characteristic time (τ) associated with it.

$$\varepsilon' = (\varepsilon_u + \frac{\varepsilon_r - \varepsilon_u}{1 + \omega^2\tau^2})\ \varepsilon_{p'} \qquad \varepsilon'' = \left(\frac{\sigma}{\omega\varepsilon_0} + \frac{\omega\tau(\varepsilon_r - \varepsilon_u)}{1 + \omega^2\tau^2} \right) \varepsilon''_p \qquad (4)$$

where ε', ε'', σ, ε_0, ω, and τ are the relative permittivity, the loss factor, the bulk ionic conductivity, the permittivity of free space, $2 \times \pi \times$ Frequency, and the dipole relaxation time, respectively. ε_r is the relaxed dielectric constant, for example, the low-frequency dielectric constant (permittivity due to induced plus static dipoles). ε_u is the unrelaxed dielectric constant, for example, the high-frequency dielectric constant (permittivity due to induced dipoles only). ε'_p is the electrode polarization term for ε' (see Ref. 10) and ε''_p is the electrode polarization term for ε'' (see Ref. 10; note that ε'_p and ε''_p are usually 1.0 except when ion conduction is very high).

Static dipoles also contribute to the loss factor as can be seen by Eq. (4). This contribution arises from viscous drag as the dipole rotates through the surrounding medium. As a result of this drag, there can be a significant phase lag between maximum applied field and maximum dipole deflection. This phase lag reaches a peak as $\omega(2\pi f)$ reaches 1/dipole relaxation time. At higher frequencies, the dipole hardly moves and so little energy is lost, while at lower frequencies the dipole can keep up with the changing field more easily and again, less energy is expended.

It should be noted that Eq. (4) is a form of the ideal Debye equation for a single relaxation time model (11). In reality, polymers often contain more than one type of

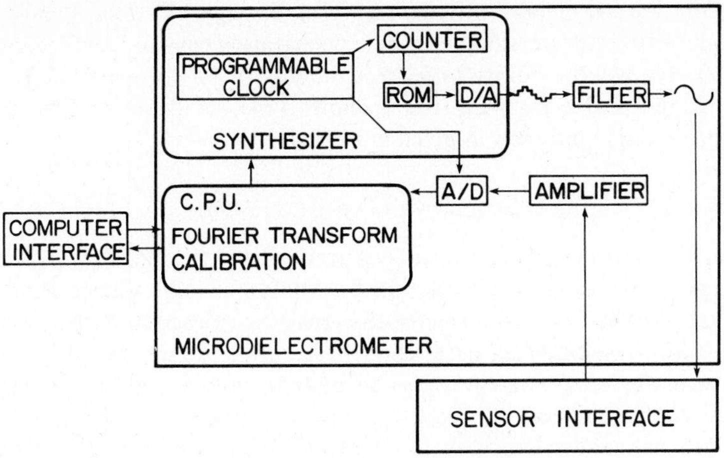

Figure 1.11: Schematic of microdielectrometer including frequency synthesizer and signal analyzer.

dipole and each of these has a distribution of relaxation times causing some differences between calculated and observed results. There has been much work at refining the model to account for distributed relaxation time systems (12). However, the Debye model [Eq. (4)] serves as a very useful template upon which dielectric data can be analyzed and interpreted.

1.3.3 Ionic Conduction

Ionic conduction is the result of current flow due to the motion of mobile impurity ions within the material under test. It is often assumed that ionic conduction is insignificant due to low mobile ion concentrations. However, it has been demonstrated that concentrations well below 1 ppm are sufficient to cause significant ionic conduction levels (13). From Eq. (4), it is seen that ionic conduction contributes only to the loss factor and does not affect the permittivity. It is very common to observe large loss factors in polymers when above T_g due to the ionic conduction contribution to the loss factor.

1.3.4 Electrode Polarization

Electrode polarization is not due to the polymer under test alone, but is the result of the resin in combination with the electrodes used to introduce the electric field. Electrode polarization occurs when ionic conduction is extremely high, causing ions to collect at the polymer/electrode interface during one-half cycle of the oscillating electric field cycle. As ions build up at the electrode in a thin boundary layer and do not exchange their charge, a large capacitance is formed (not unlike an electrolytic capacitor). This has the effect of artificially decreasing the measured values of high loss factors and increasing the measured values of permittivity. In a Cole/Cole plot or arc diagram, electrode polarization appears much like a dipole transition.

However, the measured permittivities attained through electrode polarization are usually high (>100), thus making electrode polarization easy to identify. Some commercial software has the ability to correct the loss factor for electrode polarization influence if the ε_r value from Eq. (4) is known (14). A detailed derivation of the electrode polarization influence is given in Ref. 10.

1.3.5 Monitoring Curing Reactions with Dipoles and Ions

Many changes occur within the resin system during a reaction. Typically, among these changes are a loss of dipole rotational mobility and ionic translational mobility. Both of these may be detected dielectrically and may be used as probes for following cure reactions. In monitoring cure, care must be taken not to confuse changes in mobility caused by cure with those caused by temperature and ionic concentration. Luckily, ionic concentrations for most resin systems are constant through cure except for those systems that decompose or have ion generating catalysts. However, temperature is often non-isothermal and techniques for taking thermal influence into account will be discussed later. For now, an isothermal cure of a simple thermosetting epoxy/amine system will be discussed.

Figure 1.12 shows the permittivity and loss factor during cure of a DGEBA/DDS (diglycidyl ether of bis-phenol A/diamino diphenyl sulfone) epoxy amine. The permittivity is observed to decrease from a higher plateau to a lower plateau as the cure proceeds with higher frequencies falling first and lower frequencies falling last. This behavior can be understood by illustrating a point in the cure at about 250 min where the 10 Hz permittivity curve is at about the midpoint of its falloff. The dipole relaxation time at this point is about $1/(2\pi f)$ or 0.016 s. This also means that at 10 Hz, the dipoles are at a maximum lag behind the electric field and movement is becoming difficult. At a higher frequency, for example, 10,000 Hz, the dipoles simply cannot keep up with the electric field and thus a low permittivity is measured. At a lower frequency, for example, 1 Hz, the dipoles are still able to move freely, and thus a higher permittivity is measured. As the cure progresses, the dipole relaxation time grows longer, and so eventually the dipoles cannot orient even at low frequencies.

The loss factors in Fig. 1.12 exhibit a decay and then a peak. The initially high loss factors are caused by high ion mobilities and therefore high ionic conductivities. As the mobility decreases with cure, the ionic contribution to the loss factor at some frequencies eventually falls below the dipole contribution. At this point the loss factor starts to increase again. As the dipole reaches maximum lag behind the electric field, the loss factor exhibits a peak (maximum loss of energy due to rotational viscous drag). This is precisely the same point where the corresponding permittivity is at the midpoint of its decay. Both the dipole loss factor peaks and/or the permittivity decay midpoints may be used to determine dipole relaxation times during the course of a cure.

In order to plot dipole relaxation time during the course of a cure, the times to reach the dipole peaks (or midpoints) at various frequencies must be measured. The relaxation time is simply $1/(2\pi f)$ s. Figure 1.13 shows the dipole relaxation time data derived from the dielectric data in Fig. 1.12. The data in Fig. 1.13 show that the relaxation time can change by many orders of magnitude during the course of a cure and that to adequately follow an entire cure, extremely high frequencies are

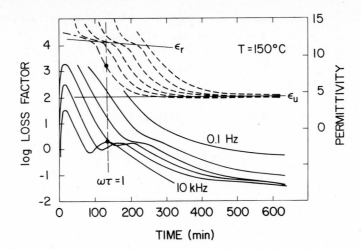

Figure   1.12:   Permittivity and loss factor during cure of DGEBA epoxy and DDS amine hardener.

required at the beginning of reaction and very low frequencies are required at the end of cure. No commercial bridge covers the entire range, however, the microdielectric sensor in combination with commercial equipment can cover eight orders of magnitude, a significant part of that range. Even if the entire frequency range could be covered, it is often the case in some resin systems that the dipole loss peaks cannot be detected due to dominance of the ionic contribution over the loss factor. Therefore it is usually more practical to monitor changes in ionic conductivity during cure since the ionic contribution can always be measured if a low enough frequency is used.

Software routines now exist (15) that automatically extract the ionic portions of the multifrequency loss factor data and calculate the ionic conductivity during the course of a cure. Figure 1.14 shows ionic conductivity data derived from the loss factor data from Fig. 1.12. With this technique, changes in the resin can be monitored continuously from the beginning to end of cure, provided low enough frequencies are used at the end of cure. To exemplify this point, the dashed line in Fig. 1.14 represents the point beyond which ionic conductivity data could not be obtained if the lowest available frequency were only 5 Hz.

The ionic conductivity response has been shown to be nearly inversely proportional to the viscosity before gelation. Dipole peaks are generally related to the T_g of the material, where low-frequency dipole relaxation peaks are often near the T_g measured by other techniques and high-frequency dipole relaxations are much higher in temperature than the actual T_g. Thus, as a cure proceeds, dipole relaxations are seen at lower and lower frequencies, meaning that the T_g is approaching the temperature at which the cure is being carried out.

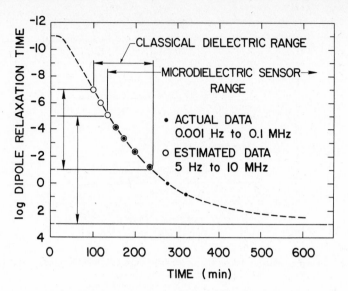

Figure   1.13:   Dipole relaxation times determined from data in Fig. 1.6.

1.3.6 Dielectric Properties and Glass Transition Temperature

Both the dipole relaxation time and ionic conductivity are related to the T_g of a polymer (3,4).   When log relaxation time or log resistivity (resistivity = 1/conductivity) is plotted against 1/temperature (degrees Kelvin) in an Arrhenius fashion, a curved result is usually obtained rather than a straight line with a single activation energy.   This behavior of the dielectric properties is similar to that of viscosity and is attributed to a WLF (Williams-Landel-Ferry) dependence.   Figure 1.15 shows such a plot for an epoxy resin.   Note that both dipole relaxation time and log resistivity start falling off steeply as the glass transition temperature is approached.   It is for this reason that low-frequency dipole relaxations or the appearance of ionic conductivity can often be used to track changes in T_g.   At temperatures well above T_g, both dipole relaxation time and ionic resistivity often change proportionally with viscosity since all three variables follow a WLF dependence (3).

In recent years there has been much work in determining WLF coefficients for various thermoset systems (16-19).   It has been found that for varying molecular weight resins and for curing systems that some of the WLF constants are, in fact, not constant.   Recent work has shown that these changes can usually be approximated with a linear dependence on T_g and that once the constants for the linear dependence are determined, the dielectric properties can be accurately determined as a function of T_g and temperature during an isothermal cure.   However, as the cure temperature is changed, the WLF constants change in other ways and work is currently under way to understand this behavior.

A recently introduced empirical model based on the near linear relationship between log conductivity and T_g during isothermal cure has been shown to adequately predict changes in T_g during nonisothermal cure for a rather simple epoxy

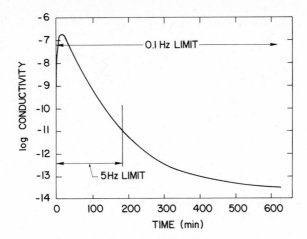

Figure 1.14: Ionic conductivity (S/cm) during cure determined from data in Fig. 1.6.

system (20). This technique requires that the resin be precharacterized through measurement of the uncured and cured T_g as well as the temperature dependencies of conductivity for the uncured and cured material. An example of predicted T_g (from dielectrically measured ionic resisitivity) compared to actual T_g is shown in Fig. 1.16 for a nonisothermal cure.

Many polymers exhibit sub T_g dipole relaxations which are related to beta or gamma transitions in the polymer. Just as in mechanical measurements, these are attributed to motions of polar side chain groups or any polar group than can move below the T_g. These transitions characteristically exhibit linear behavior when plotted on Arrhenius axes and activation energies can be extracted (3).

1.3.7 Crystallinity

As expected, the degree of crystallinity influences the dielectric response. Although there are some dipole transitions which can occur in the crystalline phase (3), dipole transitions are more often inhibited by the crystalline confines. Ionic mobilities should also be much less within the crystalline regions. The annealing process which occurs between T_g and T_m can sometimes be observed dielectrically in polymers. Low-crystallinity poly(ether-ether ketone) (PEEK) has been observed to crystallize at temperatures above T_g and well below T_m. In Fig. 1.17, the low-crystallinity material initially shows an ε_u of about 2.9 and an ε_r of about 4.1. The ε_r values are observed just after the polymer is taken above its T_g (140°C) and as the dipoles start to orient. At 155°C, ε_r starts a descent to a final value of 3.6. The decrease in ε_r during the temperature ramp is attributed to crystallization and consequently, the inhibition of dipole motion of segments within the crystalline regions. All temperature ramps thereafter show no further change in ε_r since maximum crystallinity was attained.

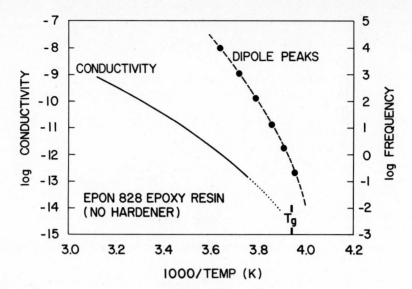

Figure 1.15: Arrhenius plot of dipole relaxation times and ionic conductivity (S/cm) as a function of temperature for pure DGEBA epoxy.

1.3.8 Moisture Uptake and Diffusion Coefficient Calculation

It has been shown that at relatively low percentages of moisture uptake, the dielectric permittivity (or dielectric constant) is proportional to the weight of water taken up (6) (Fig. 1.18). Advantage can be taken of this relationship in that rate of moisture uptake can be measured through dynamically measured changes in the permittivity. A thin organic film on a microdielectric sensor can be modeled as a thin infinite sheet where only diffusion in the z direction (perpendicular to the chip surface) needs to be considered. The traditional solution to Fick's law for one-dimensional diffusion is

$$\frac{C - C_0}{C_1 - C_0} = 1 - \frac{4}{\pi} \sum_{n=0}^{\infty} \frac{(-1)^n}{2n + 1} \exp\left(\frac{-D(2n + 1)^2 \pi^2 t}{4L^2} \right) \cos\left(\frac{(2n + 1)\pi x}{2L} \right) \quad (5)$$

where C, C_0, C_1, D, t, x, and L are the concentration at a distance x into the film, the initial uniform concentration, the concentration at the surface (in equilibrium with vapor), the diffusion coefficient, the elapsed time, the distance from the back surface of the film, and the distance from the back of the film to the surface of the film. n need only be carried to about 20 for convergence of summation. From the linear relation between permittivity and moisture concentration, the left hand of Eq. (5) can be replaced with

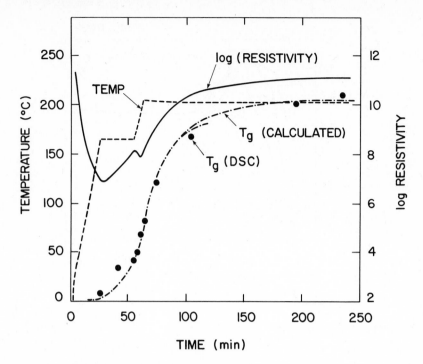

Figure   1.16:   Dielectrically determined T_g and actual T_g during nonisothermal cure for DGEBA epoxy and DDS amine hardener (20).

$$\frac{C - C_0}{C_1 - C_0} = \frac{\varepsilon' - \varepsilon'_0}{\varepsilon'_1 - \varepsilon'_0} \tag{6}$$

where ε', ε'_0, and ε'_1 are the permittivity at a distance x into the film, the initial uniform permittivity across the film, and the permittivity at the surface after equilibrated with vapor, or the final permittivity at any point after total equilibration with vapor.

The implications of Eq. (5) can be more easily visualized in Fig. 1.19 where the left edge of the figure represents the chip/polymer interface and the right edge is the polymer/vapor interface. Concentration curves for various Dt/L^2 values are plotted. Since D and L are presumed constant, these curves really represent various times. Note that at small times, the surface has equilibrated with the vapor and has a fractional concentration of 1.0. Not very far into the bulk this value decays to zero. Beyond Dt/L^2 values greater than 1.0, the entire film has nearly come to a uniform concentration in equilibrium with the vapor.

If the film thickness is large compared to the sensor electrode spacing, then the sensor can be considered to sense only concentration changes at the chip/polymer interface region. This can be modeled through Eq. (5) by setting the x variable to 0 and calculating the fractional change in concentration (or permittivity) as a function

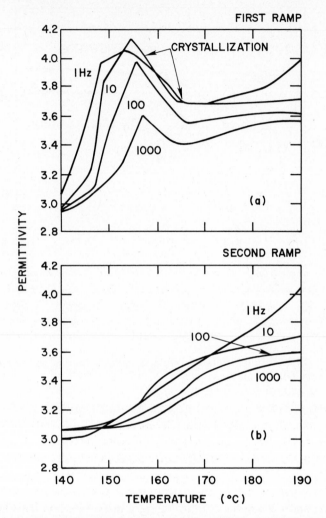

Figure 1.17: (a) Thermal ramp of low-crystallinity PEEK thermoplastic; (b) Thermal ramp of high-crystallinity PEEK thermoplastic.

of time. The only unknown is D, the diffusion coefficient. There are various ways of determining the best value of D ranging from a midpoint calculation (6) to a least-squares fit (21). Figure 1.20 shows the change in ϵ' during the exposure of a 30 μm polyimide film to 90% humidity at room temperature (solid line). The dotted line is a plot of Eq. (5) with D set to 2.0×10^{-9} cm^2/s.

This technique is useful for carrying out diffusion studies very quickly and with good sensitivity. The thin films allow for quicker equilibration times and samples need not be weighed. Figure 1.21 shows the result of a study on a DGEBA/DDS epoxy system in which the diffusion coefficient was measured as a function of cure state (or T_g). Surprisingly, there is a slight increase in diffusion coefficient with

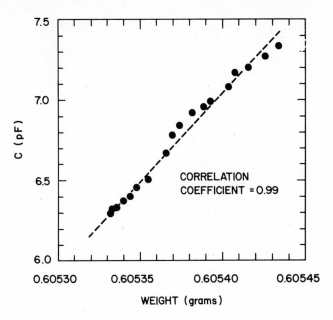

Figure 1.18:   Capacitance of polyimide as a function of moisture uptake (6). Reprinted with permission from *J. Electron. Mat.*, Vol. 14, pg. 19, a publication of The Metallurgical Society, Warrendale, PA.

degree of cure, which is in agreement with published weight uptake data on this system (22). The $\delta \epsilon'$ was also observed to increase with cure, suggesting that more moisture may enter the system with increasing degree of cure. Although this is counter intuitive, it is also in agreement with weight uptake data shown in Fig. 1.22. This behavior arises from the more cured samples contracting during cooling at less of a rate than less cured samples with lower T_g. This causes the specific volume/temperature curves to cross so that more cured material, which will have the least specific volume above T_g, will have the most specific volume at temperatures well below T_g (22).

1.3.9 Sub-Picoampere Leakage Measurement

The microdielectric sensor is well constructed for making low leakage current measurements in unusual environments such as with high humidity and temperature. Any current through the sample is integrated on an internal capacitor built in high-quality thermal oxide. The signal is converted by the FETs to high current so that once the signal leaves the sensor it is well buffered and not subject to errors caused by cable leakage. However, a drawback to DC measurements is that they can only be carried out for relatively short periods of time due to saturation of the on-chip integrating capacitor. In the AC mode, normal conductivity measurements can be made at very low frequencies. At 0.001 Hz, conductivities may be measured down to about 10^{-16} S/cm.

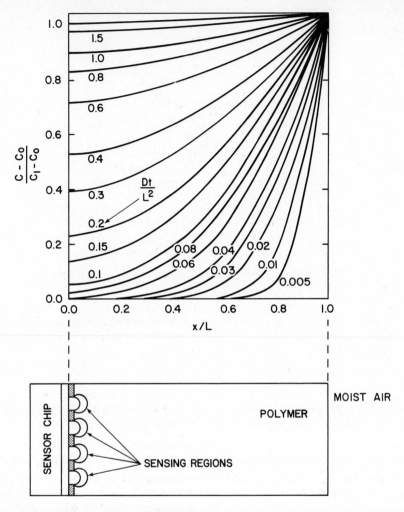

Figure 1.19: Concentration versus distance into a one-dimensional film at various intervals of Dt/L^2 (or time) for Fickian diffusion. Note that the microdielectric sensor only detects changes at the back surface of the film under test.

DC measurements may be made in two ways. The first is to simply apply a DC potential to the excitation electrode and utilizing the sensor electronics, follow the potential on the integrating capacitor. Since the equivalent circuit is an RC [the sample is R (resistance) and the integrating capacitor is C (capacitance)] an exponential decay is usually observed (Fig. 1.23). From the time scale of the decay and the known capacitor value, the sample resistance may be calculated. Sample resistivities up to 10^{18} ohm-cm have been observed with this method. The drawback of this method is that the potential across the sample is constantly changing during the decay which could cause problems if there is a nonlinear response.

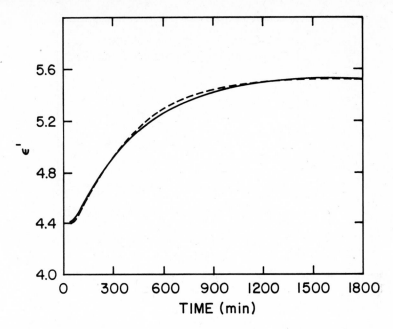

Figure 1.20: Permittivity versus time during exposure of DGEBA/DDS epoxy/amine film to 90% relative humidity.

The second method is to apply a ramped DC potential to the excitation electrode; this potential is constantly adjusted so that there is a constant difference (say, 1.0 V) between it and the sensing electrode (the potential across the integrating capacitor). In this way a constant electric field is applied to the sample and the current may be determined from the rate at which the potential increases. This method also has a drawback in that the applied potential is limited to about 10 V due to possible FET gate breakdown. Once the potential across the integrating capacitor reaches 10 V the polarity of the excitation electrode has to be switched. This effectively limits the total amount of charge that is allowed to pass through the sample. For the present microdielectric sensor design this is about 5×10^{-10} C.

1.3.10 Process Control

Dielectric measurements are particularly useful for various types of process control ranging from drying to curing phenomena. They have been shown to be particularly sensitive to change in viscosity and to end of reaction. Figure 1.24 shows an example of closed loop control for an epoxy system (used in structural composites) utilizing feedback information from the microdielectric sensor (23). There were four programmed steps during the controlled cure:

1. Heat to 250°F and hold until log resistivity is equal to 7.
2. Hold log resistivity at 7 until 350°F is reached.
3. Hold at 350°F until slope of log resistivity equals zero.

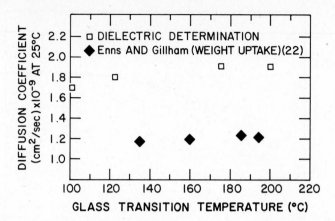

Figure 1.21: Diffusion coefficient versus T_g (or cure state) for dielectric and weight uptake experiments (22) for DGEBA/DDS epoxy amine.

4. Cool to room temperature.

The purpose of the first step is to heat the material so that volatiles are emitted. The second step monitors the viscosity (log resistivity) until a predetermined value of 7.0 is attained. At that point the viscosity is held constant through control of the temperature. As the reaction proceeds faster with temperature, the temperature must also be increased faster. At one point the temperature cannot be increased fast enough and the viscosity starts to rise. The third step is to keep the temperature constant during the final hold until the log resistivity stops changing. When that occurs, cool down is initiated and the cure is complete.

What is particularly novel about this technique is that the viscosity can be controlled through the monitoring of both the dielectric and temperature information returned from the microdielectric sensor. Typical cures of the epoxy system shown in Fig. 1.24 exhibit a second viscosity minimum as the temperature is increased to the final temperature. Through dielectric feedback, this was not only eliminated, but viscosity (resistivity) was also kept constant at a chosen value. There are many ways in which the dielectric data could be used for process control. Some commercial software packages now exist which allow the user to select a critical slope of log (conductivity), the sign of the second derivative at the critical point, the number of data points to be averaged, and finally, time and temperature windows. During the course of a reaction, as the critical point is reached, and if all user selected conditions are satisfied, the controlling computer will issue a pre-programmed analog voltage which is used to initiate any kind of process.

1.4 SUMMARY

Microdielectric sensors have the ability to make sensitive measurements on thin films. As a result of the microelectronic design, the sensors have the ability to function at very low frequencies enabling the detection of very low ionic conduction levels

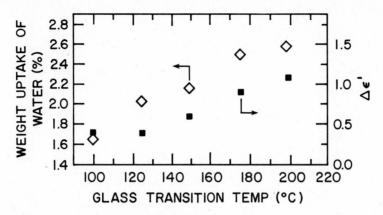

Figure 1.22:   Weight uptake and change in permittivity as a function of T_g (cure state) in DGEBA epoxy cured with DDS amine hardener.

and long dipole relaxation times. The sensors may be custom mounted to facilitate certain film deposition techniques or they may be used from within a standard polyimide flexible package. The sensors are well suited for monitoring chemical changes occurring in thin polymer films ranging from simple drying to polymerization and crosslinking. Cures may be analyzed in terms of changing dipole relaxation times or ionic conductivity. However, ionic conductivity is usually more easily extracted from multifrequency loss factor data over the entire range of cure. If enough is known about the material under test, the microdielectric sensors may be used to observe changes in T_g , crystallinity, moisture content, as well as most other chemical changes. Care must be taken in analyzing dielectric results since they can be effected by any and all changes in a material under test. Finally, the microdielectric sensor may be useful for process control operations ranging from film drying to film polymerization and curing.

1.5 REFERENCES

1.   S. D. Senturia and S. L. Garverick, "Method and Apparatus for Microdielectrometry," *U.S. Patent No. 4,423,371.*

2.   N. F. Sheppard, Jr., S. L. Garverick, D. R. Day, and S. D. Senturia, "Microdielectrometry: A New Method for In Situ Cure Monitoring," *Proc. 26th SAMPE Symp.*, 65 (1981).

3.   P. Hedvig, **Dielectric Spectroscopy of Polymers**, McGraw-Hill, New York (1955).

4.   S. D. Senturia and N. S. Sheppard, Jr., "Dielectric Analysis of Thermoset Cure," *Adv. Polym. Sci.*, 80, 1 (1986).

5.   J. Gotro and M. Yandrasits, "Cure Monitoring Using Dielectric and Dynamic Mechanical Analysis," *Proc. S.P.E. 45th Ann. Tech. Conf.*, May 4, Los Angeles, CA, 1039 (1987).

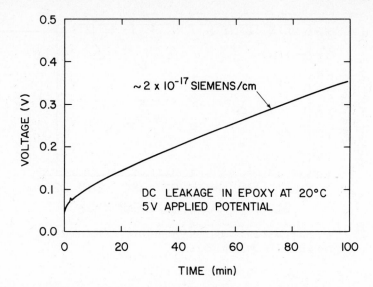

Figure 1.23:   Charging of microdielectric sensing electrode during application of 5-V DC bias at room temperature on an epoxy composite sample.

6.   D. D. Denton, D. R. Day, D. F. Priori, S. D. Senturia, E. S. Anolick, and D. Scheider, "Moisture Diffusion in Polyimide Films in Integrated Circuits," *J. Electron. Mat.*, 14, 119 (1985).

7.   C. W. Reed, in **Dielectric Properties of Polymers**, F. E. Kraus, Ed., Plenum Press, New York, 343 (1972).

8.   H. L. Lee, "Optimization of a Resin Cure Sensor," *M. S. Thesis*, MIT Dept. of Electrical Eng., Cambridge, MA (1982).

9.   Micromet Instruments, Inc., "Ultra Fast Dielectric Measurements for RIM Applications," *Applications Report # AR0156* (1988).

10.   D. R. Day, T. J. Lewis, H. L. Lee, and S. D. Senturia, "The Role of Boundary Layer Capacitance at Blocking Electrodes in the Interpretation of Dielectric Cure Data in Adhesives," *J. Adhesion*, 18, 73 (1985).

11.   P. Debye, *Polar Molecules*, New York, Chemical Catalog Co. (1929).

12.   See, for example, G. Williams and D. C. Watts, "Non-symmetrical Dielectric Relaxation Behavior Arising from a Simple Empirical Decay Function," *Trans. Farad. Soc.*, 66, 80 (1970).

13.   A. R. Blythe, **Electrical Properties of Polymers**, Cambridge University Press, Cambridge (1979).

14.   D. R. Day, in **Quantitative Nondestructive Evaluation**, Plenum Press, New York, 5B, 1037 (1986).

15.   D. R. Day, "Effects of Stoichiometric Mixing Ratio on Epoxy Cure - a Dielectric Analysis," *Poly. Eng. Sci.*, 26, 5, 362 (1986).

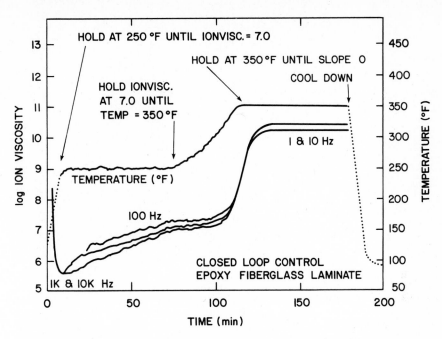

Figure   1.24:   Closed loop control on Fiberite 932 epoxy/graphite prepreg.

16.   N. F. Sheppard, Jr., "Dielectric Analysis of the Cure of Thermosetting Epoxy/amine Systems," *Ph.D. Thesis*, MIT, Cambridge, MA (1986).

17.   S. A. Bidstrup, N. F. Sheppard, Jr., and S. D. Senturia, "Dielectric Analysis of the Cure of Thermosetting Epoxy/amine Systems," *Proc. S.P.E. 45th Ann. Tech. Conf.*, May 4, Los Angeles, CA, 987 (1987).

18.   J. W. Lane and R. K. Khattak, "Correlation Between Dielectric Cure Models and Rheometric Viscosity," *Proc. S.P.E. 45th Ann. Tech. Conf.*, May 4, Los Angeles, CA, 982 (1987).

19.   W. Bidstrup, S. A. Bidstrup, and S. D. Senturia, "Correlation Between Dielectric and Structural Properties During Epoxy Cure," *Proc. S.P.E. 46th Ann. Tech. Conf.*, Atlanta, GA, April 18, 960 (1988).

20.   D. R. Day, "Dynamic Cure and Diffusion Monitoring in Thin Encapsulant Films," *Proc. 38th Elect. Comp. Conf.*, Los Angeles, CA, 457 (1988).

21.   B. K. Bhattacharyya, W. A. Huffman, W. E. Jahsman, and B. Natarajan, "Moisture Absorption and Mechanical Performance of Surface Mountable Plastic Packages," *Proc. 38th Elect. Comp. Conf.*, Los Angeles, CA, 49 (1988).

22.   J. B. Enns and J. K. Gillham, "Effect of the Extent of Cure on the Modulus, Glass Transition Temperature, Water Absorption, and Density of an Amine-cured Epoxy," *J. Appl. Polym. Sci.*, 28, 2831 (1983).

23.   D. R. Day, in **Quantitative Nondestructive Evaluation**, Plenum Press, New York, 7B, 1573 (1988).

2
BENDING-BEAM CHARACTERIZATION OF THIN POLYMER FILMS

H. M. Tong and K. L. Saenger

Thomas J. Watson Research Center
IBM Research Division
Yorktown Heights, New York

Recent advances in the speed and density of integrated circuits (ICs) have stimulated the development of multilayer, multichip packages which provide high-density interconnections for propagating high-speed signals with minimal delay and distortion. These packages, as reported in published literature, include high-density printed wiring boards with glass reinforced polymers (1), and thin film multilayer structures using polymer dielectrics such as polyimide (2-4). As a result of this technology trend, thin polymer films are finding increased use in microelectronics not only as dielectrics but also as passivation layers, barrier layers, adhesives, encapsulants, and fabrication aids (e.g., photoresists).

The selection of a polymer thin film for microelectronics fabrication is often dictated by a combination of requirements on electrical, thermal, mechanical, and processing properties. During the fabrication or operation of high interconnect density packages, a polymer film in use can be subjected to hostile conditions as diversified as thermal cycling (between room temperature and 400°C for polyimide) and repetitive exposure to process and environmental fluids (e.g., solvents and moisture). Properties indicative of a polymer's capability to withstand these conditions include not only mechanical properties such as the modulus and thermal expansion coefficient but also transport characteristics of the fluids in the polymer. Both thermal cycling and fluid transport create stress in the structure which, in the extreme case, can lead to interlayer delamination and cracking (5-7). Additionally, this stress can be accompanied by a substantial amount of structural bending which is detrimental to crucial process steps such as photoresist exposure and interlayer registration. While thermal cycling and fluid pickup often impact negatively on the behavior of the structure, positive outcome such as adhesion enhancement (8,9) can be obtained with proper understanding and control of these processes.

The determination of the stress and related properties or characteristics of a polymer bearing structure can be conveniently achieved using the bending-beam technique, which requires supported samples that can mimic the actual microelectronic structure of interest. This technique is perhaps the most commonly used technique for metal stress studies (10). In the bending-beam technique, a thin slender beam supporting the sample film (which can be a single or a multilayer coating) is clamped at one end, and the deflection of the free end of the beam is monitored in situ as a function of time after subjecting the film to practically any wet or dry experimental conditions of interest (e.g., thermal cycling or immersion in liquid). From the changes in deflection, alterations in stress can be calculated with the accuracy desired. With a suitable theory to model the physical and/or chemical processes occurring in the film during the experiment (e.g., thermal expansion mismatch, polymer curing reaction, fluid transport and/or physical aging), one can analyze the changes in the stress or end deflection to obtain the fundamental properties (e.g., mechanical properties) or characteristics (e.g., fluid diffusion coefficient) governing such changes. The task of data interpretation for two major applications of bending-beam techniques, thermal cycling, and fluid transport studies, is well supported by an abundance of beam mechanics theories (see, for instance, Refs. 10-12) and the literature of fluid transport in polymers (e.g., Refs. 13-17).

The body of this chapter is divided into three sections: Experiment, Models, and Applications. The key elements of the bending-beam technique are covered in the Experiment section. Beam materials, setups, and deflection measurement techniques are described, as well as the typical considerations that aid in their selection. The Models section describes the models that have been utilized to analyze the beam deflection data collected for single and multilayer coatings of practical importance to the electronics industry (e.g., polyimide and Cu/polyimide). The results obtained from these studies are shown in the Applications section. For simple (i.e., one component, single layer) polymer coatings, the bending-beam technique has been utilized to determine

- the stress level in the polymer (6,18-21),
- the polymer's glass transition temperature (6,22),
- the inception of a cracking or delamination process (6),
- the elastic stiffness [defined as $E_f/(1 - \nu_f)$, where E_f is the elastic modulus and ν_f is Poisson's ratio], and in-plane thermal expansion coefficient of the polymer (17),
- the modulus of the polymer (18),
- the polymer curing mechanism with the help of complementary techniques (20,21),
- Fickian and non-Fickian transport of low molecular weight penetrants in glassy polymers exhibiting either negligible or substantial swelling upon penetrant diffusion (23-25).

With respect to multilayer structures, various researchers have also studied thermomechanical behaviors such as stress levels upon thermal cycling (7,22,26,27) and exposure to fluids (28).

2.1 EXPERIMENT

In this section, we will describe the important elements of a successful bending-beam experiment and the considerations that go into their selection. This section is based in part on several excellent review articles which can be found elsewhere (10,29,30). Although polymers are our main concern here, most of these considerations also apply to other types of materials such as metals and ceramics.

2.1.1 Beam Materials

The thin beam chosen for the bending-beam experiment is usually between 3 and 50 times longer than it is wide, with a constant rectangular cross section. In general, the beam can have macroscopic or microscopic dimensions depending on which better simulate the situation of interest. While macroscopic beam dimensions are practically limitless, approaching those of structural beam supports, microbeams with thickness, width, and length of 1×10^{-4}, 2×10^{-3}, and 3×10^{-3} cm, respectively, have been fabricated using IC fabrication techniques (31). Typically, the longer the beam or the smaller the beam thickness the higher the bending-beam sensitivity (see below).

To simplify data analysis, the beam must be well characterized mechanically, with known (and preferably temperature-independent) modulus and Poisson's ratio. Ideally, the beam should remain elastic during the experiment, that is, the maximum stress in the beam should not exceed its elastic limit. If a beam material does yield during the experiment, the resulting effects must be considered in the data analysis with the use of suitable elasto-plastic or viscoelastic models (see, for instance, Refs. 12 and 22). Ideal beams should also be stable in time and initially flat, so that any observed end deflection changes may be attributed to processes occurring in the film alone. The properties of ideal beams are frequently associated with rigid materials, but stability and flatness may sometimes be obtainable only after annealing or heat treatment.

Beam materials which have been used in polymer studies include

- fused quartz (6,22,26)
- silicon (7,18),
- SiO_2 (31),
- Cu (23,32),
- Au (31),
- Mo (33),
- stainless steel (19),
- polyimide (34),
- Cu/Invar/Cu (33),
- printed circuit board (33).

A comprehensive list of substrate materials that have been used in various stress studies has been compiled by Campbell (29).

In our studies, we favor two beam materials: (i) fused quartz, because of its rigidity and commercial availability, and (ii) silicon, which permits precise and complex mechanical structures and devices to be created using established IC techniques. The fused quartz beams (from Hibshman Co., San Luis Obispo, CA 93403) are typically 3.8 cm long, 0.31 cm wide, and 8×10^{-3} cm thick. Prior to film coating, each beam must be cleaned using an approach tailored to the beam material and nature of the contaminants. For the case of fused quartz, the beams might be washed successively with a dilute aqueous solution of DeContam (from Electronic Space Product International, Westlake Village, CA), deionized water, and ethanol, followed by drying under an infrared lamp to remove the residual cleaning fluids (6). When a laser pointer (see below) is used for end deflection measurements with weakly reflecting samples, one end of the beam can be coated with a metal layer, e.g., chromium (24), to increase the reflected laser intensity.

2.1.2 Preparation of Polymer-Containing Samples

Beams or structures containing multilayer coatings are typically needed for studying the behaviors of real product structures. They are often available from the respective manufacturing processes. For more fundamental studies, one might use a single polymer coating, such as a monodispersed polymer, a photoresist, or a polymer blend. In either case, the beam thickness is usually much greater than the overall coating thickness.

Beams coated with simple polymer thin films can be conveniently prepared by spin casting the polymer solution on one side of the beam and then drying the film to drive off the casting solvent system via baking. Great care must be taken to avoid coating on both sides of the beam; otherwise stress compensation effects occur, leading to erroneous results.

To improve the film thickness uniformity, a multiple coating approach (19) or a technique involving coating on a large substrate which is subsequently cut or diced to form slender beams (33) can be adopted. Better film thickness uniformity can improve the accuracy of data analysis and interpretation. After film coating, the beams are either used immediately or kept in a vacuum desiccator until use.

2.1.3 Bending-Beam Setups

In a bending-beam experiment, the sample beam must be held in a well-controlled environment while its end deflection is monitored in situ. According to Rottmayer and Hoffman (35), it is extremely important to grip the beam using clamps such that no sliding takes place during the experiment. In certain cases, a rigid joining or gluing is recommended. Two types of bending-beam setups have been developed: one for dry experiments performed in a gas phase or vacuum; and the other for liquid-phase experiments.

Dry Setup: The dry setup can be used for studying the effects of heat treatment (e.g., thermal cycling or curing) and transport of volatiles in the film. The schematic of a dry bending-beam setup is shown in Fig. 2.1 (6). It consists of a clamping fixture for holding the sample (not shown), a Cu box for sample heating, an environmental chamber for creating the desired environment, and an optical grade quartz window for viewing end deflection. Uniform sample heating is achieved by controlling the power input to the resistance wires wound uniformly around the copper box. For the experiments to be performed in a gas other than air (e.g., dry nitrogen), the environmental chamber is first evacuated and then flushed or filled with the gas.

Effects of Temperature Nonuniformities: Tight temperature control is necessary to ensure the uniform beam temperatures required for straightforward data interpretation. For every bending-beam study, the temperature nonuniformities in the length of the beam and across the beam thickness must be assessed using thermocouples or other temperature sensing devices as necessary. The effects of temperature nonuniformities on the beam's behavior may then be assessed following the approaches of Alexander (36) and Hoffman (10) below.

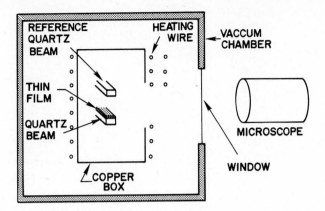

Figure   2.1:   Schematic of a dry bending-beam setup.

For a temperature difference ΔT_L established between the two ends of the beam, Alexander derived an approximate expression for the resultant stress

$$\sigma_L = \frac{E_f}{6(1 - \nu_f)} \, \Delta\alpha \, \Delta T_L \qquad (1)$$

where E_f is the elastic modulus of the film, ν_f, its Poisson's ratio, and $\Delta\alpha$ is the difference between the thermal expansion coefficients of the film and the beam. This lengthwise temperature nonuniformity creates a nonuniformly bent beam whose deflection can be estimated using Eqs. (4) and (5) below. From Eq. (1), it is clear that this temperature nonuniformity effect may be eliminated entirely if the film and substrate have identical expansion coefficients.

Compared to the temperature nonuniformity in the length of the beam, the thickness direction temperature nonuniformity is usually much smaller due to the small thickness to length ratios used. Assuming the thermal expansion strains vary linearly through the beam, and neglecting temperature nonuniformities across the film, Hoffman proposed the following expression for estimating the end deflection, δ_T, caused by the thickness direction temperature nonuniformity ΔT_T across the beam

$$\delta_T = \frac{L^2}{2t_b} \, \alpha_b \, \Delta T_T \qquad (2)$$

where L is the unclamped length of the beam, t_b the beam thickness, and α_b the thermal expansion coefficient of the beam (in the beam-length direction).

These temperature nonuniformity effects can be minimized with the use of beams which are good thermal conductors, and with a good design of the setup's thermal environment. The former, however, is not always possible as it may impact the phenomenon of concern.

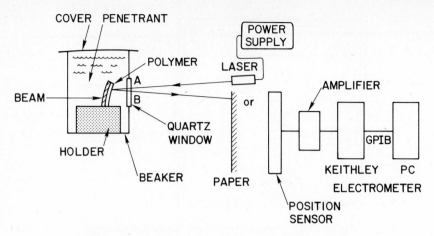

Figure 2.2:   Schematic of a wet bending-beam setup.

Wet Setup: A wet bending-beam setup for fluid transport studies is shown schematically in Fig. 2.2 (24). It consists of a jacketed beaker with an optical-grade quartz window and a fixture for holding the sample beam. Good control of the liquid penetrant temperature can be readily achieved by gently stirring the liquid penetrant and by controlling the temperature of the thermal fluid (e.g., water or silicone oil) circulating between the jacket and a constant temperature bath (not shown). For liquid-phase experiments, temperature nonuniformity effects frequently can be neglected if no self-heating of the sample occurs.

2.1.4 Deflection Measurement Techniques

The sensitivity of the bending-beam technique depends on the detection system used to observed the end deflection, as well as on beam properties and dimensions. Techniques for measuring the end deflection can be divided into six classes:

- direct observation using a microscope (6,23),
- laser pointer method (Fig. 2.2) using a low power laser (e.g., ~ 1 mW He-Ne laser at 6328 Å) with either recording paper (24) or a position sensor whose signals are continuously monitored and analyzed by a personal computer (19,22),
- restoring method, in which a known force is applied to the end of the beam to balance out the deflection of the beam (37,38),
- interferometric method using a Michelson interferometer (39),
- inductance transducer method (40),
- capacitance-change method (30).

Among these classes of detection techniques, the microscope technique is the simplest and, for many polymer related experiments, its low sensitivity is sufficient. For liquid-phase experiments, the laser pointer method is very convenient. Correction for the refractive index difference between the penetrant and air is straightforward, and an increased sensitivity to small angular deflections of the quartz beam can be achieved by increasing the distance between the sample and the recording paper or a position sensor (Fig. 2.2).   The capacitance-change and inductance transducer methods have the highest sensitivities: deflection changes as small as 2.5×10^{-7} cm (inductance method; Ref. 40) and 3×10^{-10} cm (capacitance-change method; Ref. 41) have been reported, although for practical purposes, noise and other effects may limit the working sensitivity to higher values (e.g., 5×10^{-6} cm for the inductance method; Ref. 42).   Compared to the bakeable capacitance-change method, the main disadvantage of the inductance method is that measurements have to be made at room temperature.   Sophisticated interferometric schemes for monitoring the minute deflection changes of a cantilever beam are employed in the newly developed atomic force microscope (43,44).

2.1.5 Bending-Beam Experiment

In a typical bending-beam experiment, both the coated sample beam and (optional) noncoated reference beam of the same material are clamped at one end and subsequently subjected to the experimental conditions of interest (Fig. 2.1).   The optional reference beam is useful primarily as a reference point for measuring the bending changes in the sample beam.   The collection of the deflection data proceeds until the deflection of the coated beam no longer changes with time or until the length of the data collection period is sufficient to simulate the phenomenon of interest.

2.1.6 Film Thickness Measurements

As will become clear below, knowledge of the film thickness is needed for deducing the stress related properties or characteristics from the end deflection data.   For a bending-beam experiment which does not involve significant thickness changes, one can use a stylus profilometer and/or a weighing method (which calculates the film thickness from the weight, in-plane dimensions, and density of the film) to measure the initial film thickness.   When the thickness of the film changes significantly in the course of a bending-beam experiment due to, for example, film drying or pickup of a strong swelling agent by the film, one can measure the film thickness changes by using either the ex situ profilometry technique requiring a calibration sample to be moved out of the setup every time a thickness measurement is made, or, preferably, an in situ technique such as laser interferometry (45).

2.1.7 Beam Modulus Measurement

To analyze the deflection data for practically any bending-beam study, both the elastic modulus and Poisson's ratio of the beam are needed.   Although their values can be taken from published data, it is preferable to directly measure the elastic modulus of the beam material to account for variations in beam fabrication processes.   This is particularly true for polymeric beams.   Highly constrained room temperature vulcanizing rubber has been known to develop an apparent modulus which is at least two orders of magnitude greater than its elastic modulus (46).   Berry and Pritchet (27) also found, in a study of trilayer magnetic recording tapes, that the calendering treatment increased the modulus of the (magnetic particle containing) polymer coating by over a factor of five.

For rigid beams, the simplest way to measure the beam's modulus is to hang a known weight on one end of the beam and observe the deflection of the beam (29). In this case, the equation relating the beam modulus, E_b, to the end deflection δ_F is

$$E_b = \frac{4GL^3}{wt_b^3 \delta_F}$$

(3)

where G is the weight placed on the end and w is the substrate width. More sophisticated methods include the measurements of the elastic modulus from the natural resonant frequency of flexural vibration excited by an alternating electrostatic field (30), and from the load-deflection data collected using a nanoindenter and microbeams (31). For flexible materials such as polymers, tensile tests (27) can be used. A release technique involving the measurement of the length of a polymer microbeam before and after it is released from a silicon support may also be useful (34).

2.2 MODELS

The utility of the bending-beam technique for measuring any property or characteristic hinges upon the availability of a model or criterion for analyzing the measured deflection data. For thermal cycling and fluid transport studies, thermomechanical models based on either simple coatings or multilayer structures have been developed for the calculation of mechanical properties and/or stress levels from the end deflection data, and vice versa. To analyze data collected during fluid transport, however, the time-dependent fluid concentration distribution in the film must be known. In this section, all these models are reviewed.

2.2.1 Stress Calculation for Simple Coatings

We shall begin with a stress model for the case of a simple coating on an elastic beam with parameters defined in Fig. 2.3. Under the following conditions:

- the beam shape approximates part of a circle,
- good adhesion between film and beam,
- the radius of curvature, R, of the beam is much greater than the beam thickness which, in turn, is greater than the film thickness (30),
- the width of the beam is less than half the length (38),

the formula below has been obtained for calculating the total stress, σ, in the film from the radius of curvature of the beam (10,30,47)

$$\sigma = \frac{E_b t_b^2}{6R(1 - \nu_b)t_f}$$

(4)

where ν_b is the Poisson's ratio for the beam and t_f the film thickness. By applying the Pythagorean theorem, R can be calculated exactly from the end deflection δ (relative to a typically flat reference beam with $R = \infty$) to be

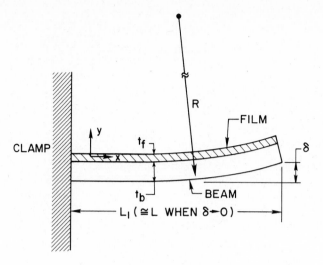

Figure   2.3:   Definition of parameters for a bilayer beam.

$$\frac{1}{R} = \frac{2\delta}{L_1^2 + \delta^2} \tag{5}$$

where the distance L_1 is the bent beam's projection on the "undeflected" beam (x-axis in Fig. 2.3). Provided that $R << L$ (the unclamped beam length) $<< \delta$, then $L_1 \rightarrow L$ and, according to Eq. (5),

$$R \sim \frac{L^2}{2\delta} \tag{6}$$

whereupon Eq. (4) becomes

$$\sigma = \left[\frac{E_b t_b^2}{3(1 - \nu_b)L^2 t_f} \right] \delta \tag{7}$$

The term $(1 - \nu_b)$ arises from the biaxial film stress and can be omitted when the deflections are less than half of the beam thickness (48). According to Eq. (7), the sensitivity of the bending-beam technique (i.e., the deflection at a given stress) is greater when the beam thickness (t_b) is smaller or when the unclamped length of the beam (L) is larger, keeping other parameters constant.

   Sign of Total Stress: The total stress σ in the film can be tensile (i.e., positive) or compressive (i.e., negative) depending on whether the film assumes a length longer or shorter than its free-standing length. A polymer film with a higher thermal expansion coefficient compared to the beam develops a tensile stress upon cooling because it is restricted from shrinking to its free-standing length. This definition comes about

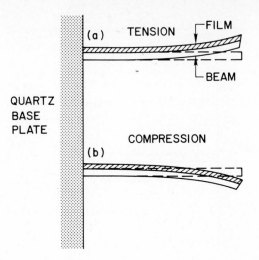

Figure 2.4:   Beam bending under (a) a tensile and (b) a compressive stress.

because the beam applies a tensile stress to the thin film to prevent it from elastically relaxing. A tensile stress will bend a thin beam so that the film is concave (Fig. 2.4a), and a compressive stress so that is convex (Fig. 2.4b).

The total stress measured using the bending-beam technique is the sum of thermal and intrinsic stresses, that is,

$$\sigma = \sigma_{thermal} + \sigma_{intrinsic} \tag{8}$$

The origin of the thermal stress is well understood and its value is usually amenable to calculation. The intrinsic stress, whose origin is much less understood for polymers, is the manifestation of all other contributions including the curing reaction and nonequilibrium processes such as film shrinkage (6) or swelling related to solvent transport. The thermal (or intrinsic) stress can be tensile or compressive, as can the total stress.

Thermal Stress: The thermal stress can arise from (10):

- the effect of differential thermal expansion between film and beam if the temperature of the sample is uniform, but changing with time,
- the effect of the temperature gradient along the length of the beam,
- the effect of the temperature gradient in the beam thickness direction.

The second and third temperature effects on beam deflection have been discussed in the Experiment section [e.g., Eq. (2)]. The first temperature effect, which is often the predominant effect in a well-designed bending-beam experiment, will now be treated. Neglecting all other effects, a temperature excursion between T_1 and T_2 produces the thermal stress

$$\sigma_{\text{thermal}} = \int_{T_1}^{T_2} \frac{E_f}{1 - \nu_f} \Delta\alpha \, dT \tag{9}$$

Here, T_1 can be taken as the temperature at which the sample is flat and stress free. For beams with glassy polymer coatings, this temperature can be the glass transition temperature. In cases where the stress changes are caused by changes in thermal stress, the temperature derivative of the total stress, that is,

$$\frac{d\sigma}{dT} = \frac{E_b t_b^2}{6(1 - \nu_b)t_f} d\left(\frac{1}{R}\right) = \frac{E_f}{1 - \nu_f} \Delta\alpha \tag{10}$$

can be used to calculate the product of the elastic stiffness, $E'_f \equiv E_f/(1 - \nu_f)$, and the thermal expansion coefficient of the film, α_f, if the beam's thermal expansion coefficient is known. Conversely, the value of E_f can be obtained from this product if ν_f and $\Delta\alpha$ are known.

Effects of Less Ideal Beam Geometries: A basic assumption of beam theory is that the stress system is one of plane stress, that is, the stresses in the beam width direction are all zero. The consequence of this assumption is that if there is a radius of curvature R in the xy plane, there is an induced radius of curvature $-R/\nu_b$ in the yz plane. For wide beams, this so-called anticlastic curvature effect is not present and the term $(1 - \nu_f)$ in the equations above must be replaced by $(1 - \nu_f^2)$ (12). The exact width (w) to thickness (t_b) ratio where this occurs is when $(w^2/R)/t_b > 1000$ (10). For most practical purposes, this wide-beam effect is not significant unless one is dealing with a beam of nearly square geometry. More detailed references to judge the effect of beam width can be found elsewhere (49,50). Discussion of the complicating effects of thick beams and nonuniform beam thicknesses is also available (see for instance, Ref. 11), but these effects are best avoided by selection of the appropriate beam materials.

Nonelastic Effects: The foregoing discussion has been based for the most part on the assumption of purely elastic action. When a beam sample is stressed beyond its elastic limit, plane sections remain plane or nearly so but unit stress is no longer proportional to strain and distance from the neutral surface. Some discussions of this effect are available (10-12,27). Because of the exceedingly small strains imposed upon thin polymer coatings on beams, the modulus E_f is expected to approximate the elastic modulus of the film in most cases. For films exhibiting significant stress relaxation or plastic deformation, Chen et al. (22) recommend replacing E_f or E'_f with an apparent modulus E_{fa}, where

$$E_{fa} = \frac{\sigma}{\varepsilon[\sigma]} \tag{11}$$

where σ and the stress-dependent strain $\varepsilon[\sigma]$ are determined from an experimental stress-strain curve.

Double-Beam Determination of Film Properties: Provided that stress variations depend only on temperature, thermal cycling data from beams of two different materials coated with the same polymer can be used [with Eq. (10)] to simultaneously determine the polymer's elastic stiffness (E'_f) and in-plane thermal expansion coefficient (α_f) at any temperature. The accuracy of this calculation can be enhanced by using two beam materials of distinctly different physical properties. This double-beam approach (19) is similar in principle to the approach of Retajczyk and Sinha (51) who obtained the elastic stiffness and thermal expansion coefficient for refractory silicides and silicon nitrides using wafer substrates. With the help of a complementary technique for estimating either E_f or ν_f, it is possible to determine the respective values of the elastic modulus, Poisson's ratio, and thermal expansion coefficient for the polymer.

2.2.2 Stress Calculation for Multilayer Coatings

When stresses are present in a multilayer film coated on a rigid substrate, the substrate will bend into a section of a sphere with a radius of curvature R. Assuming elastic behaviors, thermal origin of all stresses, and radius R much greater than the overall sample thickness, Feng and Liu (52) have derived generalized formulas for calculating the radius of curvature and layer stresses caused by the thermal strains in a general N-layer structure (Fig. 2.5). By balancing the forces, F_i (positive when tensile), and moments, M_i, in the structure, they obtained the following general formulas

$$F_i = \frac{E_i t_i}{\sum_j E_j t_j} \left(\frac{1}{R} \sum_j E_j t_j \left[\frac{t_i - t_j}{2} + \sum_{k<i} t_k - \sum_{k<j} t_k \right] + \sum_j E_j t_j \Delta T \left[\sum_{k<j} (\alpha_{k+1} - \alpha_k) - \sum_{k<i} (\alpha_{k+1} - \alpha_k) \right] \right) \tag{12}$$

and

$$\frac{1}{R} = \frac{6 \sum_i E_i t_i \left(2 \sum_{j<i} t_j + t_i \right) \sum_k E_k t_k \; \Delta T \left[\sum_{p<i} (\alpha_{p+1} - \alpha_p) - \sum_{p<k} (\alpha_{p+1} - \alpha_p) \right]}{\sum_i E_i t_i \left[\sum_j E_j t_j^3 + 3 \left(2 \sum_{j<i} t_j + t_i \right) \sum_k E_k t_k \left(t_i - t_k + 2 \sum_{p<i} t_p - 2 \sum_{p<k} t_p \right) \right]} \tag{13}$$

with i, j, k, p = 1, 2, . . ., N. Knowing R, and F_i [from Eq. (12)], the stress in the ith layer, σ_i, at the position x_i from the central line of the ith layer, can be calculated using

$$\sigma_i[x_i] = \frac{F_i}{t_i} + \frac{E_i x_i}{R} \tag{14}$$

In the above, E is the elastic modulus, α the thermal expansion coefficient, t the thickness and the subscript, i, j, k, or l, denotes the layer the property is referring to. ΔT is defined as the difference between the observation temperature and a constant reference temperature at which the sample is flat and free of stress (27). The radius of curvature R is related to the deflection δ shown in Fig. 2.5 according to Eq. (5). Lengthy but straightforward formulas have been derived (53) for the case when a constant stress-free temperature does not exist for all the layers.

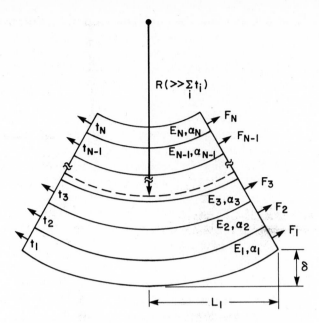

Figure   2.5:   Definition of parameters for a N-layer beam.

In order to use Eqs. (12)-(14) for situations involving biaxial and wide-beam effects, one must choose the appropriate expression for the elastic stiffness E'_i [$\equiv E_i/(1 - \nu_i)$] or "effective" elastic stiffness [e.g., $E_i/(1 - \nu_i^2)$]. It should be noted that Eq. (13) reduces to Eq. (4) assuming the applicability of the assumptions used to derive Eq. (4), constant physical properties, and $\sigma = \sigma_{thermal}$. When plastic deformation or stress relaxation occurs in a film, Chen et al. (22) suggest replacing E_i (or E'_i) with the apparent modulus E_{ia} calculated from the experimental stress-strain curve using

$$E_{ia} = \frac{\sigma_i}{\varepsilon_i[\sigma_i]} \tag{15}$$

as discussed in connection with Eq. (11).

2.2.3 Fluid Transport Models

To our knowledge, detailed fluid transport studies in polymers using the bending-beam technique have only been performed for single layer coatings. In these studies, models have been developed for the determination of the governing transport mechanism and associated transport parameters from the end deflection data. For studies of fluid transport in multilayer structures, approaches have also been suggested for estimating the stress distributions from end deflections under typically Fickian-type diffusion conditions. In this case, deduction of transport parameters from the measured deflection data is not, however, advisable due to the structural complexities involved.

Simple Polymer Coatings Exhibiting Negligible Swelling: Diffusion of a low molecular weight species into a simple polymer coating can change the stress level in the polymer, resulting in the end deflection of the beam upon which the polymer is coated. One mechanism for stress reduction is polymer relaxation in the presence of the penetrant. The end deflection data collected during fluid transport provide a detailed fingerprint of the overall diffusion process.

Berry and Pritchet (23) have pioneered the use of the bending-beam technique for studying Fickian transport, using the assumption of negligible penetrant-induced swelling. In the case of negligible thickness changes, they derived a general expression relating the curvature ($1/R$) of a beam to the penetrant concentration distribution (C)

$$\psi = \frac{\dfrac{1}{R} - \dfrac{1}{R_0}}{\dfrac{1}{R_\infty} - \dfrac{1}{R_0}} = \int_{-0.5}^{+0.5} \left(\frac{C}{C_\infty} \right) (1 + Ku)\, du \qquad (16)$$

where the subscripts, 0 and ∞, indicate initial and equilibrium properties, $K = 2a_1 (1 + a_1 K_1)/(1 + a_1)$, $a_1 = t_f/t_b$, and $K_1 = E_f(1 - \nu_b)/E_b(1 - \nu_f) = E'_f/E'_b$. According to Tong and Saenger (24), the concentration distribution which causes beam bending can be established by considering the transport process as the linear superposition of independent contributions from Fickian diffusion and first-order molecular relaxation (non-Fickian effect), that is,

$$\frac{C}{C_\infty} = \lambda F + (1 - \lambda)[1 - \exp(-kt')] \qquad (17)$$

where λ is the fraction of equilibrium uptake due to the Fickian process, F is a function of the constant diffusion coefficient, D,

$$F = 1 - \frac{4}{\pi} \sum_{n=0}^{\infty} \frac{(-1)^n}{2n + 1} \exp\left[\frac{-D(2n + 1)^2 \pi^2 t'}{4t_f^2} \right] \cos\left[\frac{(2n + 1)(u + 0.5)\pi}{2} \right] \quad (18)$$

t' is the time, and k is a relaxation rate constant. In arriving at Eqs. (16) and (17), the following assumptions have been made: (i) both polymer and beam behave elastically, (ii) D, E_f, and ν_f are independent of penetrant concentration, and (iii) the polymer's hygroscopic strain is linearly proportional to the penetrant concentration. Provided that the assumptions behind Eq. (4) are valid, the film stress change as a function of time can be estimated using Eq. (4) and the end deflection data. Based on Eqs. (16)-(18), one can determine the transport parameters (D, λ, and k) by varying their values until a good match in ψ versus $\sqrt{Dt'/t_f^2}$ (dimensionless time) data is obtained between theory and experiment.

Simple Polymer Coatings Exhibiting Substantial Swelling: Recently, Fu et al. (25) developed a general bending-beam model to study penetrant transport in both swelling and nonswelling polymer films. Both Fickian and non-Fickian transport behaviors can be studied with this model. The polymer film during fluid transport is considered to consist of a swollen (or fluid penetrated) layer of thickness λ and a glassy or unpenetrated layer of thickness δ' which are separated by a sharp boundary (Fig. 2.6). Under the assumptions of

- unidirectional fluid transport,
- plane cross sections remain plane,
- net stress over the beam surface is zero,
- the shear stress in the xy plane and the normal stress in the y direction are negligible (Fig. 2.6),
- film thickness change is much smaller than the lateral dimensions of the polymer coated beam,
- hygroscopic strain is linearly proportional to the penetrant concentration,

Fu et al. developed the formula below for analyzing the end deflection data collected during a fluid transport experiment

$$\psi[t'] = \frac{\frac{1}{R} - \frac{1}{R_0}}{\frac{1}{R_\infty} - \frac{1}{R_0}} = \left[\frac{A_\infty}{A} - \frac{9m_1 A_\infty B}{AV_1 t_b a_1 (1 + \theta)} \right] \frac{\check{C}}{C_\infty} + \left[\frac{24m_2 A_\infty B}{At_b a_1 (1 + \theta)} \right] \frac{\check{C}z}{C_\infty} \qquad (19)$$

In the above, the various terms are defined as

$$A = \frac{t_b}{2} \left(\frac{K_1 a_1^3 (1 + \theta)^3 + 1}{3[1 + a_1(1 + \theta)]} \left[\frac{1}{K_1 a_1(1 + \theta)} + 1 \right] + 1 + a_1(1 + \theta) \right)$$

$$B = \frac{a_1^2 (1 + \theta)^2 t_b}{6[1 + a_1(1 + \theta)]} [1 + K_1 a_1(1 + \theta)]$$

$$K_1 = \left(\frac{E'_g}{E'_q t_{exp}^3} \right) [\delta'^3 + K_2(t_{exp}^3 - \delta'^3)]$$

$$m_1 = \frac{8E'_s E^{(2)}}{8E^{(3)}E^{(1)} - (3E^{(2)})^2}$$

$$m_2 = \frac{8E'_s E^{(1)}}{8E^{(3)}E^{(1)} - (3E^{(2)})^2}$$

$$E^{(1)} = \frac{E'_g \delta' + E'_s \lambda}{t_{exp}}$$

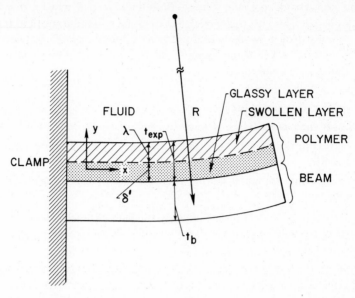

Figure 2.6: Definition of parameters for fluid transport in a polymer-coated beam.

$$E^{(2)} = \frac{E'_g\big[(\delta' - \lambda)^2 - (\delta' + \lambda)^2\big] + E'_s\big[(\delta' + \lambda)^2 - (\delta' - \lambda)^2\big]}{t_{exp}^2}$$

$$E^{(3)} = \frac{E'_g\big[(\delta' - \lambda)^3 + t_{exp}^3\big] + E'_s\big[t_{exp}^3 - (\delta' - \lambda)^3\big]}{t_{exp}^3}$$

$$\tilde{C} = \frac{1}{t_{exp}} \int_{(\delta'-\lambda)/2}^{(\delta'+\lambda)/2} C[z, t]\, dz$$

$$\tilde{C}z = \frac{1}{t_{exp}} \int_{(\delta'-\lambda)/2}^{(\delta'+\lambda)/2} zC[z, t]\, dz$$

A_∞ = value of A at equilibrium ($t \to \infty$),
E'_s = elastic modulus of the swollen layer,
E'_g = elastic stiffness of the glassy polymer,
E'_q = elastic stiffness of the beam,
K_2 = ratio of the elastic stiffness of the swollen layer to that of the glassy layer,
V_1 = partial specific volume of the penetrant,
t_{exp} = experimentally measured film thickness (using, for instance, laser interferometry; see Ref. 45),
$a_1 = t_{f0}/t_b$,

t_{f0} = initial dry film thickness,

$\theta = (t_{exp} - t_{f0})/t_{f0}$ = dimensionless swelling,

z = lab coordinate defined in Fig. 2.6.

Here, λ, δ', and t_{exp} are allowed to be time dependent with $t_{exp} = \delta' + \lambda$. With the substitutions of $\delta' = 0$, $t_{exp} = \lambda = t_f = t_{f0}$ = constant and $E'_g = E'_s$, it can be easily shown that Eq. (19) reduces to Eq. (16) for transport involving negligible polymer swelling. The change in stress caused by fluid transport can be calculated using

$$\Delta\sigma = \frac{E'_q t_b^3[K_1 a_1^3(1 + \theta)^3 + 1]\left(\dfrac{1}{R} - \dfrac{1}{R_0}\right) - 4V_1 C_\infty K_1 a_1^2 E'_q t_b^2(1 + \theta)^2 \tilde{C}z}{6a_1 t_b^2(1 + \theta)[a_1(1 + \theta) + 1]} \qquad (20)$$

When a_1 and $K_1 \ll 1$, that is, when the film is soft and is much thinner than the beam, Eq. (20) reduces to Eq. (4) with $t_f = t_{exp}$. In order to use Eq. (19) or (20) for any bending-beam study, one needs to know the mechanical properties of all the layers and have a transport mechanism for predicting the film thicknesses and concentration distributions which are needed for estimating $\tilde{C}$ and $\tilde{C}z$.

Multilayer Coatings: When the ith layer picks up or loses a fluid (e.g., moisture), it can change its weight and dimensions. This can produce a hygroscopic strain, ε_i^H, that varies linearly with the fluid concentration C_i, that is,

$$\varepsilon_i^H = \beta_i(C_i - C_0) = \beta_i \Delta C_i \qquad (21)$$

where β_i is a swelling parameter and C_0 the initial fluid concentration. Equation (21) has been assumed in deriving Eqs. (16) and (19). Comparing Eq. (20) with the linear thermal strain expression

$$\varepsilon_i^T = \alpha_i \Delta T_i \qquad (22)$$

it is clear that the following correlations exist

$$\beta_i \sim \alpha_i$$

$$\Delta C_i \sim \Delta T_i$$

Some authors have utilized these correlations to analyze their end deflection data with thermal strain equations under isothermal conditions (32).

2.3 APPLICATIONS

In this section, various bending-beam applications will be discussed in the order of the appearance of the relevant equation or equations in the Models section. We shall begin with applications involving simple polymer coatings.

2.3.1 Heat Treatments of Simple Polymer Coatings

For thin polymer films, the bending-beam technique has been used to study stress development during heat treatment (6,18-21,26). In these studies, the stresses were calculated from the end deflection data and attempts were made to relate stress development to the physical and/or chemical processes occurring in the polymer.

Stress Development During Thermal Cycling: Shown in Fig. 2.7 is the stress/temperature relationship for the poly(methyl methacrylate), that is, PMMA/chlorobenzene system during thermal cycling. Heating in this case removes the casting solvent chlorobenzene without involving any chemical reaction. The quantity $\sigma \times t_f$ instead of σ is plotted as a function of temperature up to 150°C, in the first heating period (dashed line in Fig. 2.7) because of film thickness changes due to solvent removal. Beyond this point, the PMMA film is practically free of solvent and its thickness remains fixed at 3×10^{-4} cm (= 3 μm). This thickness was used to calculate the stresses following the first heating period (solid lines in Fig. 2.7).

The stress in the as-spun PMMA/chlorobenzene film is tensile (i.e., positive) at room temperature. This is probably the intrinsic stress created as a result of the inability of the polymer to relax the stress developed during spin coating. As the temperature increases, the total stress σ decreases as a result of changing thermal and intrinsic stresses. It is completely relaxed beyond 100°C. This temperature is consistent with the glass transition temperature of PMMA determined by differential scanning calorimetry. Biernath and Soong (54) have attempted to predict the stress development in the solvent cast films undergoing thermal cycling using a linear viscoelastic (Maxwell) model which accounts for the solvent evaporation and thermal expansion effects.

In Fig. 2.7, it can be seen that cooling of the PMMA film below 100°C results in a nearly linear increase of tensile stress with decreasing temperature. The stress level reaches 2.3×10^8 dynes/cm² at room temperature. Repeated thermal cycling does not markedly alter the stress/temperature relationship corresponding to the first cooling period, indicating that the stress changes are mainly thermal in origin. From Eq. (10), the linear stress/temperature relationship, and constant values of E_f and ν_f from the literature, the average thermal expansion coefficient α_f of PMMA was calculated to be $7 \times 10^{-5}/$°C. This value compares well with published data. In Fig. 2.7 and Figs. 2.8-2.10 below, the linear stress/temperature relationship exhibits minimal hysteresis, indicating that the bending changes probably occur in the elastic regime.

Detection of Stress Relaxation Processes: Using the bending-beam technique under dynamic heating or cooling conditions, the polymer's glass transition temperature can be identified as the temperature at which the stress level becomes negligible, as has been illustrated for PMMA above. For polyimide materials (22), this has also been found to be the case.

When stress relaxation occurs as a result of other causes such as cracking and/or delamination, the inception of such a process can also be detected using the bending-beam technique. Shown in Fig. 2.8 is the stress/temperature relationship for the cooling of a solventless amide/imide polymer in air from 200 ° C. The existence of the linear stress/temperature relationship between 200 and 50°C is characteristic of thermal stress changes. Deviations from linearity at temperatures below 50°C were

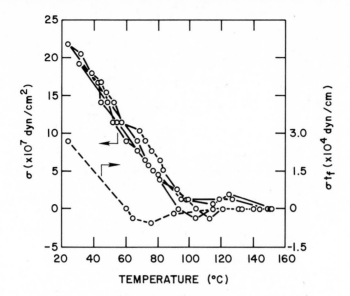

Figure   2.7:   Stress development in a PMMA sample coated on a quartz beam during thermal cycling (6).

found to be accompanied by cracking of the polymer which occurred because the tensile stress exceeded the strength of the film.   In a bending-beam study of metals (i.e., Cr/Cu/Cr; Ref. 7), high stresses were also found to cause sample cracking or delamination manifested by an abrupt change in the stress level.

Study of Curing Mechanism: Using the bending-beam technique and complementary techniques such as FTIR and tensile tests, Han et al. (20,21) have studied the curing reaction of polyamic acid in the presence of n-methyl-2-pyrrolidone (NMP) as it is converted into polyimide under different curing schedules.   In this study, the stress/temperature relationship was attributed to the processes of water sorption, solvent evaporation, material softening, imidization reaction and thermal expansion mismatch.

Shown in Fig. 2.9 is the stress development (plotted as $\sigma \times t_f$) for a partially dried polyamic acid film coated on a quartz beam during curing in air at a heating rate of 10°C/min and a cooling rate of 5°C/min.   Upon heating, the stress first increases with temperature up to 50°C due to rapid evaporation of moisture.   Between 50 and 100°C, the stress decreases with increasing temperature, a behavior dominated by the thermal expansion behavior of the polyamic acid/NMP complex.   This proceeds with no significant change in film thickness and extent of cure up to 100°C.   Marked film shrinkage caused by NMP evaporation was observed as the temperature increased from 100 to 175°C. The decrease in stress after correction for these thickness changes was attributed to the proximity of the softening temperature of the partially imidized polyamic acid.   Near 175°C, the accelerated imidization process rapidly shifts the softening temperature of the polymer and as a result the stress begins to increase until 225°C is reached.   Within this temperature range, the fast

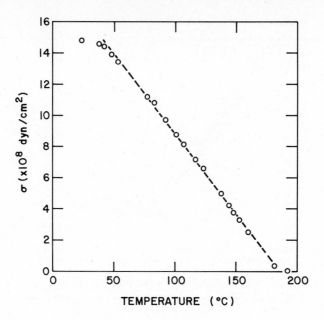

Figure 2.8: Stress development in an amide/imide polymer coated on a quartz beam during cooling (6).

imidization process leads to rapid vitrification and the concomitant changes in the mechanical properties of the film. This causes the stress to decrease with temperature beyond 225°C. Above 225°C, the curing reaction, which tends to increase the stress, and the thermal expansion mismatch effect, which tends to decrease the stress, compete, giving rise to small changes in stress between 225 and 330°C. At higher temperatures, the thermal expansion mismatch effect dominates and consequently the stress decreases with increasing temperature. Upon cooling, this effect prevails resulting in an increase in stress with decreasing temperature until moisture pickup lowers the stress at about 35°C.

Han et al. studied the effect of the heating rate on stress development during polyimide curing. Keeping the cooling rate fixed at 5°C/min, they found that the polyimide stress at 50°C increases by a factor of 1.6 as the heating rate increases from 2 to 20 °C/min. This can have significant practical implications in the optimization of the polyimide curing schedule for more reliable polyimide-containing structures.

Property Determination Using Double Beams: Assuming that stress variations with temperature are thermal in origin, we have shown previously that one can use two different beam materials coated with the same polymer in a bending-beam experiment to determine the elastic stiffness and thermal expansion coefficient of the polymer.

Using fused quartz and annealed stainless steel beams, Han et al. (19) studied the stress development in polyimide and PMMA films during cycling between 50 and 400°C in air. The stress data for the polyimide are shown in Fig. 2.10. Quartz and

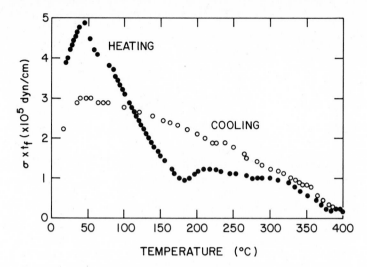

Figure 2.9: Stress development in a polyamic acid/polyimide film coated on a quartz beam during heating (at a rate of 10°C/min) and subsequent cooling (at 5 °C/min) in air (21).

stainless steel were chosen for their distinctly different thermal expansion coefficients. From the stress/temperature relationships, they calculated the elastic stiffness and thermal expansion coefficient for both polymers. For PMMA, the thermal expansion coefficient (7.8×10^{-5} /°C) and the Poisson's ratio (0.33) calculated from the elastic stiffness and the reported value of the modulus were found to agree well with the literature.

Typical Polymer Stress Levels: The room temperature stress levels in fully dried or cured polymer thin films measured using bending-beam techniques are typically of the order of 10^8 dynes/cm². Polymers studied include PMMA (6), an epoxy resin (6), an amide/imide polymer (6), and different polyimide materials (18,20-22). The polymer stress levels are low compared to the $\sim 10^8$ to 10^{10} dynes/cm² stress levels for metals (29). This is due to the much lower moduli of polymers.

2.3.2 Thermomechanical Behaviors of Multilayer Beams

The thermomechanical behaviors of multilayer structures have been investigated by various workers. From these studies, which will be reviewed herein, they were able to demonstrate the stress relief effect of a polymer placed between a metal thin film and a rigid substrate (7,22,26) and the effects of processing history on film properties and structural bending (27,53).

Metal Containing Beams: Chen et al. (7,22,26) have used the bending-beam technique to study the following thin film coatings on either quartz or silicon beams:

- polyimide,
- Cu,
- Cr,
- Cu/polyimide/beam,

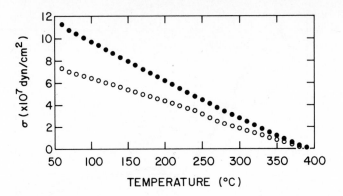

Figure 2.10: Stress/temperature relationships in a polyimide coated on quartz and stainless steel beams (19).

- Cr/polyimide/beam,
- Cr/Cu/Cr/beam,
- Cr/Cu/Cr/polyimide/beam.

The results indicated that upon thermal cycling, a blanket metal film without the cushioning polyimide underlayer tends to develop a higher stress in comparison to one containing the polyimide underlayer. The extent of stress relaxation depends upon the type of metal used and the polyimide thickness. As an example, Fig. 2.11 shows the deflection (an indication of stress) during thermal cycling of a quartz beam (0.0127 cm thick) coated on one side with a (2.9×10^{-4} cm) thick layer of polyimide before and after the deposition (on the polyimide coated side) of a (5.3×10^{-5} cm thick) Cu film. The shapes of the deflection versus temperature curves for the control sample (i.e., Cu/quartz without the polyimide) were similar to those shown for the Cu/polyimide/quartz sample. However, the room temperature beam deflection for the Cu/polyimide/quartz sample was about 25% smaller than that of the control sample. Transmission electron microscopic examinations indicated that the stress is partially released through the deformation of the polyimide near the metal/polyimide interface. To account for the stress relaxation effects in multilayer samples containing metal and polyimide films, the stress levels and distributions were calculated using apparent moduli for layers exhibiting stress relaxation and plastic deformation.

Trilayer Polymer Tapes: Berry and Pritchet (27) have studied the thermomechanical behavior of a trilayer magnetic tape which consists of a 2.3×10^{-3} cm thick poly(ethylene terephthalate) support, a 4×10^{-4} cm thick front coat of magnetic CrO_2 particles in a polymeric binder, and a back coat of nominally similar composition and thickness. From tensile tests, they found that the frontcoat which had received a calendering treatment had a modulus five times larger than that of the backcoat which did not receive the calendering treatment. This finding indicates the importance of the process history in determining material properties. The elastic asymmetry existing in the trilayer tape rather than the small

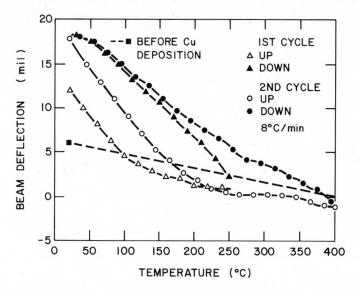

Figure   2.11:   Deflections of Cu/polyimide and polyimide coated beams as a function of time during thermal cycling (26).

difference in the thermal expansion coefficients of the front and back coatings was found to be largely responsible for the thermal curling behavior of the tape.

2.3.3 Fluid Transport Studies

The fluid transport studies to be reviewed here are Fickian-like and non-Fickian moisture transport in PMMA (24), and non-Fickian transport of NMP in polyimide films (25). While no significant swelling of the PMMA is involved, the polyimide films can swell as much as 20% by volume.

Studies Involving Negligible Polymer Swelling: Using the bending-beam equations, Eqs. (16)-(18), with $\lambda = 1$, Berry and Pritchet (23) have studied Fickian-type vapor transport in epoxy samples exposed to moisture. By allowing for a non-Fickian molecular relaxation effect, their model has been extended by Tong and Saenger (24) to the study of both Fickian-like and non-Fickian moisture diffusion in PMMA films upon exposure to water. To examine the effects of film thickness and sample thermal history on the transport behavior, different PMMA thicknesses (7 to 12 $\times$ 10^{-4} cm) and cooling rates (following a 120°C PMMA drying step) were used in the experiments. For each experiment, the curvature of a polymer coated beam exposed to liquid water was monitored as a function of time by a low power laser pointer and the end deflection data were analyzed using Eqs. (16)-(18).

For all PMMA samples, sorption into a penetrant-free PMMA is initially dominated by a rapid Fickian-like diffusion step, characterized by a diffusion coefficient of 2 to 4 $\times$ 10^{-9} cm^2/s. Figure 2.12 shows representative ψ versus $\sqrt{Dt'/t_f^2}$ (square root of dimensionless time) data for a slowly cooled ($\sim$ 6°C/h), 1.13 $\times$ 10^{-3} cm thick PMMA film after immersion in water at t=0. The early time data are well fit

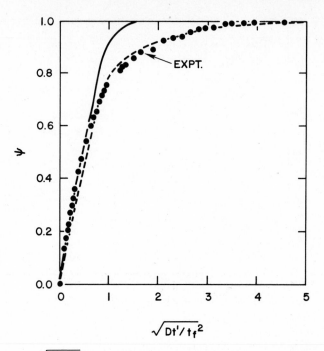

Figure 2.12: ψ - $\sqrt{Dt'/t_f^2}$ relationships from experiment (dots) and models (Fickian: solid line; non-Fickian: dashed line) for a slowly cooled PMMA film immersed in water at room temperature (24).

by a Fickian model (solid line) with a diffusion constant of $\sim 2 \times 10^{-9}$ cm^2/s. At later times, significant relaxation contributions lead to non-Fickian diffusion behavior, an effect that is more pronounced as the film thickness or sample cooling rate decreases. The better match between the non-Fickian theory (dashed line) and experiment suggests that moisture transport in PMMA is actually non-Fickian, with a diffusion coefficient of $\sim 2 \times 10^{-9}$ cm^2/s, a relaxation rate constant, k, of 3 $\times 10^{-4}$ s^{-1}, and an equilibrium fraction of uptake due to Fickian diffusion (λ) of 0.82. A mechanism involving free volume and molecular relaxations was proposed to account for the observed transport behaviors. In addition, sorption of water was found to reduce the film stress (initially tensile at $\sim 10^8$ dynes/cm^2) at an initial rate that increases with sample cooling rate.

Studies Involving Significant Polymer Swelling: The bending-beam technique has been applied by Fu et al. (25) to study the swelling of polyimide films in NMP at 20 and 80°C. From their time-dependent ψ data and previously reported thickness data measured by interferometry (55), they determined that NMP transport is best described as intermediate between Fickian-like and Case II at both temperatures. This is consistent with a previous laser interferometric study (55). In Fig. 2.13 their 20°C ψ data and fits are plotted versus $D't'/t_{f0}$, where t_{f0} is the initial (dry) film thickness and D' is the effective diffusion coefficient characteristic of intermediate transport (55).

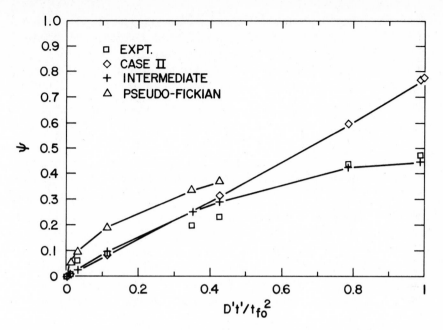

Figure 2.13: ψ - $D't'/t_{fo}^2$ relationships from experiment and Eq. (19) based on three transport mechanisms for a polyimide film immersed in NMP at 20°C (25).

2.4 SUMMARY

The utility of a bending-beam technique has been explored for monitoring the stress and related properties of polymer thin films (i) during curing, (ii) during thermal cycling, and (iii) during exposure to gaseous or liquid penetrant fluids. The advantages of the bending-beam technique are as follows: (1) it is in situ and simple, (2) it uses samples naturally compatible with actual structures, and (3) it possesses sensitivities adequate for studying polymer thin film properties not accessible by other techniques. The information obtained via the bending-beam technique can have a significant impact on the choice of new polymeric materials and processing strategies.

2.5 ACKNOWLEDGMENTS

The authors would like to thank Drs. P. Ho, C. J. Durning, B. S. Berry, C. Feger, and C. C. Gryte for many useful discussions.

2.6 REFERENCES

1.   J. R. Bupp, L. N. Challis, R. E. Ruane, and J. P. Wiley, "High-density Board Fabrication Techniques," *IBM J. Res. Develop.*, 26, 306 (1982).

2.   R. J. Jensen, "Polyimides as Interlayer Dielectrics for High-performance Interconnections of Integrated Circuits," in *Polym. High Tech.*, M. J. Bowden and S. R. Turner, Eds., ACS Symposium Series 346, 1987.

3.   C. W. Ho, "High-performance Computer Packaging and the Thin Film Multi-chip Module," in N. G. Einspruch, Ed., **VLSI Electronics: Microstructure Science**, Vol. 5, Ch. 3, Academic Press, New York, 1982.

4.   J. Hagge, "Ultra-reliable Packaging for Silicon- on-Silicon WSI," *Proc. 38th Electron. Comp. Conf.*, Los Angeles, California (1988).

5.   A. C. M. Yang and H. R. Brown, "Solvent-Induced Local Deformation Zones in Polyimide Films Adhered to a Rigid Substrate," *IBM Res. Rep.*, RJ 5555 (1987).

6.   H. M. Tong, C. K. Hu, C. Feger, and P. Ho, "Stress Development in Supported Polymer Films During Thermal Cycling," *Polym. Eng. Sci.*, 26, 1213 (1986).

7.   C. H. Yang, D. Y. Shih, S. T. Chen, H. Yeh, and P. S. Ho, "The Study on the Adhesion Strength and Thermal Stress of Cr/Cu/Cr/polyimide TFM Packaging Materials," private communication.

8.   S. P. Kowalczyk, Y.-H. Kim, G. F. Walker, and J. Kim, "Polyimide on Copper: The Role of Solvent in the Formation of Copper Precipitates," *IBM Res. Rep.*, RC 12894 (1987); Y.-H. Kim, G. F. Walker, J. Kim, and J. Park, "Adhesion and Interface Studies Between Copper and Polyimide," *IBM Res. Rep.*, RC 12602 (1987).

9.   K. L. Saenger, H. M. Tong, and R. D. Haynes, "Improved Polyimide/Polyimide Adhesion Via Swelling Agent Enhanced Interdiffusion," *J. Polym. Sci., Polym. Lett. Ed.*, in press.

10.   R. W. Hoffman, "Mechanical Properties of Non-metallic Thin Films," in **Physics of Nonmetallic Thin Films**, C. H. S. Dupuy and A. Cachard, Eds., Plenum, New York (1976).

11.   R. J. Roark and W. C. Young, **Formulas for Stress and Strain**, 5th Ed., McGraw-Hill, New York (1982).

12.   J. Williams, **Stress Analysis of Polymers**, 2nd Ed., Halsted Press, New York (1980).

13.   J. Crank and G. S. Park, Eds., **Diffusion in Polymers**, Academic Press, New York (1968).

14.   J. Crank, **The Mathematics of Diffusion**, Clarendon Press, Oxford (1968).

15.   G. E. Zaikov, A. L. Iordanskii, and V. S. Markin, **Diffusion of Electrolytes in Polymers**, Utrecht, The Netherlands (1988).

16.   A. R. Berens and H. B. Hopfenberg, "Diffusion and Relaxation in Glassy Polymer Powders - 2. Separation of Diffusion and Relaxation Parameters," *Polym.*, 19, 489 (1978); "Induction and Measurement of Glassy-state Relaxations by Vapor Sorption Techniques," *J. Polym. Sci., Polym. Phys. Ed.*, 17, 1757 (1979).

17.   G. Astarita and G. C. Sarti, "Class of Mathematical Models for Sorption of Swelling Solvents in Glassy Polymers," *Polym. Eng. Sci.*, 18, 388 (1978).

18.   J. H. Jou, J. Hwang and D. C. Hofer, "In Situ Measurement of Temperature Dependence of Internal Residual Stress in Polyimide Films Coated on Silicon Substrate," *IBM Res. Rep.*, RJ 5984 (1987).

19. B. Han, H. M. Tong, K. L. Saenger, and C. C. Gryte, "Mechanical Property Determination for Supported Polymer Films Using Double Bending-Beams," *Mat. Res. Soc. Symp. Proc.*, 76, 123 (1987).

20. B. J. Han, "A Study of Cure and Cure-Related Topics of Thin Polyimide Films," Ph.D. Thesis, Columbia University, 1988.

21. B. Han, C. Gryte, H. Tong, and C. Feger, "Stress Buildup in Thin Polyimide Films During Curing," *Proc. 46th Annual Tech. Conf.*, Society of Plastics Engineers, pp. 994 - 996 (1988).

22. S. T. Chen, C. H. Yang, F. Faupel, and P. Ho, "Stress Relaxation During Thermal Cycling in Metal/polyimide Layered Films," *J. Appl. Phys.*, 64, 6690 (1988).

23. B. S. Berry and W. C. Pritchet, "Bending-Cantilever Method for the Study of Moisture in Polymers," *IBM J. Res. Develop.*, 28, 662 (1984).

24. H. M. Tong and K. L. Saenger, "Bending-Beam Study of Water Transport by Thin Poly(methyl methacrylate) Films," *J. Appl. Polym. Sci.*, in press.

25. T. Z. Fu, C. J. Durning, and H. M. Tong, "A Simple Model for Swelling Induced Stresses in a Supported Polymer Thin Film," submitted to *J. Appl. Polym Sci.*

26. S. T. Chen, C. H. Yang, H. M. Tong, and P. Ho, "Temperature Dependence of Stress Relaxation at the Copper/polyimide Interface," *Mat. Res. Soc. Symp. Proc.*, 79, 351 (1987).

27. B. S. Berry and W. C. Pritchet, "Elastic and Viscoelastic Behavior of a Magnetic Recording Tape," *IBM J. Res. Develop.*, 32, 682 (1988).

28. H. Bouadi and C. T. Sun, "Hygrothermal Effects on the Stress Field of Laminated Composites," *J. Reinforced Plastics Composites*, 8, 40 (1989).

29. D. S. Campbell, "Mechanical Properties of Thin Films," in **Handbook of Thin Film Technology**, L. I. Maissel and R. Glang, Eds., McGraw-Hill, New York (1969).

30. J. D. Wilcox and D. S. Campbell, "A Sensitive Bending-Beam Apparatus for Measuring the Stress in Evaporated Thin Films," *Thin Solid Films*, 3, 3 (1969).

31. T. P. Weihs, S. Hong, J. C. Bravman, and W. D. Nix, "Mechanical Deflection of Cantilever Microbeams: A New Technique for Testing the Mechanical Properties of Thin Films," *J. Mater. Res.*, 3, 931 (1988).

32. R. A. Susko and P. A. Engel, "Novel Methods for In Situ Measurement of Coefficients of Expansion for Polyimide Films," *Extended Abstracts*, vol. 87-2, 656, Fall Meeting, The Electrochemical Society, October 18-23 (1987).

33. H. M. Tong and K. L. Saenger, unpublished data.

34. M. Mehregany, R. T. Howe, and S. D. Senturia, "Novel Microstructures for the *In Situ* Measurement of Mechanical Properties of Thin Films," *J. Appl. Phys.*, 62, 3579 (1987).

35. R. E. Rottmayer and R. W. Hoffman, "Boundary Conditions for Stress Measurement Using a Cantilevered Beam," *J. Vac. Sci. Tech.*, 7, 461 (1970).

36. P. M. Alexander, *AEC Tech. Rept. 85*, Case Western Reserve University, Cleveland, Ohio (1974).

37. E. Klokholm, "A Sensitive Device for Measuring Residual Stresses in Thin Films," *IBM Res. Rep.*, RC 1352 (1965).

38. H. S. Story and R. W. Hoffman, *Proc. Roy. Phys. Soc.*, B70, 950 (1957).

39. A. E. Ennos, "Stresses Developed in Optical Film Coatings," *Appl. Optics*, 5, 51 (1966).

40. H. Blackburn and D. S. Campbell, "The Development of Stress and Surface Temperature During Deposition of Lithium Fluoride Films," *Phil. Mag.*, 8, 823 (1963).

41. R. V. Jones, *Bull. Inst. Phys.*, 18, 325 (1967).

42. R. Carpenter and D. S. Campbell, "Stress in Alkali Halide Films," *J. Mat. Sci.*, 2, 173 (1967).

43. G. Binning, C. F. Quate, and C. Gerber, "Atomic Force Microscope," *Phys. Rev. Lett.*, 56, 930 (1986).

44. D. H. Reneker, "Prospects for Examination of Polymer Molecules With the Scanning Tunneling Microscopy and the Atomic Force Microscope," this volume.

45. K. L. Saenger and H. M. Tong, "Laser Interferometry: A Measurement Tool for Thin Film Polymer Properties and Processing Characteristics," this volume.

46. J. L. Dais and F. L. Howland, "Fatigue Failure of Encapsulated Gold-Beam Lead and TAB Devices," *IEEE Trans. Components, Hybrids, Manuf. Technol.*, CHMT-1, 158 (1978).

47. A. Brenner and S. Senderoff, "Calculation of Stress in Electrodeposits from the Curvature of a Plated Strip," *J. Res. NBS.*, 42, 105 (1949).

48. S. P. Timoshenko, "Analysis of Bi-metal Thermostats," in *Collected Papers*, McGraw-Hill, New York (1953).

49. D. G. Ashwell and E. D. Greenwood, *Engineering*, 170, 51 (1970).

50. D. G. Bellow, G. Ford, and J. S. Kennedy, *Exp. Mech.*, 227 (1965).

51. T. F. Retajczyk and A. K. Sinha, "Elastic Stiffness and Thermal Expansion Coefficients of Various Refractory Silicides and Silicon Nitride Films," *Thin Solid Films*, 70, 241 (1980).

52. Z. Feng and H. Liu, "Generalized Formula for Curvature Radius and Layer Stresses Caused by Thermal Strain in Semiconductor Multilayer Structures," *J. Appl. Phys.*, 54, 83 (1983).

53. J. H. Jou, "General Formula for Internal Stress and Curvature Radius in Multilayer Structures - Consideration of Processing Condition," *IBM Res. Rep.*, RJ 6058 (1988).

54. R. Biernath and D. Soong, "Stress Development in Thin Films of Polymeric Interlayer Dielectrics and Packaging Materials," *Extended Abstracts*, vol. 87-2, 658, Fall Meeting, The Electrochemical Society, October 18-23 (1987).

55. H. M. Tong, K. L. Saenger, and C. J. Durning, "A Study of Solvent Diffusion in Thin Polyimide Films Using Laser Interferometry," *J. Polym. Sci., Polym. Phys. Ed.*, 27, 689 (1989).

3
POLYMER FILM STRESS MEASUREMENT
BY X-RAY DIFFRACTION

L. T. Nguyen

Philips Research Laboratories
Signetics Corporation
Sunnyvale, California

Residual stresses generated in polymer films on silicon substrates are of special interest in microelectronic packaging. For instance, during the curing and postcuring stages, polymer films form three-dimensional networks, and in the process shrink to varying extent due to reduced molecular conformations. The induced stresses can reach sufficient magnitude to cause drifts of electrical parameters, move metal lines, impart wire bond fatigue, or even crack the silicon die, as reported in Chapter 8, "Surface Sensors for Moisture and Stress Studies." Thus, an understanding of the origins and nature of these stresses is required for the optimization of the whole film package.

Various techniques have been introduced to characterize intrinsic stresses in polymers. A concise but by no means complete list may include, for instance, photoelasticity with birefringent materials or coatings, laser interferometry, bilayer bending beams, ultrasound, strain gages embedded in bulk parts, or semiconductor surface transducers. X-ray diffraction is a recent complement to this list, although it has been used for some time to characterize residual stresses in metals.

This chapter deals with methods of X-ray diffraction analysis adapted to the study of polymer film stresses. General techniques of X-ray diffraction are mentioned first, followed by a discussion of variants of the basic method to analyze passivation and encapsulation films. Stresses are derived from the strain imparted to a crystalline lattice from the shrinking polymer. To observe bulk stresses, the crystalline material is usually embedded into the polymer matrix undergoing cure or thermal treatment. An example of stresses in copper filler particles imparted by the shrinking polymeric matrix is given. On the other hand, surface stresses from a polymer film are monitored from the strain within the crystalline backing substrate. Examples of silicone and polyimide passivation films are also described, together with thick films of heavily filled encapsulants. Finally, some models of film stresses are reviewed within the context of stresses measured by X-ray diffraction.

3.1 BACKGROUND INFORMATION

Theoretical aspects of X-ray analysis are thoroughly covered in the literature, and only pertinent points relevant to the technique are briefly reviewed in this section.

X-rays reside within the region 0.1 to 4.5 Å of the electromagnetic spectrum. Under the proper conditions, a given substance emits X-rays which are discretely distributed in groups. User convention assigns letters to these groupings, with the most intense being labeled as K, followed by L, M, N, O, and P. The K group occurs when sufficient incident energy, in the form of photons, strikes the K shell of an element and ejects one of its electrons. The resulting vacancy must be filled by an electron from the next shell, the L shell, or one further out. The transition from a high potential energy to a lower energy level appears as emitted radiation. Since the energies within the shells of each atom are well defined, each electronic switch yields a monochromatic emission line characteristic of the element. Thus, for instance, when an L shell electron fills a vacancy in the K shell, the resulting emission lines are called $K\alpha_1$ and $K\alpha_2$. These lines are close doublets since the transition can occur from two possible configurations differing only slightly in energy. Similarly, $K\beta_1$ and $K\beta_2$ lines describe the drop of an M electron into the K shell. $K\alpha$ is always more intense and occurs at a longer wavelength than $K\beta$ since the energy differential between the K and L shells is less than that between the K and M shells. Although K emission lines have discrete wavelengths, they vary continuously in a stepwise manner from one atomic number to the next higher one. Thus, lithium exhibits $K\alpha$ of 240 Å while that for fermium resides at 0.104 Å. $K\alpha$ lines of elements between these extremes have decreasing wavelengths from lithium to fermium. By developing suitable detection methods, materials of different atomic numbers can be characterized.

To describe the crystal structure of a material, on the other hand, a diffraction technique is needed. In this case, a monochromatic X-ray beam is necessary for meaningful data. The necessary condition for diffraction is described by the well-known Bragg law

$$n\lambda = 2d \sin\theta \qquad (1)$$

where n, λ, d, and θ represent the order of the spectrum, the wavelength of the X-rays, the lattice spacing of the crystal, and the angle of incidence, respectively. For stress work, usually, n = 1. λ is determined by the target material used in the diffraction tube. Typical target materials include molybdenum, copper, nickel, cobalt, and iron. The corresponding $K\alpha$ lines are 0.71, 1.54, 1.65, 1.79, and 1.93 Å, respectively. Although each material has its own specific application, general usage is based on a copper target tube. X-ray diffraction offers useful information on the structure of a sample. Amorphous structures with poorly defined characteristics yield diffuse patterns, compared to the distinct intensity peaks obtained with crystalline materials.

The typical basic setup used for diffraction experiments is illustrated in Fig. 3.1, and involves a high voltage source, an electronic radiation detector, a goniometer, and a curved crystal monochromator. The latter, made of graphite, nickel, or LiF, eliminates excess background radiation and undesirable $K\beta$ emission. The crystal functions as a focusing spectrometer with the receiving slit acting as the X-ray source. Such a system requires that the source, crystal, and detector reside within a circular trajectory such that

$$D = R_C \sin \theta \qquad (2)$$

with D, R_C, and θ being the distance between the X-ray source and the crystal, the radius of the curved crystal, and the Bragg angle for a particular wavelength λ, respectively.   The collimated primary X-ray beam strikes the sample and the diffracted lines collected by the crystal are focused on the solid state scintillation detector. Scattered intensity measurements are usually carried out by pulse height analysis and counting techniques. By rotating the fixture containing the receiving slit and crystal at the same rate as the detector, the goniometer allows a recording of the whole spectrum of emission lines. Naturally, the process of data collection can be automated so that the diffraction pattern of angular position and relative intensities can be obtained within 30 to 60 min. In the particular arrangement depicted in Fig. 3.1, strains perpendicular to the sample surface can be detected.

3.2 BASIC TECHNIQUES

3.2.1 General Methods

The principle underlying the use of X-ray diffraction for residual stress measurement is the fact that when a crystalline material is placed under stress, the lattice shows a corresponding elastic strain. The stress can be either applied externally or residual from some previous processing operation.   As long as the yield strength is not exceeded, such stress is taken up by the interatomic strain. X-ray diffraction measures this interatomic spacing, which is indicative of the macrostrain experienced by the sample. A knowledge of the elastic constants would provide an estimation of the stress exerted, with the usual assumption of proportionality between stress and strain.

With X-ray diffraction, the basic equation governing stress/strain relations is given by (1)

$$\varepsilon_{\phi\psi} = \left(\frac{1 + \nu}{E} \right) \sigma_\phi \, \sin^2\psi - \frac{\nu}{E} (\sigma_1 + \sigma_2) \qquad (3)$$

where σ_ϕ is the stress within the plane of the sample at an angle ϕ with a principal stress direction, and σ_1 and σ_2 are the principal stresses. ψ is the angle between the normal to the surface and the normal to the crystallographic planes which diffract an X-ray peak. $\varepsilon_{\phi\psi}$ is the strain in the direction defined by the two angles ϕ and ψ , while ν is the Poisson's ratio and E is the modulus of the material. In terms of interatomic spacing of crystal planes parallel to the sample surface, d_1, Eq. (3) can be rewritten as

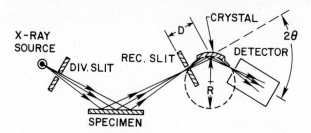

Figure 3.1: Schematic of a typical X-ray diffraction setup with high voltage source, radiation detector, and goniometer. The receiving slit, curved crystal, and detector rotate at the same rate. The typical material for the receiving slit is graphite, nickel, or LiF.

$$\sigma_\phi = \left(\frac{d_{\phi\psi} - d_1}{d_1} \right)\left(\frac{E}{1 + \nu} \right)\left(\frac{1}{\sin^2\psi} \right) \qquad (4)$$

with $d_{\phi\psi}$ as the interatomic spacing of crystal planes whose normal is defined by the angles ϕ and ψ. The difference between $d_{\phi\psi}$ and d_1 is rather small, so that from Bragg's law, $\Delta d/d = -\cot\theta\,(\Delta 2\theta/2)$ and Eq. (4) can be rewritten as

$$\sigma_\phi = (2\theta_1 - 2\theta_2)\left(\frac{\cot\theta_1}{2} \right)\left(\frac{E}{1 + \nu} \right)\left(\frac{1}{\sin^2\psi} \right)\left(\frac{\pi}{180} \right) \qquad (5)$$

where θ_1 and θ_2 are the Bragg angles diffracting at ψ_1 and ψ_2, respectively. This is depicted in Fig. 3.2 which illustrates the diffraction technique for a single exposure. Equation (5) indicates that as θ_1 increases, a given σ_ϕ would produce a larger difference $(2\theta_1 - 2\theta_2)$. Thus, X-ray peaks in the far back reflection range and θ values near 90° are preferred because of the greater sensitivity to stress. In Fig. 3.2, β is the angle between the normal to the surface and the incident beam. η_1 and η_2 are the angles between the incident beam and the diffracting plane normals (1 and 2). S_1 and S_2 are measured parameters depicting the distances between a known reference point and the incident and diffracted beams. Finally, ΔR_0 is the measured error in the specimen-to-detector radius, R_0.

An X-ray beam incident on a sample is diffracted in a cone of reflections as shown in Fig. 3.3. This is the underlying principle behind the *single exposure* technique. An unstressed specimen will project a circle on a plane intersecting perpendicularly to the cone axis. However, under stress, the circular projection becomes ellipsoidal.

Three basic techniques for determining residual stresses exist, namely, the *single exposure method* (SE), the *double exposure method* (DE), and the *multiple exposure* or $\sin^2\psi$ method (2). The angle referred to is that between the incident X-ray beam and the normal to the sample surface. All methods assume plane stress conditions within the shallow depth of penetration of the beam, typically around 0.01 mm.

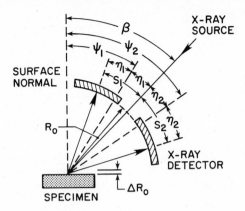

Figure 3.2: Schematic of single exposure diffraction method. Shown are the angles between the normal to the diffracting planes and the sample surface normal, ψ_1 and ψ_2. β is the angle between the incident beam and the surface normal. R_0 is the distance from sample to detector and ΔR_0 is the potential displacement error during a double exposure measurement (4).

This deviation can be translated into stress (strain) exerted on the sample. Detectors are generally placed 180° apart, as illustrated previously in Fig. 3.2, to measure the interatomic strain. The stress is described by the following equation

$$\sigma_{SE} = \left(\frac{E}{1 + \nu} \right) \left(\frac{1}{4R_0 \sin^2\theta \sin 2\beta} \right) (S_2 - S_1) \qquad (6)$$

where R_0 is the distance between specimen and detector, β is the angle between the normal to the surface and the incident beam, and both S_1 and S_2 are related to the Bragg angles θ_1 and θ_2.

On the other hand, the *double exposure* method requires the measurement of the interatomic spacing of two planes with different orientations to the surface. As depicted in Fig. 3.4, two typical orientations are $\psi_1 = 0°$ and $\psi_2 = 45°$. The stress in any given direction ϕ is given by Eq. (4). For R_0 as the distance separating the sample from the detector and $(S_2 - S_1)$ as the angular differential shown in Fig. 3.4, the equation describing the method is given by

$$\sigma_{DE} = \left(\frac{E}{1 + \nu} \right) \left(\frac{\cot \theta_0}{2R_0 \sin^2\psi_2} \right) (S_2 - S_1) \qquad (7)$$

Finally, the *multiple exposure* or $\sin^2 \psi$ method is similar to the double exposure technique, except that multiple readings are needed. This is required for specimens displaying markedly different texture. The plane containing the incident and

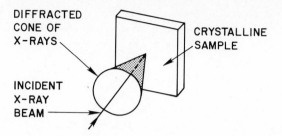

Figure 3.3: Diffracted X-rays from a beam incident on a polycrystalline material form a cone. A plane intersecting the cone axis gives rise to a circle for an unstressed material. Upon stressing, the circle becomes an ellipse. Such deformation can be used to estimate the stress exerted on the sample.

diffracted beam is kept the same for all ψ angles. The line of coincidence of this plane and the plane of the specimen surface is the direction of measured stress. Generally, when the interatomic spacings are plotted against $\sin^2 \psi$, a straight line will result with an intercept of $\nu/E\,(\sigma_1 + \sigma_2)$ and a slope m described by

$$\sigma_\phi = m \left(\frac{E}{1 + \nu} \right) \tag{8}$$

m is the slope of the line $(d - d_0)/d_0$ plotted against $\sin^2 \psi$ where d_0 is the spacing in the unstressed material. This technique is quite time consuming although a high degree of statistical confidence can be reached. When time is not a major consideration, the multiple exposure is desirable due to the superior accuracy obtainable. Constraints on the number of measurements within a certain period, however, generally force a choice between the single and double exposure methods. It has been shown theoretically that the latter is more accurate than the former (3). The sensitivity of these two techniques can be expressed by the following ratio (4)

$$\frac{\text{DE sensitivity}}{\text{SE sensitivity}} = \left(\frac{\sin^2 \psi_2}{\cot \theta_0} \right) \left(\frac{1}{2 \sin^2 \theta_0 \, \sin 2\beta} \right) \tag{9}$$

For typical values of θ_0, ψ_2, and β of 78, 45, and 30° substituted into Eq. (9), a ratio of 1.42 is obtained. Thus, under a given experimental stress condition, the double exposure is about 40% more sensitive than its single exposure counterpart.

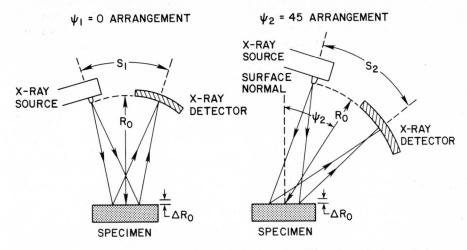

Figure 3.4:   Schematic of double exposure technique for diffraction. ψ is the angle between the normal of the diffracting planes and the surface normal. Two angular measurements are shown, namely, $\psi_1 = 0°$ and $\psi_2 = 45°$ (4).

3.2.2 Macrostresses and Microstresses

In X-ray diffraction, residual stresses are classified in terms of the volume of material affected, for example, a few mm^3 down to $\mathring{A}^3$. Regardless of the size of the region affected, however, strain compatibility must always hold. These stresses are conventionally organized into three classes, namely, first, second, and third order (5). First-order stresses affect a group of crystalline grains, and the resulting global strain will be revealed by a shift in the diffraction peak. Second-order stresses influence only individual grains, creating a mosaic distribution of strains in each grain. Crystalline materials with large grains produce diffraction spots that are moved from the mean diffraction ring. Only interatomic distances govern third-order stresses which show up as fluctuations around the mean lattice parameter. In this case, small grained materials will exhibit a broadening of the diffraction ring, while larger grained substrates will witness a spreading of the diffraction spots. The residual stresses observed are generally the physical manifestation of a combination of these three types of stresses.

Macrostresses are generally determined by the $\sin^2 \psi$ technique discussed previously. In carrying out the measurements, the following assumptions must be kept in mind: (1) There is no shear stress in the irradiated volume. (2) There are no stress gradients, for example, the macroscopic state of stress and strain must be homogeneous throughout.   (3)   All the crystallites behave in a linear isotropic manner. These assumptions are also the intrinsic limitations of the technique. Complications generally happen at the diffraction level which requires small domains for continuous diffraction. Furthermore, nonlinearities in the $\sin^2 \psi$ law can happen from the existence of texture or the influence of a triaxial stress state, leading to large errors (6,7). Several techniques for studying microstresses from diffraction data include, for instance, Fourier analysis on at least two orders of the same reflection, Fourier analysis on a single profile, or application of the integral breath method on a few orders of the same reflection. This was reviewed critically with a discussion on recent advances in theory and technology for measuring residual stresses (5).

3.3 WAFER CURVATURE MEASUREMENT

X-ray diffraction is part of the growing inventory of methods devised to measure the curvature of silicon wafers. The bending of the substrates observed is caused by macrostresses imparted during various processing steps. For instance, an initially flat wafer will be elastically deformed into a convex shape after deposition of a layer of silicon dioxide. This results from the smaller coefficient of thermal expansion (CTE) of the SiO_2 compared to that of silicon. As a result, the oxide film is under compression upon cooling. Similarly, when a polymeric film is spun coated and cured, the ensuing shrinkage will also be transmitted to the supporting substrate. Thus, the rapid progress in microelectronics has made it more imperative to understand the physics of both metal and polymer thin films. Multilayer structures of metal and oxide are typically deposited on silicon wafers, which then undergo various thermal processes. Stresses arising from lattice misfit during deposition, nonequilibrium treatment conditions, or thermal mismatch account for subsequent manufacturing defects (8). Knowledge of the radius of curvature and film thickness would give a good estimate of the mean stress within the film.

Substrates used in semiconductor technology have a high degree of perfection, and therefore, X-ray topography methods can be used to gather information on properties such as film adhesion, film integrity, and both macro and microstresses. The topography terminology refers to the fact that on an X-ray film, the image of a dislocation in an otherwise perfect crystal appears as a dark line against a light background. This is due to the attenuation by absorption and primary extinction of transmitted and diffracted X-ray beams from perfect crystal regions. On the other hand, beams diffracted from imperfect regions, for example, near dislocations are less intense since they suffer only from normal absorption.

For instance, edges of film stripes have high-strain gradients which may be the cause of faults and dislocations. A quantitative estimate of this condition was obtained by measuring the local curvature of silicon wafers with stripes and etched oxide patterns. An X-ray topography camera operating at 45 kV and 16 mA with Mo $K\alpha_1$ radiation was equipped with a set of autocollimator and mirror to record the diffracted intensity along the wafer (9). Elastic distortion of the wafer under stress, however, alters the Bragg position as the wafer moves across the primary incident beam. Efforts have been devoted to devise means for automatically controlling the Bragg position, and will be discussed in more detail later (10,11).

Two common topography techniques for studying film stresses are the *single crystal* and *double crystal* spectrometry (12,13). Both are variations of the single exposure technique discussed previously. Figure 3.5 illustrates schematically the principle underlying the *single crystal* method with an X-ray diffraction topography camera. The X-ray incident on the coated wafer diffracts at a known angle. As long as the wafer remains unstressed, this angle stays constant even when the wafer is translated along the x direction. Residual stresses or external loading, however, deform the wafer into either a concave or convex shape. The crystalline spacing becomes distorted which changes the diffraction angle. By rotating the specimen a

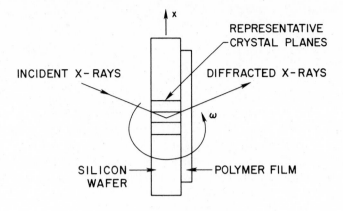

x = TRAVERSE POSITION
ω = ROCKING ANGLE

Figure 3.5:   Single crystal measurement scheme for a coated wafer exposed to an incident X-ray beam. The diffracted angle remains constant even when the sample is translated in the x direction, as long as the substrate is unstressed. However, stressing the sample alters the interatomic spacing and affects the diffraction angle. Only by rocking the sample by a certain angle ω that the beam can be brought back to its original position. Knowledge of x and ω would give a measure of the stress applied or residual within the sample.

certain angle ω, the diffracted beam can be returned to its original position facing the detector. The rocking angle $\Delta\omega$ and the travel length Δx can be related to the radius of curvature, R, by (14)

$$R = 57.295 \, \frac{\Delta x}{\Delta \omega} \tag{10}$$

The derivative of the lattice rotation, $\omega(x)$ yields the local curvature, $1/R(x)$. The mean stress within the film is given by the relation

$$\sigma_f = \left(\frac{1}{R} \right) \left(\frac{t_s^2}{t_f} \right) \left(\frac{E}{6(1 - \nu)} \right) \tag{11}$$

where t_f, t_s, E, and ν are the thickness of the film, the thickness of the substrate, the modulus, and the Poisson's ratio of the substrate, respectively. Material constants for typical substrates such as silicon, germanium, gallium arsenide, and gallium phosphide are readily available in the literature (15).

The *double crystal* method in the transmission/transmission mode is depicted in Fig. 3.6. An unstrained crystal in position two is initially aligned to obtain Bragg

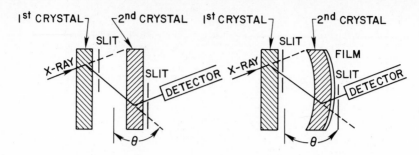

Figure 3.6: Schematic for double crystal X-ray topography method. The principle is similar to that of the single crystal technique, except that two crystals are used. The diffracted intensity changes only when the second crystal is strained by a film.

diffraction from a set of planes. The Bragg angle is constant across the crystals, and translation would not alter the double diffracted X-ray intensity. The corresponding divergence is only about a few seconds of arc. When the second crystal is replaced by one with a curvature from a thin film, translation of the set produces a sharp decrease in the diffracted intensity due to the lattice curvature. The Bragg angle has to be adjusted continuously during translation for maximum diffracted intensity. The radius of curvature is also given by Eq. (10). Similar in concept, both techniques can give quantitative stress measurements to within 4% of each other (12). Generally, the double crystal method is slightly more sensitive than the single crystal method.

To regulate the Bragg reflection angle in both methods as the setup travels across the primary beam, an automatic feedback control was introduced (10). This early version, known as ABAC (automatic bragg angle control), uses analog circuits to maintain the wafer at the maximum of the rocking curve ω during translation. The diffracted intensity, I, is highest at $\omega = 0$. The shape of the rocking curve does not depend on the direction of the diffraction vector. Similarly, a small change in the lattice parameter should not alter its shape or width, although the Bragg angle will be shifted. Since the rocking curve is a symmetric function, two values of ω correspond to any single intensity value. However, the output of the typical sensor used in conventional control systems is a monotonic function of the error between the deviation and the operating point. Thus, it is difficult from one measurement of the intensity to decide which direction ω should be adjusted. The feedback control unit bypasses this issue by dealing with the angular derivative of the rocking curve, $dI/d\omega$. Thus, optimal positioning is obtained by controlling ω so that $dI/d\omega = 0$. This scheme was reported to be quite successful in mapping impurity segregation and dislocations in silicon slices.

A more advanced strategy for measuring the local curvature of wafers was reported recently (11). In this single crystal approach, the rocking curve is digitized above a certain level (e.g., 30 to 85% of the peak intensity), and the corresponding peak position and intensity are determined by least-squares fitting. By keeping track of all motions, the digital approach eliminated the need for cumbersome mirrors and potentiometers. The curvature of the wafers is measured with a modified Elliot

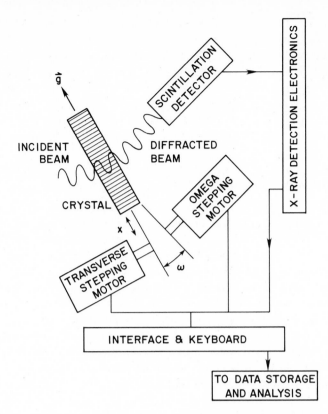

Figure 3.7: Block diagram of computer controlled Lang topography camera (11).

Bristol Lang topography camera, shown schematically in Fig. 3.7. In a typical run, the wafer is mounted on a foam rubber base to eliminate artifacts from external stresses such as clamping or bonding. Bragg reflection in transmission by lattice planes is set as the wafer traverses the primary beam in steps of 0.01 mm. The angular position and the peak intensity are determined at each step with a precision considerably better than 0.001°. Measurements are susceptible to temperature variations and linearity of the transverse motion. For instance, drift of the peak position is induced by temperature changes, and can be as high as a few 0.001° of ω per degree temperature. To check for the linearity of the transverse motion, a silicon wafer with a uniform film of SiO_2 on one side is first measured. The wafer was then turned 180° around the goniometer axis and remeasured. The nonlinearity of the travel motion, expressed as the difference between the two measurements, was less than 0.001° over a 30 mm range. This is equivalent to a probable relative error in the radius of curvature of $\Delta R/R = R/(1700 \text{ m})$.

3.4 APPLICATIONS

With the use of least-squares for on-line peak search, the precision of the rocking angle measurement in the single crystal method described earlier, <0.001°, approximates that of the double crystal technique. All the stress data reported in this chapter (16-19) were collected with this method. A 6-kW Rigaku rotating anode with a Mo target was used. Beam dimensions were approximately 0.2 mm x 2 mm with the shorter length in the transverse direction. Examples are given for one-sided encapsulation (16), passivation (17), and stresses within filled resin systems (18,19).

3.4.1 Polymeric Encapsulation

Filled encapsulant systems used in plastic packages undergo both curing and thermal stresses. Such stresses can affect the performance and long-term reliability of the package. Steps commonly taken to alleviate this problem involve either redesigning the whole package configuration, or reducing the inherent stress within the encapsulant. Since altering the whole structure is an expensive solution applicable only to specific cases, most research efforts are devoted to either flexibilizing the material, or lowering its coefficient of thermal expansion (20). Curing stresses arise from the combined effect of resin shrinkage during polymerization and differential thermal contraction after cooling from high-temperature postcure treatment. The latter thermal stresses occur due to the mismatch in the CTEs of the resin, the fillers, and the backing substrate. The induced microstresses cannot be easily relieved.

The following experiment was initiated as part of a systematic study of the behavior of molding compounds used in low-end packaging. The low-end terminology refers mainly to inexpensive plastic packages, compared to the high-end which deals only with high-cost hermetic ceramic modules. Plastic packages, in turn, exist in different configurations, from the familiar dual-in-line package (DIP) to the newer tape automatic bonding (TAB) technology. The actual silicon device in the DIP is molded, and becomes totally surrounded and protected by its plastic shell. As will be mentioned in the next section on film stress models, such configuration balances the stress profile across the package. Inherent compressive stresses from the curing reaction still exist, however, and are described in more detail in the chapter on "Surface Sensors for Moisture and Stress Studies." On the other hand, a harsher stress profile exists in the one-sided configuration of TAB. Only the circuit side of the device is protected by a coating in this case. This sets up a stress imbalance which can be sufficiently high to crack the die. Due to the severity of the situation, efforts in studying packaging stress were directed at the one-sided stucture. X-ray diffraction was carried out to measure the radii of curvature of silicon wafers coated with different thicknesses of encapsulant. Results are compared with those obtained from scanning the back of the wafers with a Dektak profilometer.

Several encapsulants were investigated and included filled, unfilled, and flexibilized epoxies, and solvent-based and solventless silicone-modified epoxies. Typical stress results described here are for two epoxies (16,18,19). One is a two-component blend of bisphenol-A and cycloaliphatic epoxy cured with anhydride. It is available commercially under the trade name ES-4322 (Hysol Dexter, Industry, CA). About 65 wt % of silica fillers with an average particle size of 20 μm and 3 wt % of carbon black are compounded to the resin. Typically, the resin becomes nontacky after 10 min at 121°C. Postcuring at 150°C for 6 h is needed to reach full properties. The second resin studied is a flexibilized cycloaliphatic epoxy, referred to as CAEF for CycloAliphatic Epoxide Formulation. It is a blend of cycloaliphatic epoxide, ERL-4221 (Union Carbide, Danbury, CT), and an aliphatic epoxide modifier, ERL-4350 (Union Carbide). Flexibilizing is required since the neat resin is inherently quite brittle. Methyl hexahydrophthalic anhydride was the flexibilizer, and the curing reaction was initiated with an amine catalyst. Curing requires 2 min at 140°C, followed by 2 h at 185°C (21).

One-inch (25.4-mm) silicon [001] wafers served as substrates. A cylindrical stainless steel mold capped with two plungers was used to coat the wafers. The following procedure was initiated for each wafer. Teflon sheets were initially inserted underneath the wafer for extra support and more uniform redistribution of molding pressure. A calibrated volume of resin was then added to the mold cavity to cover the wafer. More sheets of Teflon were subsequently added to cap the resin snugly and prevent flashing. Finally, the second plunger was inserted and loaded to a relatively low molding pressure of 0.14 MPa. Curing and postcuring were conducted within the mold, and the coated wafer was allowed to cool down to room temperature before removal.

Each coated wafer was positioned so that maximum diffracted intensity is obtained with the detector. The wafer was translated and rotated at the same time around the camera axis to reposition the diffracted beam for maximum intensity. Once the procedure was repeated a number of times, the curvature was estimated from Eq. (10). Figure 3.8 illustrates a typical trace of the rotation needed to maintain the peak of the intensity within the detector. The uniform curvature of the wafer translates into a linear variation of the rotation angle with translation. Curvatures measured along two perpendicular directions agreed within 5% of each other.

Bending of the coated wafers from shrinkage and thermal stresses varied with the thickness of the encapsulant. This can be seen in the measured radii of curvature illustrated in Fig. 3.9. Copper particles were added to the CAEF system to provide an estimation of the shrinking matrix on the fillers, as will be discussed later. Wafers coated with unfilled ES-4322 showed the lowest radii of curvature, and correspondingly, higher stress profiles. This is not surprising considering that unfilled resins typically have much higher coefficients of thermal expansion than their filled counterparts (19). Interestingly, thin coatings (e.g., <0.06 mm) did not impart much stress on the substrates. Within this range, the behavior of the laminate was dominated by the stiffer silicon. As the coating was gradually increased to about three to four times the thickness of the wafer, stresses built up resulting in low radii of curvature. Thicker coatings, however, produced an opposite effect since the ensuing stiffening of the composite countered any bending influence of the curing shrinkage. Moderate curvature was found with coatings of the filled ES-4322 and the CAEF resin. No significant differences in curvature were found between the two systems, although the corresponding stress profiles can vary dramatically, as will be discussed in Section 3.4.3.

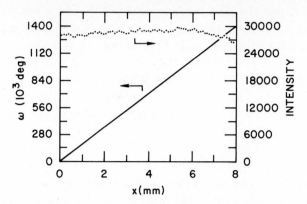

Figure 3.8: Lattice rotation ω and intensity I versus traverse position for a [001] silicon wafer coated on one side with a film of epoxy encapsulant. Thickness of the substrate, $t_0 = 0.28$ mm; thickness of the epoxy coating, $t_1 = 0.27$ mm (16).

3.4.2 One-Sided Passivation

Aside from encapsulants, X-ray diffraction was also carried out on wafers coated with passivation materials such as polyimide and silicone. The former is often used as interlevel dielectric in microelectronic circuits, while the latter protects packages as conformal coatings.

Polyimide is usually spun coated, sometimes in multiple runs to build up to the desired thickness, and cured at temperatures reaching up to 400°C. The final properties of the polyimide depend on the degree of imidization, the specific structure of the polymer, and the amount of residual solvent. For instance, solvent evolution from the film can cause significant volumetric shrinkage, sometimes reaching as much as 20%. As a result, induced residual stresses are locked into the composite laminate. Furthermore, recent work indicated that stresses also arise from the differential mismatch in CTE between the silicon ($2.5 \times 10^{-6}/$°C) and that of the film (50×10^{-6} /°C). The uneven contraction after cooling down from the high-temperature cure induces residual stresses at the film/substrate interface. The magnitude of these stresses, however, was found to be negligible (13).

Thermal stresses are not the only factor governing the mechanical and electrical properties of the polyimide. Rapid increase in the viscosity of the resin during cure leads to the entrapment of solvent bubbles, and ultimately, to the formation of pinholes (22,23). Thick layers are likely to aggravate the problem even further. The typical film would then behave as a two-phase material, with voids as the characteristic second phase. Interpretation of the stress profile would be complicated by the distribution and size of these microvoids. Indeed, cure shrinkage would be somewhat compensated by the presence of these voids.

On the other hand, silicone used as conformal coatings does not cause such problems. Shrinkage stresses are not generated since most silicones react at room temperature, and require only high-humidity conditions for fast cure. Furthermore, the low rubbery modulus does not generate undue stresses on the die. The main drawback is

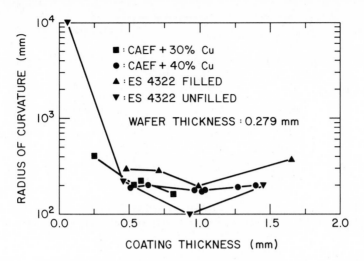

Figure 3.9:   Variation in the radii of curvature for [001] silicon wafers coated on one side with four epoxy systems: CAEF + 30 vol % of copper fillers; CAEF + 40 vol % of Cu; ES-4322 unfilled; and ES-4322 filled with 65 wt % of fused silica. Copper fillers have an average diameter of 10 μm. The fused silica particles have a broader distribution centering around 20 μm (19).

the material high CTE ($\sim$ 300 to 500 x 10^{-6} /°C in the range from 0 to 200°C), which upon thermal cycling can cause wire bond fatigue.

Two polyimides and two silicones were used in this diffraction study. All films were spun coated to the desired thickness and cured (17). Polyimides were obtained from solutions of pyromellitic dianhydride-oxydianiline (PMDA-ODA) polyamic acid/NMP/xylene and hexafluorodianhydride-oxydianiline (HFDA-ODA) polyamic acid/NMP/xylene (Dupont, Wilmington, DE). Both polyimides showed good adhesion to the silicon substrates. Curing of PMDA-ODA follows the normal schedule of 85°C for 10 min, 150°C for 30 min, 200°C for 50 min, 300°C for 50 min, and 400°C for 1 h. HFDA-ODA, on the other hand, needs 85°C for 10 min, 150°C for 30 min, 225°C for 30 min, and 350°C for 30 min. The heating step at 150°C was carried out under nitrogen, while all the other high-temperature phases required vacuum. These curing profiles are summarized in Fig. 3.10, which also depicts the solvent evolution from the films and the corresponding weight loss measured by thermogravimetry.

Both silicones, 3-6550 and 1-2577 (Dow Corning, Midland, MI), are normally used as primary coatings or overcoats of electronic packages before molding. The low viscosity of 3-6550 (1300 cps) allowed spin coating of films which cured at ambient relative humidity and temperature after 72 h. However, 1-2577 required dilution with xylene to enhance flow coverage. Full cure, in this case, was achieved after 24 h at room temperature to produce films of moduli slightly higher than those of 3-6550 films.

Figure 3.11 shows the radii of curvature of wafers with one-sided coats of different thickness of PMDA-ODA. Certainly, compared to Fig. 3.9, much less stress is trans-

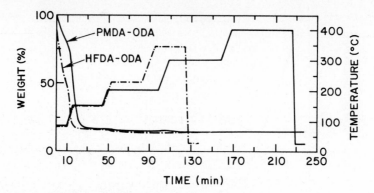

Figure 3.10: Curing profiles of PMDA-ODA and HFDA-ODA. Also shown are the corresponding weight losses of the polymers during imidization as monitored by thermogravimetry (17).

mitted to the silicon substrate by the polyimide than by the encapsulation materials. Nevertheless, the following trend is observed. Between 2 to 10 μm, the curvature decreases steeply with increasing film thickness. However, the rate of change becomes less dramatic for thicker films (>15 μm). Silicone films, on the other hand, had negligible influence on the substrate. As illustrated in Fig. 3.12, the wafers with silicone overcoats remained essentially flat since the rubbery films were more yielding.

To monitor the development of the stress profile within the polyimide films during the various curing steps, diffraction was conducted after each phase. This is shown in Figs. 3.12 and 3.13 for PMDA-ODA and HFDA-ODA, respectively. With PMDA-ODA films less than 10 μm thick, the curvature relaxed after exposure to successive higher temperatures (e.g., 150, 200, and 300°C), and increased after the final treatment at 400°C. This behavior was inconsistent with results reported previously (13,24), and was due to the presence of entrapped solvent bubbles within the film (17). Voiding from solvent removal was not an issue with the silicone films. Curing proceeds over an extended period of time, and during that period of solvent evaporation so critical with polyimide films, the silicone remained sufficiently fluid to heal itself.

The development of instrinsic stresses in the polyimide films during bake was studied recently. With a polyimide made from benzophenone tetracarboxylic acid diethyl ester and 4,4'-methylene diamine (BTDA-MDA), the wafer curvature changed reversibly and linearly with each curing step (13). This polyimide was found to have a glass transition close to the highest temperature subjected to the sample. Each time the BTDA-MDA reached a higher temperature, the film relaxed all stresses. Upon cooling, it became rigid, and stress rised with the mismatch in CTE. Similar results were reported with an amide/imide system (AI-lite from Amoco Chemicals Co.) (24). Using a bending-beam method, the stresses were found to decrease near the T_g of the film. Stress would also be reduced if cracking or delamination of the film occurred during thermal treatment. Voids from air entrapment during molding or casting imperfections would yield similar end results

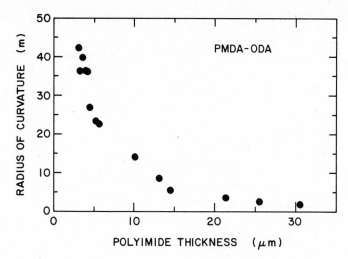

Figure 3.11: Radii of curvature of wafers coated with varying thickness of cured PMDA-ODA (17).

(16).   The HFDA-ODA films of Fig. 3.13 exhibited behavior similar to those reported, where the radii of curvature decreased (or, correspondingly, the residual stresses increased) with the temperature setting.

X-ray diffraction topography can also be used to monitor the effect of moisture on polyimides. For instance, Fig. 3.14 shows the plasticization of PMDA-ODA and HFDA-ODA films after a 24-h exposure to 80°C and 85% relative humidity. The ensuing relaxation of the coatings translated into slightly higher radii of curvature, and correspondingly, lower stresses. The effects observed were somewhat small, although the two polyimides displayed distinctly different characteritics. Under similar exposure conditions, HFDA-ODA seemed to absorb more moisture, and as a result, relaxed more than its counterpart, PMDA-ODA.   This variation in permeability certainly depends on chemical factors such as morphology, chain orientation, degree of crystallinity, density, chain orientation, or extent of crosslinking. Plasticization of the polymer usually translates into a temporary (reversible after bakeout) lowering of the modulus and the glass transition temperature.   This behavior was also observed in other types of polymides subjected to moisture (25-27), and even harsher gaseous environments such as pure anhydrous ammonia and pure sulfur dioxide (28).

3.4.3 Microstresses Within Fillers

To minimize the residual stresses within encapsulation compounds, both the molding process or the chemical formulation can be modified. Process changes include, for instance, slow heating and cooling profiles or lower temperature cure. On the other hand, the compounding can either be a decrease in the effective modulus of the matrix by incorporating flexibilizers, or lowering the CTE of the material. Various types of fillers have been incorporated into the polymeric matrix, and include metals

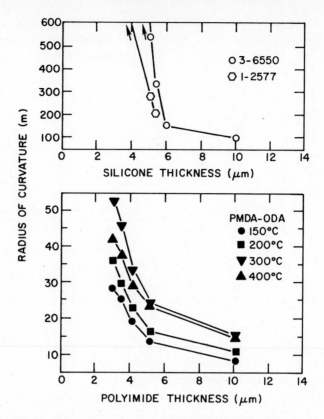

Figure 3.12: Radii of curvature of wafers coated with different thickness of silicone (3-6550 and 1-2577). Also illustrated is the history of the PMDA-ODA film stress during the various steps of the curing phase (17).

(Al, Au, Ag, and Cu), metal oxides (Al_2O_3, Sb_2O_3, and MgO), ceramic (crystalline and fused silica), SiC, and glass fibers (29,30).

Compounding such fillers at predetermined loading can lower the CTE, but raises the effective modulus of the composite, which affects the magnitude of the stresses exerting on the die. With the current packaging tendency toward larger dies, thinner coatings are required to minimize the compressive curing stresses. This is a serious issue to consider, especially in one-sided coating technology such as tape automatic bonding. For instance, when the thickness of the coating approaches the diameter of the filler particle, film cracking can occur readily (16). Therefore, an understanding of the role played by fillers is a prerequisite to the design of low-stress encapsulating materials.

This section discusses some typical results on the stresses imparted to fillers embedded within a polymeric network. Copper particles with an average diameter of 10 μm were added to the CAEF resin. No special interfacial coupling agent was used to enhance the adhesion between fillers and matrix (18,19).

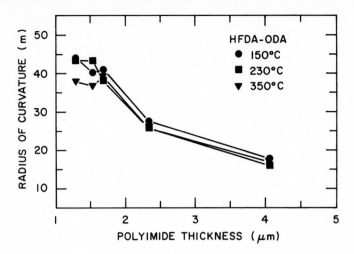

Figure 3.13: Radii of curvature of wafers coated with HFDA-ODA during the various curing steps (17).

Both the macrostress on the coated wafer and the microstress on the copper particles were recorded. The former was measured by X-ray diffraction with the Lang topography camera described earlier, and by scanning the wafer backside with a Sloan Dektak profilometer. Expressions for the stress profile as a function of radii of curvature depend on the film stress model selected and its underlying assumptions. Stresses within the copper fillers, on the other hand, were determined with the standard single exposure, back reflection mode. A horizontal Picker diffractometer with a Cu tube, operating at 40 kV and 10 mA, was selected. The sampling beam size was approximately a 4-mm-square spot. The standard $\sin^2 \psi$ method was used. Peak diffracted intensities at [240] were recorded at 6 ψ tilts by curve fitting to the top 15% of the Bragg peak line with background substraction. The lattice parameters were then computed, and a least-squares line was fitted to the "d" versus $\sin^2 \psi$ spacing. Figure 3.15 shows a representative profile for a wafer with one coat of CAEF and 40 vol % of copper fillers. The slope m of the curve is described by Eq. (8). Repeated measurements were conducted for improved accuracy, and good correlation between experimental data and predictions was obtained. To check on the accuracy of the technique, the state of zero stress in copper fillers prior to compounding was also determined. This value was determined to be 5 ± 10 MPa, which is indicative of the experimental variation.

Stresses within copper particles are described in Fig. 3.16 for CAEF systems with both 30 and 40 vol % of fillers. A combination of resin shrinkage and differential thermal contraction between the polymeric matrix and the metallic fillers accounted for the existence of these stresses. The typical analogy of the physical interaction between a single particle and the matrix can be compared to embedding a sphere of radius $r + \delta r$ into a cavity of diameter r. In such case, the volumetric shrinkage is proportional to δr^3. However, microstresses can also be tensile. This tends to occur across the polymer/filler interface for resin pockets trapped between closely packed

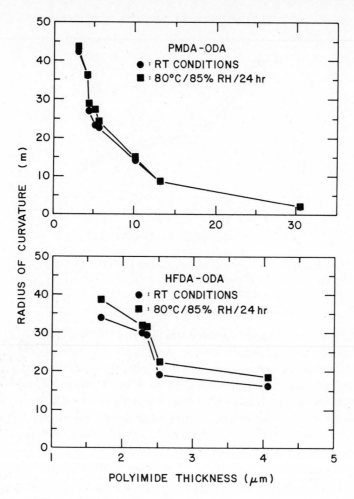

Figure 3.14: Effect of moisture on the radii of curvature of wafers coated with PMDA-ODA and HFDA-ODA. Exposure conditions: 80°C, 85% RH, 24 h (17).

fillers at high volume loading. In this case, fillers can refer to inorganic or metallic particulates, short glass fibers, or long polymeric or metallic fibers. The existence of this stress profile has been confirmed by studies with electrical resistance strain gages (31,32), Moire fringe methods (33), photoelasticity (34), holographic interferometry (35), and X-ray diffraction (36,37).

All stress values acting on the fillers were within ±10 MPa. This was within the range of experimental error (±7 MPa) reported in the X-ray diffraction study on fillers (36). In this case, Ag, Nb, and CdO particles (1 to 40 μm) were dispersed in small amounts (<2 mg/cm²) between the first and second plies of 6-ply unidirectional graphite epoxy laminates. Elastic strains in the fillers were proportional to the stress applied in the direction of the fibers, and expressions describing the three principal strains parallel to the orthotropic axes of the composite were pre-

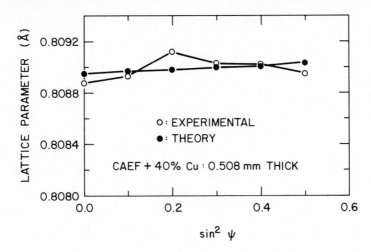

Figure 3.15:   Lattice spacing versus Exposure conditions: 80°C, 85% RH, 24 h (17). $\sin^2 \psi$ of a silicon wafer coated with CAEF and 40% (volume) of copper fillers (19).

sented (38). Residual stresses were also estimated from the particle X-ray strains, assuming that all the fillers were isotropic. Although none of the fillers were treated with adhesion promoters, there was surprisingly no interfacial adhesion problems. However, this was not the case with the copper fillers compounded into the CAEF resin. The stress in the average copper particle was less than the stress in the wafers coated with CAEF calculated from curvature data. In one instance, the average stress for fillers in the 40% CAEF filled formulation even dropped down to zero. This strongly suggests that crazing may have occurred in the vicinity of the fillers. Consequently, full load transfer from the CAEF network to the fillers was not achieved. Such interfacial delamination was probable, since the copper particles were not treated.

3.5 FILM STRESS MODELS

Thus far, the discussion on X-ray diffraction topography and related radii of curvature did not dwell into the stress aspects of the films or substrates. The stress profile obtained depends on the film stress model selected and the underlying assumptions governing the validity of the model. The proper theory should predict the stresses arising from the thermal mismatch between materials either adhesively bonded or soldered, and subjected to uniform cooling or heating. Both normal stresses responsible for the material strengths and the shear (or peeling) stresses accountable for the cohesive strength of the interfacial regions must be addressed.

Several models describing the stress and strain distribution in both films and backing substrates exist. The earliest model goes back to the analysis of the tension in electrolytically deposited metal films (39). For instance, poorly adhering nickel films would curl up once a certain critical thickness was reached. The expression for the film tension derived, however, was based on inherently thin films with the same elastic moduli. Another classical model was postulated for bimetallic thermostats

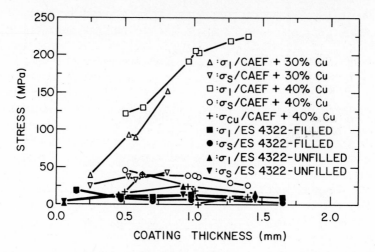

Figure 3.16: Stresses at the interface between substrate and encapsulant (σ_I) and at the outer surface of the encapsulant (σ_S). Silicon substrates were coated on one side with two systems: (a) cycloaliphatic epoxy and two volume loadings (30 and 40%) of copper fillers. Average particle size of 10 μm; (b) ES-4322 (19).

(40). This theory described the deformation of metal strips during heating (or cooling) and the resulting internal stresses. Widely used in various areas of electronic packaging, this theory determined only the normal stresses within the components, assuming that the stresses were uniformly distributed over the entire assembly length. For smaller assemblies such as chip/substrate joints, the stresses tended to be overestimated (41).

Other models have since been formulated to consider different variations of the coating issue. Stresses were estimated from the deformation of the substrates through various means such as bending beam, bending plate, optical interference, or X-ray diffraction (14,42). However, these models suffer drawbacks from simplifying assumptions such as equal film and substrate moduli and Poisson's ratios (43), isotropic properties (44), and homogeneous stresses (45). Fracture mechanics and Griffith's energy criterion were applied in another model which analyzed the adhesion of thin brittle films ($\sim \mu$m range) to low modulus polymeric substrates (46,47). The interfacial stresses were addressed as a function of film thickness, material constants, and external loading conditions. This theory was subsequently extended to layered composite media where viscoelastic effects were important (48).

3.5.1 Two-Sided Coating Configuration

Polymeric passivation and encapsulation are generally thicker than the evaporated metal layers. Thus, a state of plane strain is more likely to exist in the composite rather than the plane stress with thin films. Viscoelastic relaxation does not occur because the typical operating conditions, even during qualification tests, rarely exceed the glass transition temperature of the coating. For instance, this limit is around 400°C for most polyimide resins and 150°C for epoxy encapsulation. A generalized model for films and substrates of different material constants was presented recently (16). Both film and substrate were assumed to follow the usual thermoelastic laws. Two commonly encountered cases were considered, namely, two-sided coating and one-sided coating. Two-sided encapsulation can be found in the normal plastic package configuration, where the resin completely covers the die on all sides. For simplicity, only the top and bottom layers were incorporated into the model. This configuration is also sometimes introduced into the processing of wafers during the deposition of various top metallurgy. Back coating with the appropriate film counters the stresses from metal deposition, reduces wafer bending, and helps keep line registration for subsequent photolithographic steps.

Figure 3.17(a) illustrates an idealized silicon substrate of characteristic length L_0 coated with layers of encapsulation on both sides. The deposited resin flows out to cover an area of dimension L_1 and thickness t_1. Plane strain is assumed, or in other words, stresses parallel to the thickness direction in both films are smaller than the normal stresses. Stresses build up in the substrate as curing starts and the polymer network is formed. Volumetric shrinkage leads to a final structure where both films and substrate have the same length L_f but different thickness $(t_1)_f$ and $(t_0)_f$, respectively, as shown in Fig. 3.17(b). The corresponding free-body diagram is depicted in Fig. 3.17(c) with tensile (F_1) forces acting on the films and compressive forces (F_0) exerting on the silicon, and the ensuing bending moments, M_1 and M_2. Equilibrium requires that no net force or torque exists, so that (16)

$$\left[\sum F_i = 0\right] \quad 2F_1 - F_0 = 0 \tag{12}$$

$$\left[\sum M_i = 0\right] \quad F_1\left(t_0 + \frac{t_1}{2}\right) - F_1\left(t_0 + \frac{t_1}{2}\right) = M_1 + M_2 \tag{13}$$

The moment balance in Eq. (13) is taken at the midpoint A on the neutral plane $\phi\phi'$ of the laminate. The requirement that $M_1 + M_2 = 0$ indicates that the structure covered on both sides with the same thickness of material does not undergo any bending. The maximum stress in the encapsulant, $(\sigma_1)_{xx}$, is given by

$$(\sigma_1)_{xx} = \frac{F_1}{(t_1)_f L_f} \tag{14}$$

where $(t_1)_f$ is the final thickness of the film. After some manipulation, Eq. (14) yields the following expression

$$(\sigma_1)_{xx} = E_1 \frac{\Delta L}{L_1} \tag{15}$$

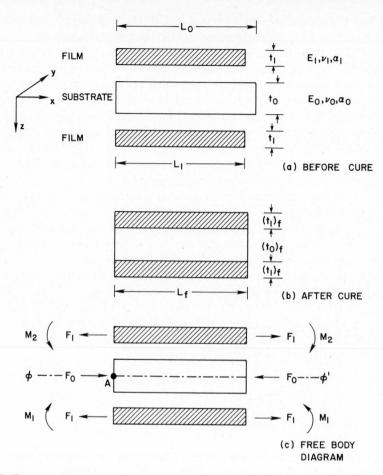

Figure 3.17: (a) Idealized two-sided coating configuration of silicon substrate before curing of the resin; (b) Configuration after curing; (c) Free-body diagram showing forces and moments exerting on the laminate (16).

where E_1 is the film modulus. Similar considerations for the substrate gives

$$(\sigma_0)_{xx} = 2E_1 \frac{t_1}{t_0} \frac{\Delta L}{L_1} \qquad (16)$$

The previous two equations raise three important implications for two-sided coating configurations. First, for a given shrinkage, the stress in each film is independent of the coating thickness. Stress relief may occur, however, especially in thin films since the film dimensions can approach those of the filler particles. The sharp corners of these fillers act as stress concentrators, which ultimately lead to minute crazing or cracking in the films. In real packages, the compressive stresses can be sufficiently large to force fillers against the surface of the devices causing bit errors (20). Second, as thicker coatings are deposited, stresses within the substrate increase linearly. This underlines the role of polymeric overcoats on devices prior to molding. Before the advent of low-stress encapsulation compounds, overcoats are necessary to redistribute the hydrostatic compression, and relieve the stresses within the silicon substrates. And, finally, the theoretical radius of curvature of the laminate is zero for similar thickness of the same coating material. In practice, however, this is harder to achieve due to molding variations.

3.5.2 One-Sided Coating Configuration

Models for one-sided film configurations can be found in diverse situations. For instance, to determine the residual stresses in sheet metals, a chemical etching method was developed, where thin layers were successively removed and the resulting increasing radius of curvature of the sheet was measured by optical microscope (49). The situation was analogous to film coating, except that in this case, layers were milled away rather being added to the substrate. After the removal of a given number of layers, the average stress in the metal was described by a recursive expression of the substrate thickness and radius of curvature. A more generalized version of the stress distribution in thin films produced by evaporation, sputtering, or electrolytic deposition was later presented (50). Such films, typically 1000 Å thick, required the assumption of plane stress for the solution of inhomogeneous and anisotropic stress conditions. Experimental results obtained from bending plate methods correlate well with predictions using linear approximations. However, better results would be obtained if local variations in stresses were taken into account. This would require minimizing the elastic energy using the compatibility equation as a boundary condition.

Some simple expressions for describing the deformation of a substrate of thickness t_0 from an adhering film t_1 thick can be advanced (51). When the substrate deforms under the intrinsic and thermal stresses, Hooke's law yields the following maximum stress in the substrate occurring at the interface

$$(\sigma_0)_{max} = \frac{t_0}{R} \frac{E_0}{\alpha(1 - \nu_0)} \tag{17}$$

where R is the radius of curvature, α is a constant, and E_0, ν_0 are the modulus and Poisson's ratio of the substrate material, respectively. For elastic deformation, biaxial stresses in the film are given by

$$\sigma_1 = \frac{1}{R} \frac{t_0^2}{t_1} \frac{E_0}{6(1 - \nu_0)} \tag{18}$$

Two assumptions underline Eq. (18). First, the distribution of stress in the film is neglected since $t_1 \ll t_0$. And, second, the neutral plane is located at $2/3\, t_0$ away from the film/substrate interface. This derives from balancing forces on any cross section of the laminate, and results in $\alpha = 3/2$. The ensuing maximum stress in the substrate is then described by

$$(\sigma_0)_{max} = \frac{2}{3} \frac{t_0}{R} \frac{E_0}{1 - \nu_0} \tag{19}$$

The existence of plastic deformation requires some modifications to Eq. (18). In this case, equilibrium of forces in both substrate and film requires after some manipulation

$$\alpha = \left(\frac{1}{2} + \frac{t_1 R}{t_0^2} \frac{1 - \nu_0}{E_0} \sigma_1(R) \right)^{-1} \tag{20}$$

As the film stress exceeds the elastic limit, the neutral plane shifts from the initial $2/3\ t_0$ position from the interface to the $t_0/2$ location. $(\sigma_0)_{max}$ can be subsequently calculated from Eqs. (17) and (20) and a known stress/strain curve. Generally, however, a constant plastic stress is assumed for simplicity, for example, $\sigma_1(R) = \sigma_Y$ where σ_Y is the yield strength of the film. Since once plastic deformation is initiated, the neutral plane shifts from $2/3\ t_0$ to $t_0/2$, it has been suggested that positioning the neutral plane arbitrarily at the midpoint would be acceptable (51). This resulting equation

$$(\sigma_0)_{max} = \frac{7}{12} \frac{t_0}{R} \frac{E_0}{1 - \nu_s} \tag{21}$$

yields a maximum error of about 15%.

The thin films in the previous models were assumed to be in a state of plane stress. The following model does not make any *a priori* assumption on the film thickness, and is found to give good predictions for encapsulated substrates (16). It is an extension of an analysis of the stress dependency on the density of misfit dislocations in heteroepitaxial structures (52). This was subsequently amended in another version (53). Figure 3.18 depicts the idealized structure of such one-sided configuration before (a) and after curing of the film (b). Upon cooling, the encapsulant will be under tension due to a CTE higher than that of the silicon supporting substrate. Equilibrium again dictates that

$$\left[\sum F_i = 0 \right] \quad F_1 - F_0 = 0 \tag{22}$$

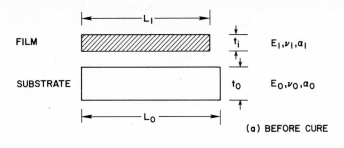

FILM

SUBSTRATE

(a) BEFORE CURE

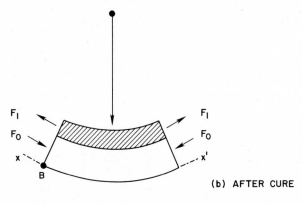

(b) AFTER CURE

Figure   3.18:   (a) Idealized one-sided coating configuration of silicon substrate before resin shrinkage; (b) Bending configuration after curing (16).

$$\left[\sum M_i = 0\right] \quad F_1\left(t_0 + \frac{t_1}{2}\right) - F_0 \frac{t_0}{2} = M_0 + M_1 \tag{23}$$

where the moments M_0 and M_1 can be found from classical beam theory as

$$M_0 = \frac{E_0}{(1 - \nu_0)R} \int_{-t_0/2}^{t_0/2} (z - \delta)^2 L_f \, dz \tag{24}$$

$$M_1 = \frac{E_1}{(1 - \nu_1)R} \int_{t_0/2}^{(t_0/2) + t_1} (z - \delta)^2 L_f \, dz \tag{25}$$

where δ is the shift of the neutral axis during bending. Solving Eqs. (22) and (23) gives the following expressions for the stresses in the substrate and the film, $(\sigma_0)_{xx}$ and $(\sigma_1)_{xx}$, respectively

$$(\sigma_0)_{xx} = \frac{2C_1}{t_0(t_0 + t_1)R} \tag{26}$$

and

$$(\sigma_1)_{xx} = \frac{2C_1}{t_1(t_0 + t_1)R} \tag{27}$$

where

$$C_1 = \left(\frac{E_0}{1 - \nu_0}\right)\left[\frac{t_0^3}{12} + \delta^2 t_0\right] + \left(\frac{E_1}{1 - \nu_1}\right)\left[\frac{t_1^3}{12} + t_1\left(\delta - \frac{t_1 + t_0}{2}\right)^2\right] \tag{28}$$

Given experimental values of the radius of curvature, Eqs. (26) and (27) can give the magnitude of the stresses within the substrate and film. Conversely, the radius of curvature of the composite can, in turn, be described from coherency conditions at the interface by the relation

$$R = \left(\frac{1}{L_1 - L_0}\right)\left[\frac{L_1 t_1 + L_0 t_0}{2} - \frac{2C_1 C_2}{t_1(t_0 + t_1)}\right] \tag{29}$$

with

$$C_2 = \left[\frac{1 - \nu_1}{E_1} + \frac{t_1}{t_0}\left(\frac{1 - \nu_0}{E_0}\right)\right] \tag{30}$$

Thus, given material constants such as E_0, E_1, ν_0, ν_1, and physical characteristics of the composite structure such as L_0, L_1, t_0, and t_1, the curvature of a given system can readily be predicted. Equations (26) and (27) only describe the stresses averaged over the corresponding substrate and film. To truly describe the stress profile throughout the composite structure, the component in the z direction perpendicular to the interface must also be known. For $-t_0 \leq z \leq 0$, z being defined from the interface (53), the stress within the substrate is

$$\sigma_0(z) = (\sigma_0)_{xx} + E_0\left(\frac{z - \frac{t_0}{2}}{R}\right) \qquad (31)$$

A similar expression for the stress within the film is given for $0 \leq z \leq t_1$

$$\sigma_1(z) = (\sigma_1)_{xx} + E_1\left(\frac{z - \frac{t_1}{2}}{R}\right) \qquad (32)$$

The clear disadvantage with one-sided coating configurations is the state of stress imbalance. All processing operations have to ensure that the strains from the curing shrinkage reside below the critical strain for die fracture. Considering the epoxy encapsulation case presented earlier, stresses within the one-sided configurations can now be estimated from Eqs. (26) and (27). This is shown in Fig. 3.19, which illustrates the stresses imparted to the silicon wafer by different layers of ES-4322 encapsulant. For instance, a layer of epoxy 0.13 mm thick on a 0.28-mm-thick wafer generates compressive stresses about -10 MPa within the silicon, and tensile stresses in the film of approximately 20 MPa. When the film is as thick as the wafer, almost the same absolute stress level is found in the silicon (-22.8 MPa) and the epoxy (23.4 MPa). Thicker coatings exert higher stresses on the substrate, although the film stress tends to remain constant. The sudden drop at the thickest coating is not an experimental anomaly. X-ray micrographs of the encapsulated samples indicate the presence of voids (due most likely to air entrapment during molding) which relax the state of stress.

The stress profile through the coated wafer can be visualized in Fig. 3.20 with Eqs. (31) and (32). Three film thicknesses are depicted, namely, 0.13, 1.01, and 2.31 mm. Only tensile stresses exist in the films, decreasing linearly toward the outer surface. In contrast, more interesting patterns exist in the substrates. The shrinking polymer films cause the wafers to curl into a convex shape similar to that of a slightly deployed parachute. With the thin film (0.13 mm), the highest tensile stresses (~11 MPa) are at the outermost fiber of the substrate. These stresses decrease to zero at the neutral plane, and become compressive up to the interface where they reach almost -30 MPa. As thicker coatings are applied, the stiffness of the epoxy starts to dominate, putting the entire die into compression.

Stress profiles for the CAEF system are shown in Fig. 3.16. The elastic moduli of the composite were calculated using the Hill approximation, which is the average of Reuss (constant stress) and Voigt (constant strain) limits of the elastic constants. Although the radii of curvature did not vary dramatically between the two resins, ES-4322 and CAEF, as depicted in Fig. 3.9, their stress profiles can be quite different. Similar effects are found between the two filler volume loadings of CAEF. In this case, even though the curvature are comparable, the stress at the interface for CAEF + 40% Cu is more severe than that of CAEF + 30% Cu. Such effect results

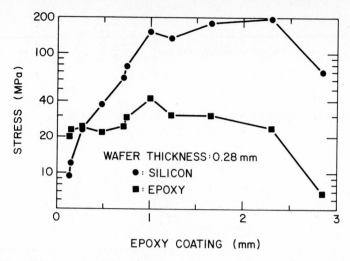

Figure 3.19: Stress in [001] silicon wafers with one-sided coatings of ES-4322. The thickness of the wafers was 0.28 mm. Stress was calculated at $t_0/2$ for silicon and $t_1/2$ for the epoxy (16).

from the increase in modulus overcoming the decrease in the composite CTE. Indeed, loading the resin with an extra 10% (from 30 to 40%) produces a compound CTE only 11% lower, while stiffening the laminate by almost 31%. Furthermore, interfacial stresses increase with thicker films, while stresses at the encapsulant surface stay approximately constant. Thus, any stress-induced failures would tend to occur at the interface.

Stresses in one-sided passivation can also be estimated. An example is given in Fig. 3.21 for the four curing steps of PMDA-ODA spun coated on silicon substrates. The stresses decrease after the first three consecutive steps, and increase after the final imidization at 400°C. Such behavior clashes with other findings (13,24), until it is found that voids from solvent trapped during cure account for the anomaly. Under polarized light, the films reveal a large distribution of voids ($\sim 5 \times 10^5/cm^3$). The TGA thermogram in Fig. 3.10 indicates that 25% of the solvent escapes during the first 10 to 15 min exposure to 85°C. As the temperature is raised to 150°C and imidization proceeds to make the films more rigid, another 50% is lost. After 200°C, not much residual solvent remains, suggesting that the final void distribution observed is obtained during the initial heating steps. Voids affect stress readings in several ways. First, solvent voiding partially counters the shrinkage from curing, reducing any residual stresses locked in from the curing. Second, the CTE of a polymer/void composite is smaller than that of a defect-free film. Third, voids also act as stress concentrators which may induce some extent of delamination at the interface and a corresponding reduction in stress level. And, finally, the effective modulus of the film obtained is less than that of void-free film. Thus, for a given strain, as indicated in Eqs. (26) and (27), lower stress is produced. There is no reported systematic investigation of the effect of solvent-induced voids on the effective modulus of the film. Nevertheless, using a volume fraction average, those cases

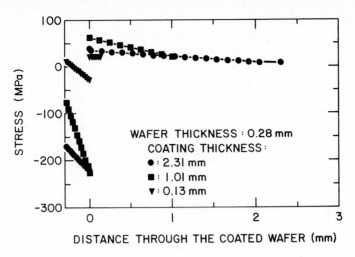

Figure 3.20:   Calculated stress profile through a silicon substrate encapsulated with ES-4322. Three different thicknesses are depicted: 0.13 mm; 1.01 mm; and 2.31 mm (16).

where voids are uniformly dispersed throughout the film, the modulus change can be as much as 20%. Thus, translating radii of curvature measurements either from Dektak profiling or X-ray diffraction into stresses can lead to erroneous values. Only those values from thin films which exhibit few defects can be considered as accurate. Others can only be used comparatively to discern trends.

3.5.3 Multilayered Coating Configuration

Complications arise in multilayered structures where it is mandatory to understand the physical interactions between films and substrate. Typically, most models are formulated for a hypothetical composite medium made up of layers with different elastic and thermal properties. The different rates of thermal expansion lead to elastic, and sometimes plastic, strains in the confined layers. Naturally, equilibrium conditions dictate that no resultant end forces or bending moments exist.

   A three-layer model was first proposed to describe the stress relief introduced by a transition zone between two crystals grown by epitaxial deposition (54). The analysis followed the elastic analysis of the bimetallic thermostats (40), but assumed equal moduli in all three layers. However, only qualitative results can be obtained, since there was a large uncertainty in the compositional dependency of the CTE at high temperatures. Furthermore, plastic deformation also occurred at high temperatures compounding further errors. Another approach based on the elastic behavior of classical laminated plate theory was presented (31,32). Residual strains and curvatures were predicted for unidirectional and multidirectional composites such as boron/epoxy and glass/polyester systems. Swelling effects from moisture absorption were also taken into account. The method decoupled the total strain into thermal and mechanical components. Thermoelastic behavior was assumed, that is, all

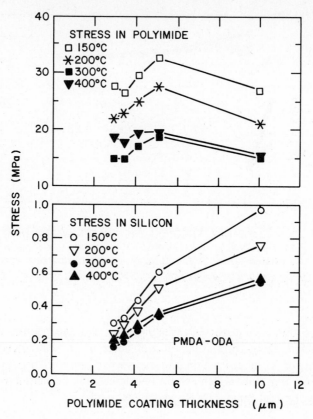

Figure 3.21: Stresses in PMDA-ODA films and in silicon substrates during each step of the curing process (17).

thermal strains and elastic compliances were allowed to fully depend on the temperature. Similarly, a two-dimensional model for laminated plates was deduced from the three-dimensional continuum theory of stratified media (55). The model, however, was geared toward developing governing plate stress equations of motion, stress/strain relations, and boundary conditions to study the propagation of harmonic waves. By extending the classical beam theory to stratified media, another model described the effect of transverse shear on a laminated plate (56). The plate was considered to be of infinite length in the z (out of plane) direction, and was simply supported along the edges. A shear correction factor accounts for nonlinear in-plane displacements. More refinement led to the inclusion of thermal stresses and warping (57). By expanding the solutions of warping and thermal stress into a power series, the theory degenerates into the classical plate model. A thermal shear correction factor can then be introduced to solve for thermal expansion problems.

Closed form expressions for the stress in individual layers of heteroepitaxial structures were obtained (58). However, equal and isotropic elastic constants were again assumed for simplicity. This was found to introduce errors on the order of the difference in elastic constants. The actual error may vary with the thickness of each film and can be slightly less. Applications to $Al_xGa_{1-x}As/GaAs$ laser diodes showed a drastic change in stress with the incorporation of Al into the active GaAs region. Qualitative correlation existed between the sign reversal of the stress with the life-times of the devices. Another model based on classical plate theory considered thermoelastic loading of multilayered anisotropic plates (59). High-temperature structural changes leading to film shrinkage were included together with thermal dimensional changes.   Radii of curvature of two- (Si/SiO_2) and three-layer $(Si-SiO_2$-polycrystalline Si) structures measured by the double crystal X-ray diffraction method agreed well with predicted values.

An attempt at simplifying the complex stress expressions typical of multilayered configurations was made recently (60). The approach considered a long and narrow structure amenable to one-dimensional stress analysis. If the experimental sample does not have this exact shape, a correction factor had to be applied to the substrate modulus. Key parameters to the radius of curvature, strain, and stress formulas were the effective layer to substrate thickness ratio, and the average differential con-traction. These expressions were found to agree well with special cases.

More recently, generalized closed form expressions for the layer stresses and radii of curvature were derived based on classical elastic analysis of beam theory (61), and on elastic surface compliance (62). Both gave good agreement between pred-ictions and experimental data. The latter model, aside from the familiar normal stresses responsible for film ultimate and fatigue strenth, also described the shear stresses which account for blistering and peeling tendencies. The normal stress in the ith film layer is described by the following relations

$$\sigma_i = E_i^0 \, t_i \, \Delta\alpha_i \, \Delta T \tag{33}$$

where $E_i^0 = E_i/1 - \nu_i$, $\Delta\alpha_i = \alpha_i - \alpha_s$. E_i, ν_i, t_i, and α_i are the modulus, Poisson's ratio, thickness, and CTE of the film, respectively. α_s is the CTE of the substrate and ΔT is the temperature gradient imposed on the structure. σ_i is uniform across the surface of the typical film, and only drops to zero near the edges. Equation (33) reveals that normal stresses can be reduced by a combination of either smaller thermal processing steps or softer materials. On the other hand, shear stresses between the ith and ith + 1 layer of a structure m layers thick are given by

$$\tau_i = k_i \sum_{j=i+1}^{m} t_j\sigma_j \tag{34}$$

where k_i is a stiffness constant such that

$$k_i^2 = \frac{3}{4} \frac{\displaystyle\sum_{j=0}^{i+1} \lambda_j}{\displaystyle\sum_{j=0}^{i+1} t_j^2 \lambda_j} \tag{35}$$

and $\lambda_i = 1/(E_i^0 t_i)$. The corresponding peeling forces between the ith and i + 1 layer are

$$P_i = \frac{1}{2} k_i \tau_i \sum_{j=i+1}^{m} t_j \tag{36}$$

As expected, the interfacial stresses are predicted to be highest near the edges of the structure. These stresses depend on the normal stress level, and increase with the thickness of the layers and the interfacial stiffness, k. Thus, properly slanting the film toward the edges would be one good option to minimize peeling stresses.

3.6 CONCLUSION

X-ray diffraction is a versatile method of determining stress levels in crystalline structures. It measures the strain in a given lattice under either thermal or mechanical loading. Stresses can thus be derived knowing the elastic constants of the material, provided that the behavior does not exceed the elastic limit. Instrumentation for X-ray diffraction ranges from portable diffractometers to laboratory devices equipped with special position sensitive scintillation detectors. Three basic methods for diffraction studies were discussed, namely the single exposure, double exposure, and multiple exposure or $\sin^2 \psi$ method. The latter was described in an application to determine stress levels in fillers compounded to encapsulants. Large potential still remains untapped in this case, since the method can provide solutions to the type of fillers used, the aspect ratio and volume loading incorporated, and the effect of interfacial coupling agents for increased adhesion. Such information would ultimately be useful for the design of low-stress molding compounds.

X-ray topography gives rapid scanning of the state of strain in a coated substrate. The method is enjoying wide acceptance as a reliable way to obtain radii of curvature of wafers. Unfortunately, this is where most agreement ends. Stresses obtained from radii of curvature data can vary depending on the model selected. One-sided coating models are relatively simple, and well-established expressions are available for describing the profile throughout the laminate. This was shown for silicon wafers coated on one side with layers of passivation and encapsulation. Models for multilayered structures, however, are still subject to debate. As long as the underlying assumptions are well understood, these models can provide good means for comparing qualitatively different processes.

3.7 ACKNOWLEDGMENTS

The author would like to thank A. Segmuller, J. Angilello, and J. Karasinski for their kind permission to use the Lang topography camera unit, and I. C. Noyan for the long nights laboring over the X-ray measurements.

3.8 REFERENCES

1.   "Residual Stress Measurement by X-Ray Diffraction," *S.A.E. 784 a*, S.A.E., Warrendale, PA (1971).

2.   B. D. Cullity, **Elements of X-Ray Diffraction**, Addison-Wesley, Reading, MA (1978).

3.   J. T. Norton, "X-Ray Stress Measurement by the Single Exposure Technique," *Adv. X-Ray Anal.*, 11, 401 (1968).

4.   C. O. Ruud and D. J. Snoha, "Displacement Errors in the Application of Portable X-Ray Diffraction Stress Measurement Instrumentation," *J. Metals*, 32, February (1984).

5.   G. Maeder, J. L. Lebrun, and J. M. Sprauel, "Present Possibilities for the X-Ray Diffraction Method of Stress Measurement," *N.D.T. Int.*, 235, October (1981).

6.   I. C. Noyan and L. T. Nguyen, "Oscillations in Interplanar Spacing *vs.* $\sin^2 \psi$: A FEM Analysis," *Adv. X-Ray Anal.*, 31, 223 (1988).

7.   I. C. Noyan and L. T. Nguyen, "Effect of Plastic Deformation on Oscillations in "d" *vs.* $\sin^2 \psi$ Plots: A FEM Analysis," *Adv. X-Ray Anal.*, 32 (1989).

8.   M. Karnezos, "Effects of Stress on the Stability of X-Ray Masks," *J. Vac. Sci. Technol.*, 4, 226 (1986).

9.   E. S. Meieran and I. A. Blech, "X-Ray Extinction Contrast Topography of Silicon Strained by Thin Surface Films," *J. Appl. Phys.*, 36, 3162 (1965).

10.   L. J. van Mellaert and G. H. Schwuttke, "Feedback Control Scheme for Scanning X-Ray Topography," *J. Appl. Phys.*, 43, 687 (1972).

11.   A. Segmuller, J. Angilelo, and S. J. La Placa, "Automatic X-Ray Diffraction Measurement of the Lattice Curvature of Substrate Wafers for the Determination of Linear Strain Patterns," *J. Appl. Phys.*, 51, 6224 (1980).

12.   E. W. Hearn, "Stress Measurements in Thin Films Deposited on Single Crystal Substrates Through X-Ray Topography Techniques," *Adv. X-Ray Anal.*, 20, 273 (1977).

13.   C. Goldsmith, P. Geldermans, F. Bedetti, and G. A. Walker, "Measurement of Stresses Generated in Cured Polyimide Films," *J. Vac. Sci. Technol.*, 1, 407 (1983).

14.   R. J. Jaccodine and W. A. Schlegel, "Measurement of Strains at Si-SiO$_2$ Interface," *J. Appl. Phys.*, 37, 2429 (1966).

15.   W. A. Brantley, "Calculated Elastic Constants for Stress Problems Associated with Semiconductor Devices," *J. Appl. Phys.*, 44, 534 (1973).

16.   L. T. Nguyen and I. C. Noyan, "X-Ray Determination of Encapsulation Stresses on Silicon Wafers," *Polym. Eng. Sci.*, 28, 1013 (1988).

17. I. C. Noyan and L. T. Nguyen, "Residual Stresses in Polymeric Passivation and Encapsulation Materials," *Polym. Eng. Sci.*, 28, 1026 (1988).

18. L. T. Nguyen and I. C. Noyan, "Macro and Microstress Distributions in Filled Epoxy Systems," *Adv. X-Ray Anal.*, 31, 223 (1988).

19. L. T. Nguyen and I. C. Noyan, "Encapsulation Stresses in Filled Epoxy Systems," *S.P.E. Reg. Tech. Conf.: Eng. Polym. Compos.*, Chicago, IL, September 23-24, 311 (1987).

20. P. V. Robock and L. T. Nguyen, "Plastic Packaging," Chapter 8, in **Microelectronics Packaging Handbook**, R. R. Tummala and E. J. Rymaszewski, Eds., Van Nostrand-Reinhold (1988).

21. L. T. Nguyen, J. C. Poler, S. L. Buchwalter, and C. A. Kovac, "Physical Properties of Cycloaliphatic Epoxide Formulations," *S.P.E. 45th Ann. Tech. Conf.*, 391, Los Angeles, CA, May 4-7 (1987).

22. T. P. Russell, "Concerning Voids in Polyimide," *Polym. Eng. Sci.*, 24, 345 (1984).

23. N. F. Chugunova, V. M. Startsev, G. M. Bartenev, I. I. Bardyshev, and A. G. Skvortsov, "Kinetic Study of Free Volume Changes Occurring in Films During Thermal Imidization of Polyamic Acid," *Polym. Sci. U.S.S.R.*, 26, 2147 (1984).

24. H. M. Tong, C. K. Hu, C. Feger, and P. Ho, "Stress Development in Supported Films During Thermal Cycling," *Polym. Eng. Sci.*, 26, 1213 (1986).

25. E. Sacher and J. R. Susko, "Water Permeation of Polymer Films. I. Polyimides," *J. Appl. Polym. Sci.*, 23, 2355 (1979).

26. E. Sacher and J. R. Susko, "Water Permeation of Polymer Films. III. High Temperature Polyimides," *J. Appl. Polym. Sci.*, 26, 679 (1981).

27. R. Ginsburg and J. R. Susko, "High Temperature Stability of a Polyimide Film," *I.B.M. J. Res. Develop.*, 28, 735 (1984).

28. L. R. Iler, R. C. Laundon, and W. J. Koros, "Characterization of Penetrant Interactions in Kapton ® Polyimide using a Gravimetric Sorption Technique," *J. Appl. Polym. Sci.*, 27, 1163 (1982).

29. B. Jordan, "Lower Stress Encapsulants," *I.E.E.E. Proc. Electron. Comp. Conf.*, 130 (1981).

30. H. W. Rauhut, "Low Stress and Live Device Performance of Experimental Epoxy Encapsulants," *S.P.E. Reg. Tech. Conf.*, Toronto, Canada, March (1983).

31. H. T. Hahn and N. J. Pagano, "Curing Stresses in Composite Laminates," *J. Compos. Mat.*, 9, 91 (1975).

32. H. T. Hahn, "Residual Stresses in Polymer Matrix Composite Laminates," *J. Compos. Mat.*, 10, 266 (1976).

33. I. M. Daniel, R. E. Rowlands, and D. Post, "Strain Analysis of Composites by Moire Methods," *Exp. Mech.*, 13, 246 (1973).

34. R. B. Pipes and J. W. Dalley, "On the Birefringent Coating Method of Stress Analysis for Fiber Reinforced Composite Laminates," *Exp. Mech.*, 12, 272 (1972).

35. R. C. Sampson, "Holographic Interferometry Applications in Experimental Mechanics," *Exp. Mech.*, 10, 313 (1970).

36. P. K. Predecki and C. S. Barrett, "Stress Measurement in Graphite/Epoxy Composites by X-Ray Diffraction from Fillers," *J. Compos. Mat.*, 13, 61 (1979).

37. C. S. Barrett and P. K. Predecki, "X-Ray Diffraction Evaluation of Adhesive Bonds and Stress Measurement with Diffracting Paint," *Adv. X-Ray Anal.*, 24, 231 (1981).

38. C. S. Barrett and P. K. Predecki, "Measuring Triaxial Stresses in Embedded Particles by Diffraction," *Adv. X-Ray Anal.*, 21, 305 (1978).

39. G. G. Stoney, "The Tension of Metallic Films Deposited by Electrolysis," *Proc. Roy. Soc. London*, A82, 172 (1909).

40. S. Timoshenko, "Analysis of Bi-Metal Thermostats," *J. Opt. Soc. Am.*, 11, 233 (1925).

41. E. Suhir, "Calculated Thermally Induced Stresses in Adhesively Bonded and Soldered Assemblies," *I.S.H.M. Int. Symp. Microelectron.*, 383 (1986).

42. K. Kinosita, "Recent Developments in the Study of Mechanical Properties of Thin Films," *Thin Solid Films*, 12, 17 (1972).

43. P. Chaudhari, "Mechanisms of Stress Relief in Polycrystalline Films," *I.B.M. J. Res. Develop.*, 13a, 177 (1969).

44. N. N. Davidenkov, "Measurement of Residual Stresses in Electrolytic Deposits," *Sov. Phys.: Solid State*, 2, 2595 (1961).

45. R. W. Hoffman, in **Physics of Nonmetallic Thin Films**, C. H. S. Dupuy and A. Cachard, Eds., 273, Plenum Press, New York (1976).

46. T. S. Chow, C. A. Liu, and R. C. Penwell, "Direct Determination of Interfacial Energy Between Brittle and Polymeric Films," *J. Polym. Sci.: Polym. Phys.*, 14, 1305 (1976).

47. T. S. Chow, C. A. Liu, and R. C. Penwell, "Strains Induced upon Drying Thin Films Solvent-Cast on Flexible Substrates," *J. Polym. Sci.: Polym. Phys.*, 14, 1311 (1976).

48. R. C. Penwell and T. S. Chow, "Minimizing Curl Induced upon Drying Layered Solvent Coated Films," *Polym. Eng. Sci.*, 25, 367 (1985).

49. F. Hospers and L. B. Vogelesang, "Determination of Residual Stresses in Aluminum Alloy Sheet Material," *Exp. Mech.*, 107, March (1975).

50. K. Roll, "Analysis of Stress and Strain Distribution in Thin Films and Substrates," *J. Appl. Phys.*, 47, 3224 (1976).

51. G. E. Henein and W. R. Wagner, "Stresses Induced in GaAs by TiPt Ohmic Contacts," *J. Appl. Phys.*, 54, 6395 (1983).

52. S. N. G. Chu, A. T. Macrander, K. E. Strege, and W. D. Johnston, Jr., "Misfit Stress in InGaAs/InP Heteroepitaxial Structures Grown by Vapor Phase Epitaxy," *J. Appl. Phys.*, 57, 249 (1985).

53. I. C. Noyan and A. Segmuller, "Comment on 'Misfit Stress in InGaAs/InP Heteroepitaxial Structures Grown by Vapor Phase Epitaxy'," *J. Appl. Phys.*, 60, 2980 (1986).

54. R. H. Saul, "Effect of a GaAs$_x$P$_{1-x}$ Transition Zone on the Perfection of GaP Crystals Grown by Deposition onto GaAs Substrates," *J. Appl. Phys.*, 40, 3273 (1969).

55. C. T. Sun, "Theory of Laminated Plates," *J. Appl. Mech.*, 231, March (1971).

56. T. S. Chow, "Theory of Unsymmetric Laminated Plates," *J. Appl. Phys.*, 46, 219 (1975).

57. T. S. Chow, "Thermal Warping of Layered Composites," *J. Appl. Phys.*, 47, 1351 (1976).

58. G. H. Olsen and M. Ettenberg, "Calculated Stresses in Multilayered Heteroepitaxial Structures," *J. Appl Phys.*, 48, 2543 (1977).

59. V. I. Lavrenyuk and V. A. Makara, "Thermal Stresses in a Multilayer Semiconductor Structure," English translation of *Problemy Prochnosti*, Kiev University, 10, 116 (1983).

60. J. Vilms and D. Kerps, "Simple Stress Formula for Multilayered Thin Films on a Thick Substrate," *J. Appl Phys.*, 53, 1536 (1982).

61. Z. C. Feng and H. D. Liu, "Generalized Formula for Curvature Radius and Layer Stresses Caused by Thermal Strain in Semiconductor Multilayer Structures," *J. Appl. Phys.*, 54, 83 (1983).

62. E. Suhir, "An Approximate Analysis of Stresses in Multilayered Elastic Thin Films," *J. Appl. Mech.*, 143, March (1988).

4

LASER INTERFEROMETRY: A MEASUREMENT TOOL FOR THIN FILM POLYMER PROPERTIES AND PROCESSING CHARACTERISTICS

K. L. Saenger and H. M. Tong

Thomas J. Watson Research Center
IBM Research Division
Yorktown Heights, New York

A large number of techniques for thickness and/or index measurements utilize optical interference. An excellent review by Pliskin and Zanin (1) covers many of them, including CARIS (constant-angle reflection interference spectroscopy), VAMFO (variable angle monochromatic fringe observation), and ellipsometry (also known as polarimetry or polarization spectroscopy).

Conventional thickness measurement techniques are not well suited for real-time measurements of rapidly evolving films. Profilometers relying on physical contact are obviously inappropriate for wet films, and inconvenient at best for films immersed in solvent. CARIS-based, noncontact film thickness analyzers (e.g., Leitz MPV-SP), which determine the thickness from the wavelength dependence of sample reflectance, are often slow (due to the scanning time required) compared with the rates of thickness change observed during drying or etching. These techniques do, however, provide *absolute* thickness (or index) measurements of static films, in contrast to the single wavelength, fixed polarization, single angle of incidence techniques that are the subject of this chapter. The latter techniques are ideally suited to real-time measurements of thickness and/or refractive index *changes*. The convenience and availability of lasers as sources of coherent, monochromatic light has also broadened the appeal of these techniques. This chapter surveys recent advances in their use in the area of polymer science and technology.

Types of Interferometers: Interferometers can generally be classified as belonging to one of two categories: wavefront-splitting interferometers and amplitude-splitting interferometers (2). Our focus here will be on amplitude-splitting interferometers, in which a light wave is split by a partially reflecting surface into two separate beams which are recombined before arriving at a detector. There are numerous designs for amplitude splitting interferometers, the following four of which are illustrated in Fig. 4.1: Michelson (I), Simple (II), Fabry-Perot type (III) and Modified Jamin (IV).

For all of these interferometers, the detected light intensity depends on the optical path length difference between the two or more recombined reflections. In interferometers II-IV, the optical path length difference is proportional to $2nL$, where L is the film thickness and n is the refractive index of the film or air gap. If

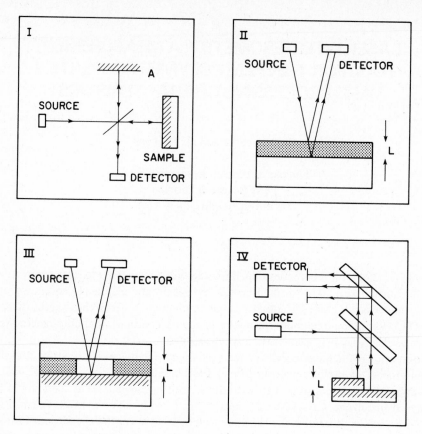

Figure 4.1: Four interferometers: I. Michelson; II. Simple; III. Fabry-Perot; IV. Modified Jamin. The quantity L is the polymer thickness.

the optical path length difference is monotonically increased (or decreased), the reflected light intensity will pass through maxima and minima as the amplitudes of the reflections originating from the top and bottom surfaces of the film or air gap constructively and destructively interfere. The difference between two successive minima or maxima is known as a fringe. The resolution (the smallest thickness change that can be accurately measured) is roughly half a fringe (typically ~ 100 nm), although much smaller fractions of a fringe can be observed (3,4).

Interferometers for Polymer Applications: In the applications of laser interferometry described here, *changes* in polymer film thickness and/or refractive index are detected from changes in reflected light intensity. Polymer properties and processing characteristics that are directly related to changes in film thickness and/or refractive index include solvent uptake/swelling and/or dissolution, thermal expansion coefficient, structural stability, and shrinkage rates during drying/curing. In choosing an interferometer for a particular application, the following factors should be considered.

Interferometers I and IV share the disadvantage that their two interfering beams travel large distances before being recombined. The physical separation of the beams makes the interferometers subject to noise from air currents which can perturb each beam differently. Interferometers II, III, and IV are immune to sample vibration since the two beam paths are equally affected by sample translation. Interferometer I, although subject to vibration noise, has the advantage that the initial optical path length difference can be adjusted independently of sample parameters (by translating reflector "A"). In sum, interferometers II and III are inherently noise-free since the optical path length difference in the two reflected beams results only from the additional excursion of one of the beams back and forth through the film (in II) or enclosed air gap (in III).

A potential limitation of the simple interferometer II arises from the fact that it is sensitive to the *product* of n and L, that is, the polymer thickness and index changes are not separable. For polymer etch rate measurements where the polymer index is fixed while only the thickness changes, the simple interferometer II is ideal. For situations where both the thickness and index of the film change during the experiment (e.g., sorption of strong swelling agents), the simple interferometer II is acceptable if a model is available to relate the film's index to its thickness. A linear model, in which the film's index is the volume-weighted average of the indices of the solvent and dry polymer, is frequently adequate, and will be discussed in the next section. For the sake of completeness, we mention another potential limitation of interferometer II: there will be no reflection at an interface with perfectly matched refractive indices. This is not a particularly severe constraint, if one considers the number of solvent/polymer and polymer/substrate pairs whose indices do *not* exactly match. The dependence of the interference signal on the reflectivities of each interface will be quantitatively treated in a later section.

For thermal expansion measurements, the simple interferometer II may not be acceptable if the polymer's refractive index varies with temperature. For this application, interferometers such as III and IV are more appropriate, in that the optical paths do not pass through the polymer.

The remainder of this chapter will primarily deal with interferometers of designs II and III, with interferometer II being used in the majority of applications we discuss.

4.1 GENERAL PRINCIPLES

4.1.1 Expressions for Reflected Intensity

We now present expressions for the amplitude and intensity (proportional to $|amplitude|^2$) of light reflected from isotropic, layered media typically encountered in simple polymer-containing systems. Throughout this section, the reflected amplitudes will be characterized by reflection coefficients (normalized to incident amplitude) and the reflected intensities will be characterized by reflectances or reflectivities (normalized to incident intensity). Our treatment roughly follows that of Heavens (5) and is applicable to interferometers II and III. A similar treatment can be found in Ref. 6.

Reflection from a Single Interface: We first consider a single interface between adjacent semi-infinite media j and k with refractive indices n_j and n_k. At the j/k interface, the ratio of the reflected electric field component, E_j^-, to the incident electric field component, E_j^+, is given by the Fresnel reflection coefficients

$$r_{j,k}^p(\theta_j) = \frac{E_{jp}^-}{E_{jp}^+} = \frac{n_j \cos\theta_k - n_k \cos\theta_j}{n_j \cos\theta_k + n_k \cos\theta_j} \tag{1}$$

$$r_{j,k}^s(\theta_j) = \frac{E_{js}^-}{E_{js}^+} = \frac{n_j \cos\theta_j - n_k \cos\theta_k}{n_j \cos\theta_j + n_k \cos\theta_k} \tag{2}$$

where n_j is the refractive index of medium j, p and s denote "p-polarization" and "s-polarization", that is, the direction of the electric field vector as parallel (p) or perpendicular (s) to the plane of incidence, and θ_j is the angle of incidence at the j/k boundary. The θ_j (see Fig. 4.2) obey Snell's law:

$$n_j \sin\theta_j = n_k \sin\theta_k \tag{3}$$

Typically, laser interferometry experiments are done at or close to normal incidence ($\theta_0 = 0$). At normal incidence, $\theta_j = 0$ for all j [from Eq. (3)]. In this case, the plane of incidence is no longer defined and Eqs. (1) and (2) reduce to

$$r_{j,k} = r_{j,k}^s = r_{j,k}^p = \frac{E_j^-}{E_j^+} = \frac{n_j - n_k}{n_j + n_k} \tag{4}$$

The single interface reflectances (for the j/k interface) are defined as the ratio of reflected to incident intensity (i.e., power), and are given by the square of the Fresnel reflection coefficients, $|r_{j,k}^s(\theta)|^2$ or $|r_{j,k}^p(\theta)|^2$, for s or p polarized light, respectively. The single interface reflectance for unpolarized light is the average of $|r_{j,k}^s(\theta)|^2$ and $|r_{j,k}^p(\theta)|^2$ (Ref. 2).

Reflection from a Multilayer System: A prototype three-medium system is shown in Fig. 4.2a. Such a system would be encountered during dissolution of a polymer layer of refractive index n_1 immersed in a solvent of index n_0 on a substrate of index n_2, providing that the solvent/polymer interface is a sharp one and that there is no reflection from the back side of the substrate. There are now reflections from two interfaces, and interference effects must be taken into account in calculating the amplitude and intensity of the reflected light. The *composite* reflection coefficient $\tilde{R}$, defined as the ratio of the electric field component reflected from the 0/1 interface, E_0^-, to the electric field component incident on the 0/1 interface, E_0^+, is now given by the complex quantity

$$\tilde{R} = \frac{E_0^-}{E_0^+} = \frac{r_{01} + r_{12}e^{-i\phi_1}}{1 + r_{01}r_{12}e^{-i\phi_1}} \tag{5}$$

where $i = \sqrt{-1}$ and ϕ_1 is the phase lag or phase shift,

$$\phi_1(\theta_1) = \left(\frac{2\pi}{\lambda}\right) 2n_1 L_1 \cos\theta_1 \tag{6}$$

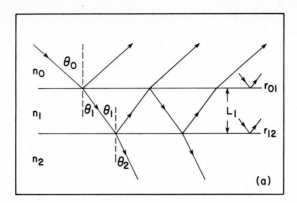

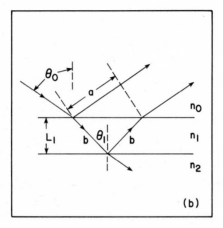

Figure 4.2:   (a) Schematic of a three-medium system; (b) Expanded view of optical path length difference for a non-normal angle of incidence.

with L_1 being the thickness of medium 1, and λ being the laser wavelength in vacuum. In the limit where multiple reflections are unimportant (i.e., $|r_{01}|$, $|r_{12}|$ $<<1$), $\tilde{R}$ is given by the numerator of Eq. (5); deviations of the denominator of Eq. (5) from unity are due to the effect of multiple reflections.

The composite reflectance $\mathscr{R}$ is the product of the composite reflection coefficient $\tilde{R}$ and its complex conjugate $\tilde{R}^*$:

$$\mathscr{R}(r_{01}, r_{12}) \equiv \frac{|E_0^-|^2}{|E_0^+|^2} = \frac{r_{01}^2 + r_{12}^2 + 2r_{01}r_{12}\cos\phi_1}{1 + r_{01}^2 r_{12}^2 + 2r_{01}r_{12}\cos\phi_1} \qquad (7)$$

For polarized light at non-normal incidence, $\mathscr{R}^s = \mathscr{R}(r_{01}^s, r_{12}^s)$ is the reflectance of the s-polarized component, and $\mathscr{R}^p = \mathscr{R}(r_{01}^p, r_{12}^p)$ is the reflectance of the p-polarized component. As before, the reflectance for unpolarized light is the average of $\mathcal{R}^s$ and $\mathcal{R}^p$; the reflectance for polarized light is the incident-intensity-weighted average of $\mathcal{R}^s$ and $\mathcal{R}^p$ (Ref. 6).

For normal incidence, ϕ_1 is $2\pi/\lambda$ times the optical path length difference $(2L_1n_1)$ between the beams reflected from the two surfaces of medium 1. The quantity $2L_1$ is the distance of a round trip through medium 1. For non-normal incidence, the optical path length difference becomes $2L_1n_1\cos(\theta_1)$. This is the difference between the optical path length traveled by the beam traversing medium 1 [$n_1 \times 2 \times$ distance "b" in Fig. 4.2b = $2L_1n_1/\cos(\theta_1)$] and that traveled by the beam reflected from the 0/1 interface ($n_0 \times$ distance "a" in Fig. 4.2b) before the two beams recombine. A "fringe" is obtained when ϕ_1 changes by 2π. The values of the reflectivity $\mathcal{R}$ for $\cos(\phi_1) = \pm 1$ are denoted by $\mathcal{R}_+$ and $\mathcal{R}_-$. For normal incidence we find [with the aid of Eq. (4)]

$$\mathcal{R}_+ = \left(\frac{n_0 - n_2}{n_0 + n_2} \right)^2 \tag{8}$$

and

$$\mathcal{R}_- = \left(\frac{n_0 n_2 - n_1^2}{n_0 n_2 + n_1^2} \right)^2 \tag{9}$$

The maximum and minimum values of the reflectance, $\mathcal{R}_{max}$ and $\mathcal{R}_{min}$, occur when $r_{01}r_{12}\cos(\phi_1)$ is equal to $+|r_{01}r_{12}|$ or $-|r_{01}r_{12}|$, respectively. In other words, if r_{01} and r_{12} are both negative or both positive, $\mathcal{R}_{max}$ and $\mathcal{R}_{min}$ are equal to $\mathcal{R}_+$ and $\mathcal{R}_-$, respectively. (For near-normal incidence, this occurs when the value of n_1 is between the values of n_0 and n_2.) If r_{01} and r_{12} are of opposite sign, $\mathcal{R}_{max}$ and $\mathcal{R}_{min}$ are equal to $\mathcal{R}_-$ and $\mathcal{R}_+$, respectively.

The quantities $\mathcal{R}_{max}$ and $\mathcal{R}_{min}$ are experimentally observable. The reflectance $\mathcal{R}_+$ is identical to the reflectance of the bare substrate. Equations (8) and (9) can be used to determine n_1 from $\mathcal{R}_+$, $\mathcal{R}_-$, n_0, and n_2, providing the signs of $r_{01}r_{12}$ and a quantity $q \equiv (n_0n_2 - n_1^2)/(n_0 - n_2)$ are known. The former sign determines whether $\mathcal{R}_-/\mathcal{R}_+ = \mathcal{R}_{min}/\mathcal{R}_{max}$ or $\mathcal{R}_{max}/\mathcal{R}_{min}$. The sign of q determines the appropriate sign in the expression below for n_1 (upper (lower) for positive (negative) q):

$$n_1 = (n_0 n_2)^{1/2} \left(\frac{[(n_0 + n_2)/(n_0 - n_2)] \mp (\mathcal{R}_-/\mathcal{R}_+)^{1/2}}{[(n_0 + n_2)/(n_0 - n_2)] \pm (\mathcal{R}_-/\mathcal{R}_+)^{1/2}} \right)^{1/2} \tag{10}$$

At large angles of incidence, the ratio $\mathcal{R}_-/\mathcal{R}_+$ deviates considerably from its value at normal incidence, and the differences between its values for s and p polarizations are very apparent, as shown in Ref. 6.

For the four-medium system shown in Fig. 4.3, the composite reflection coefficient $\tilde{R}$ is given by (7)

$$\tilde{R} = \frac{E_j^-}{E_j^+} = \frac{r_{01} + r_{12}e^{-i\phi_1} + r_{23}e^{-i(\phi_1 + \phi_2)} + r_{01}r_{12}r_{23}e^{-i\phi_2}}{1 + r_{01}r_{12}e^{-i\phi_1} + r_{01}r_{23}e^{-i(\phi_1 + \phi_2)} + r_{12}r_{23}e^{-i\phi_2}} \tag{11}$$

with

$$\phi_j(\theta_j) = \left(\frac{2\pi}{\lambda} \right) 2n_j L_j \cos \theta_j$$

where θ_0 is again the angle of incidence, and the $r_{j,k}$ are defined in Eqs. (1), (2), or (4). Such a system might be encountered during sorption with an idealized, perfectly sharp front between the penetrant-saturated and dry polymer layers; n_0 through n_3 would then correspond to the refractive indices of the penetrant, penetrant-saturated polymer film, dry polymer, and substrate, respectively. The algebra required to generate an expression for the composite reflectance ($\mathcal{R} = \tilde{R}\tilde{R}^*$) in the four-layer system is straightforward, but tedious. The minimum and maximum values of $\mathcal{R}$ are best obtained numerically with the aid of a computer.

4.1.2 Fringe Calibration

In the three-medium system shown in Fig. 4.2a, a fringe is obtained when the phase shift $\phi_1 = (2\pi/\lambda) 2n_1 L_1$ changes by 2π. In this section we discuss the dependence of ϕ_1 on the various experimental parameters. Again, n_1 is the index of medium 1 and L_1 is its thickness. For brevity of notation, we assume normal incidence; for incidence at angle θ_0, L_1 in the equations below should be replaced by $L_1 \cdot \cos(\theta_1)$. Interferometer II is assumed unless stated otherwise.

ΔL_1 /**Fringe: Fixed Refractive Index**: When the polymer's thickness changes while its index $n_1 = n_{poly}$ remains constant, the fringe calibration, ΔL_1/fringe, is

$$\Delta L_1/\text{fringe} = \left(\frac{d\phi/dL_1}{2\pi} \right)^{-1} = \frac{\lambda}{2 n_{poly}} \qquad (12)$$

This is the case for polymer etching (dry or wet), providing that (1) any etching-induced index changes in the polymer are limited to an interfacial layer between the film and the etchant, and (2) the interfacial layer has reached its steady-state thickness. These conditions are not satisfied if solvent penetration alters the index within the bulk of the film. With $\lambda = 632.8$ nm and n_{poly} typically around 1.6, ΔL_1/fringe is typically ~ 200 nm.

ΔL_1 /**Fringe: Time-Varying Refractive Index**: When medium 1 is a polymer/solvent solution whose refractive index is changing due to a changing solvent concentration (e.g., during solvent uptake or evaporation where the dry polymer mass is conserved) the fringe calibration, ΔL_1/fringe, is (8)

$$\Delta L_1/\text{fringe} = \left(\frac{d\phi/dL_1}{2\pi} \right)^{-1} = \frac{\lambda}{2 n_{solvent}} \qquad (13)$$

This expression indicates that ΔL/fringe depends on the solvent's refractive index $n_{solvent}$, rather than on the refractive index of the *solution*, $n_{solution}$. The result is based on a linear model in which the film is a binary mixture of polymer and solvent whose index of refraction is the volume-weighted average of the indices of its components; that is,

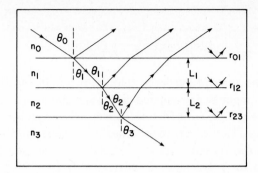

Figure   4.3:   Schematic of a four-medium system.

$$n_{solution} = \frac{(L_{poly}n_{poly} + L_{solvent}n_{solvent})}{(L_{poly} + L_{solvent})} \tag{14}$$

where L_{poly} and $L_{solvent}$ are the integrated thicknesses of the solvent and polymer, respectively. [The integrated thickness of a solution component is its (depth-dependent) volume fraction integrated over the (solution's) total film thickness; equivalently, it may be thought of as the thickness of the neat layer that would be produced if the solution separated into its component parts (while keeping a constant cross-sectional area).] The total film thickness $L_1 = L_{poly} + L_{solvent}$; L_{poly} is fixed, and $L_{solvent}$ changes during the experiment. With the appropriate substitutions, we find $(\phi/2\pi) = (2/\lambda) \times [L_1 n_{solvent} + L_{poly}(n_{poly} - n_{solvent})]$, which after differentiation yields Eq. (13). The thickness variation per fringe is thus a constant, $\lambda/2n_{solvent}$, rather than the time-dependent $\lambda/2n_{solution}$. This intuitively makes sense since the thickness change is entirely due to the changing solvent content.   With $\lambda = 632.8$ nm and $n_{solvent} \sim 1.4$, ΔL_1/fringe is ~ 230 nm.

ΔT /**Fringe: Temperature Effects**: The dependence of the optical path length difference on temperature leads to

$$\Delta T/fringe = \left(\frac{d\phi/dT}{2\pi}\right)^{-1} = \frac{\lambda}{2n_1 L_1(\alpha_1 + \beta_1)} \tag{15}$$

where $\alpha_1 \equiv (1/L_1)(dL_1/dT)$ is the thermal expansion coefficient (TEC) of the polymer and $\beta_1 \equiv (1/n_1)(dn_1/dT)$ is the temperature coefficient of the polymer's refractive index.   A measurement of ΔT/fringe (with $\lambda = 632.8$ nm) for a 25-μm-thick polyimide film on a silicon substrate (see Section 4.3.3 for approximate curing schedule) gave about $90°$/fringe. With $n_1 = n_{polyimide} = 1.72$, this gives $(\alpha_1 + \beta_1) \sim 80 \times 10^{-6}/°C$, about twice the literature value (9) of α_1 for Kapton polyimide films.   This indicates that α and β may be of similar magnitude. Since interferometer II is sensitive only to the sum of α and β, rather than to α alone, the potentially non-negligible size of β is clearly a drawback for the use of interferometer II to determine α.

Interferometer III: $\Delta T/\textbf{Fringe}$ and ΔL /Fringe: For interferometer III, n_1 is replaced by the index of the air gap (=1.00) and β_1 is zero. Equation (15) then reduces to

$$\Delta T/\text{fringe} \;=\; \frac{\lambda}{2L_1\alpha_1} \tag{16}$$

The change in polymer thickness per fringe (e.g., due to thermal expansion) is given by Eq. (12) with n_{poly} replaced by the index of air:

$$\Delta L/\text{fringe} \;=\; \frac{\lambda}{2} \tag{17}$$

4.1.3 Transition Layer Effects on Reflectance

The expressions for the Fresnel reflection coefficients [$r_{j,k}$ in Eqs. (1), (2), or (4)] and the reflectivities presented so far have been for the case of clearly defined, sharp boundaries between homogeneous media with uniform refractive indices. During dissolution and/or solvent penetration, however, the homogenous media of indices n_j and n_k may be separated by an inhomogeneous transition layer whose index varies through its thickness, typically from approximately n_j to n_k. In this section, we modify the preceding expressions to account for the transition layer, and show how the modified expressions might be used in conjunction with the experimental data to determine the transition layer thickness. A schematic refractive index profile n(z) is shown in Fig. 4.4 for (a) a three-medium system with a transition layer and (b) a four-medium system with a transition layer. The former profile (a) pertains to dissolution and the latter profile (b) pertains to sorption; for both applications, interferometer II would be the design of choice.

The presence of a transition layer at the j/k boundary can substantially reduce the reflection over what it would be if the boundary between n_j and n_k were sharp. In dissolution studies (6,10-12), this reduced reflection manifests itself as an offset between the reflectivity maxima (or minima, if one of the Fresnel coefficients is negative) during etching, and the reflectivity of the bare substrate (given by Eq. (8)) after the etching is complete. This offset is illustrated in Fig. 4.6.

A number of mathematical approaches are available to calculate the reflection coefficients of inhomogeneous films (13). Recently, in a very accessible treatment, Krasicky et al. (11) modeled the reflectivity of the transition layer in connection with their interferometric studies of poly(methyl methacrylate) (PMMA) dissolution in methyl ethyl ketone (MEK) (10). They found an expression for a complex quantity, $\widetilde{F}$, whose argument or complex phase ϕ_t is the phase lag of light passing through the transition layer and whose magnitude, f, is the "amplitude reduction factor." The effective (single interface) reflection coefficient for a diffuse boundary between media with indices j and k with an intervening transition layer (e.g., such as in Fig. 4.4a or 4.4b) is the product of the amplitude reduction factor, f, and $r_{j,k}$, where $r_{j,k}$ is the Fresnel reflection coefficient given by Eqs. (1), (2), or (4) for a sharp boundary. The reflectance for the system in Fig. 4.4a with the transition layer between medium 0 and medium 1 is thus given by the appropriate modification of Eq. (5):

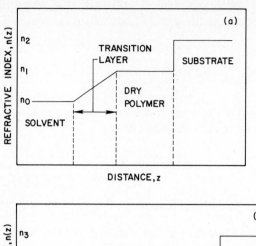

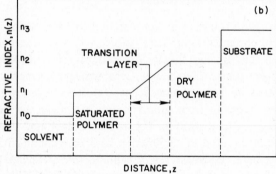

Figure 4.4: (a)Schematic refractive index profile of a three-medium system with a transition layer. (b) Refractive index profile of a four-medium system with a transition layer. Systems (a) and (b) are applicable to dissolution and sorption, respectively.

$$\mathscr{R}(f) = \frac{f^2 r_{01}^2 + r_{12}^2 + 2\,f\,r_{01}r_{12}\cos(\phi)}{1 + f^2 r_{01}^2 r_{12}^2 + 2\,f\,r_{01}r_{12}\cos(\phi)} \tag{18}$$

where $\phi = \phi_1 + \phi_t$. A detailed justification for this expression can be found in Ref. 6. In the limit of a sharp boundary, $\phi_t = 0$ and $f = 1$.

The expression for $\tilde{F}$ is

$$\tilde{F}(\gamma) = \int_0^\infty e^{i\gamma u}\,\frac{-dg(u)}{du}\,du \tag{19}$$

where u is the scaled thickness, that is, the ratio of the coordinate in the thickness direction, z, and the transition layer thickness L_t,

$$u \equiv \frac{z}{L_t} \tag{20}$$

g(u) is a scaled index,

$$g(u) \equiv \frac{n(u) - n_0}{n_1 - n_0} \tag{21}$$

$$\gamma \equiv \left(\frac{2\pi}{\lambda} \right) 2 \, \overline{n}_t \, L_t \tag{22}$$

and

$$\overline{n}_t = \frac{n_0 + n_1}{2} \tag{23}$$

For a linear profile, $g(u) = 1 - u$, and $dg(u)/du = -1$ for $0 \leq u \leq 1$ and 0 elsewhere. This yields

$$\phi_t = \frac{\gamma}{2} = \left(\frac{2\pi}{\lambda} \right) \overline{n}_t \, L_t \tag{24}$$

and

$$f = |F| = \frac{|\sin(\gamma/2)|}{(\gamma/2)} \tag{25}$$

Equations (24) and (25) relate the amplitude reduction factor, f, to the transition layer thickness. The value of f can be experimentally determined from the ratio of the value of $\mathcal{R}_+(f)$ measured during dissolution to the value of $\mathcal{R}_+(f = 1)$ measured from the bare substrate after dissolution, providing that the indices n_0, n_1, and n_2 are known. We find

$$f = \frac{r_{12} - \rho r_{02}}{r_{01} \, (\rho r_{12} r_{02} - 1)} \tag{26}$$

where we define

$$\rho \equiv \left(\frac{\mathcal{R}_+(f)}{\mathcal{R}_+(f = 1)} \right)^{1/2} \tag{27}$$

A simplified expression for f as a function only of the experimentally measured parameters $\mathcal{R}_{max}(f)$, $\mathcal{R}_{min}(f)$, and $\mathcal{R}_+(f = 1)$ for the case where $|r_{01}| \ll |r_{12}|$ is given by (11)

$$f = \frac{\mathcal{R}_{max}(f) - \mathcal{R}_{min}(f)}{2\mathcal{R}_+(f = 1) - (\mathcal{R}_{max}(f) + \mathcal{R}_{min}(f))} \tag{28}$$

The reflection coefficient (E_0^-/E_0^+) for the four-medium system in Fig. 4.4b with a transition layer between medium 1 and medium 2 is given by the appropriate modification of Eq. (11):

$$\tilde{R}(f) = \frac{r_{01} + f \, r_{12} e^{-i\phi_1} + r_{23} e^{-i(\phi_1 + \phi_2)} + f \, r_{01} r_{12} r_{23} e^{-i\phi_2}}{1 + f \, r_{01} r_{12} e^{-i\phi_1} + r_{01} r_{23} e^{-i(\phi_1 + \phi_2)} + f \, r_{12} r_{23} e^{-i\phi_2}} \tag{29}$$

where the ϕ_i are given by $(2\pi/\lambda)(2n_iL_i) + \phi_t$, and the subscripts 0 and 1 in Eqs. (21) and (23) are replaced by the subscripts 1 and 2. As before, the reflectivity is given by the product of $\tilde{R}$ and $\tilde{R}^*$.

4.2 INTERFEROMETER CONSTRUCTION

4.2.1 Apparatus

Setup: The typical experimental arrangement for interferometer II is shown in Fig. 4.5a for an air environment, and in Fig. 4.5b for a liquid environment. The typical experimental arrangement for interferometer III (in air) is shown in Fig. 4.5c. (The details of construction for interferometer II can be found in Refs. 6 and 14; for interferometer III, they will be discussed in Section 4.3.5.)

A typical reflectivity trace is shown in Fig. 4.6 for the case of polymer etching. All rate measurements are performed by measuring the fringes evolved per unit time, and converting that to a thickness decrease per unit time with the aid of the appropriate expression for $\Delta L/fringe$ [Eq. (12), (13), or (17)]. Transition layer data are extracted from the observed values of $\mathcal{R}_{max}(f)$ and/or $\mathcal{R}_{min}(f)$, and $\mathcal{R}_+(f = 1)$ (all indicated in Fig. 4.6), with the aid of Eq. (26) or (28). Thermal expansion coefficients are extracted from the observed $\Delta T/fringe$ values (using interferometer III) with the aid of Eq. (16) and the known film thickness.

Light Source/Detector: The prototypical light source for laser interferometry experiments is a low power ($\sim$ 1 mW) red HeNe laser at a wavelength of 632.8 nm. These lasers have highly stable output powers (typically exhibiting a drift of <2% after a warm-up time of 10-60 min), beam diameters d_0 of $\sim$ 0.6 mm, and angular divergences Θ of $\sim$ 1 mrad. The beam diameter d at a distance z from the laser is given by (15) $d(z) = d_0[1 + z^2(\Theta/d_0)^2]^{1/2}$. With these specifications, the beam diameter will typically be about 1.2 mm at z= 1 m. A number of other laser wavelengths are available as well, although usually at lower output power and higher cost per mW. They include a yellow HeNe at 594.1 nm , a green HeNe ("GreNe") at 543.5 nm, a blue HeCd at 441.6 nm, and an ultraviolet (uv) HeCd at 325 nm. Shorter wavelengths give better resolution (i.e., a smaller $\Delta L/fringe$, which is proportional to $1/\lambda$). However, one's choice of λ may be limited by the constraint that the polymer (and/or solvent) be transparent to the laser.

The typical light detector for laser interferometry experiments is a silicon photodiode, which can be used either in the photovoltaic (unbiased) or photoconductive (reverse biased) mode. The photodiode area should be large enough to collect all the reflected laser light. One can use a large area photodiode ($\sim$ 1 cm²), or a smaller one at the focal point of a collection lens. Both are easy to align and insensitive to small drifts in beam position. The light reaching the photodiode can be filtered by a narrow (1 to 10 nm) bandpass filter centered around the laser wavelength to selectively reject room light. The photodiode output is then displayed on a chart recorder and/or digitized and stored by a microcomputer.

4.2.2 Basic Constraints

Angle of Incidence: The angle of incidence for laser interferometry experiments is typically close to normal; the equations for normal incidence are usually sufficiently accurate for θ up to $10°$ (6). It is important to keep in mind, however, that the angle

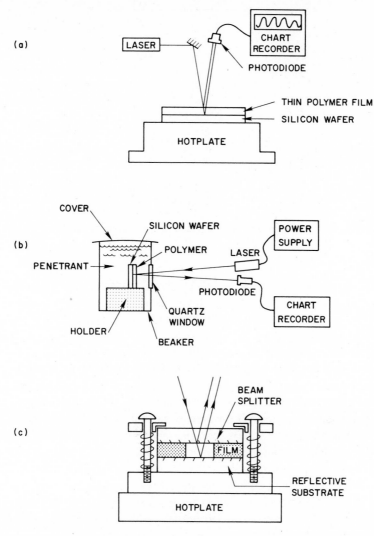

Figure   4.5:   Generic setups: (a) Interferometer II on a hot plate in air (from Ref. 8); (b) Interferometer II in liquid (from Ref. 14); (c) Interferometer III on a hot plate in air.

of incidence can influence the degree to which the interfering beams overlap; if the beams do not overlap there will be no interference.   Figure 4.7 illustrates the situation for the case of a thin film on a thick transparent substrate where the laser beam diameter is large compared to the film thickness.   The reflections from the top and bottom of the film overlap substantially.   In practice, typical beam diameters are of the order of 0.4 to 1 mm, and film thicknesses are less than 10 μm, so this is not a problem.   The reflection from the back surface of the substrate shown in Fig. 4.7 is spatially separated from the beams reflected from the polymer, and does not contribute to the interference.   This fact can be exploited experimentally to eliminate

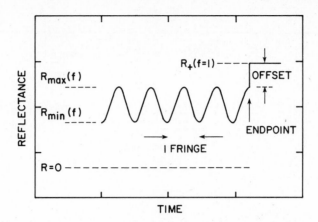

Figure   4.6:   Generic reflectivity trace for constant-rate polymer etching.

interference from the back-surface reflections of transparent substrates by using thick substrates and a slightly oblique angle of incidence. Back-surface reflections from transparent substrates can also be avoided by using wedge-shaped (as opposed to parallel-sided) substrates.

Surface Quality: The typical sample for interferometer II (or III) is an optically smooth, flat polymer film of uniform thickness bonded to an optically smooth substrate. If either the substrate or top reflector is rough, the intensity of the specular reflection from that surface will be reduced by diffuse scattering. If the film thickness has nonuniformities over the area of the incident laser beam (i.e., local variations in optical path length difference), the modulated intensity ($\mathscr{R}_+ - \mathscr{R}_-$) can decrease due to an averaging of the various local reflectances. If the film is not flat (i.e., nonparallel surfaces), the two reflections may be spatially separated before reaching the detector, thus eliminating the interference signal. In principle, the roughness effect could actually be used to good advantage: the specular reflection from the substrate's back surface could be reduced by deliberately roughening that surface (e.g., by sandblasting).

Adhesion: It is important that the film adhere well to the substrate; otherwise, spurious signals can result if the film delaminates during the experiment. A number of adhesion promoters are available, if needed. A layer of Cr ($\gtrsim 50$nm) sometimes suffices, as does an application of "A1100" (a dilute aqueous solution of a silane primer from Ohio Valley Specialty Chemical). An interesting interferometric experiment by Dijkkamp et al. (16), however, illustrates that the spurious signals in one experiment can sometimes be the desired signal in another. They studied the (248-nm) excimer laser-induced delamination of uv-transparent polymer films ($0.1 - 1$ μm thick) originally bonded to uv-absorbing substrates. Their time-resolved reflectivity measurements indicated the presence of an expanding gas layer between the substrate and bottom polymer surface, with the observed 8-ns period oscillations resulting from interference between the substrate and the (moving) delaminated polymer surface.

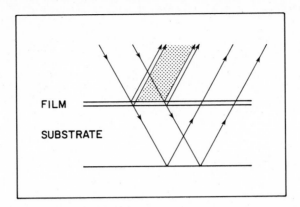

Figure   4.7:   An illustration of the overlapping reflections from a thin film on a thick transparent substrate for a beam of finite diameter and oblique angle of incidence.

4.2.3 Reflector/Materials Selection

The reflection coefficients of the front and back surfaces of the interferometer can have a dramatic effect on the interferometer's sensitivity to the signal of interest. In the following sections, two examples will be discussed in detail. In the first, we will show how interferometer II's sensitivity to a transition layer can be enhanced by choosing a substrate whose index is close to that of the polymer, that is, by minimizing the reflection coefficient of the polymer/substrate interface. In the second example, we will show how interferometer III's sensitivity to polymer thickness changes can be vastly increased by increasing the interferometer's reflection coefficients.

Sensitivity   to   Transition   Layer   (Interferometer   II):   The effect of polymer/substrate index matching on interferometer II's sensitivity to a transition layer is illustrated here for two related phenomena: dissolution (Fig. 4.8) and sorption (Fig. 4.9). In both cases the polymer film is PMMA (n = 1.489, Ref. 12) and the substrates to be compared are quartz, with a similar index of 1.457 (15), and silicon, with a much higher index of 3.8 (12). The laser light is normally incident with a wavelength of 632.8 nm.

Figure 4.8 gives the simulated reflected intensities as a function of time [calculated with Eq. (18)] for a case of constant-rate dissolution studied experimentally in Refs. 10 and 11. The solvent is MEK (n_0 = 1.380, Ref. 17), and the schematic concentration profile is given in Fig. 4.4a. The transition layer is taken to have an f value of 0.8, corresponding to a thickness of 80 nm [using Eq. (24) which is valid for a linear concentration profile]. The upper trace (a) is for a silicon substrate. The dotted lines are the values of $\mathscr{R}_{max}(f = 1)$ and $\mathscr{R}_{min}(f = 1)$, that is, the maximum and minimum reflectances expected for a transition layer of zero thickness. In this case, $\mathscr{R}_{max} = \mathscr{R}_+$ and $\mathscr{R}_{min} = \mathscr{R}_-$ because n_1 = 1.489 is between n_0 = 1.380 and n_2 = 3.8. The values of $\mathscr{R}_{max}(f)$ and $\mathscr{R}_{min}(f)$ during dissolution are only moderately sensitive to the presence of an 80-nm-thick transition layer. However, the reflected intensity signal is quite strong and well suited to measurements of etch rate. The fact that

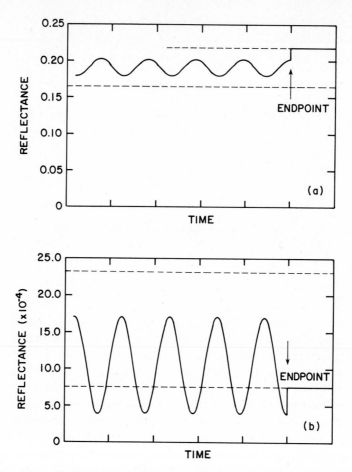

Figure 4.8: The effect of substrate index on reflectance and sensitivity to the transition layer calculated for the constant-rate dissolution of a 1-μm film of PMMA in MEK. The transition layer is characterized by an f value of 0.8 (L_t = 80 nm). Trace (a) is for a silicon substrate; trace (b) is for a quartz substrate. The dotted lines correspond to the maximum and minimum reflectivites expected for a transition layer of zero thickness.

the silicon substrate is opaque is a further advantage, in that no back-surface reflections from the substrate need be considered.

The lower trace (b) in Fig. 4.8 is for quartz. (In this figure, we assume that the substrate is configured to avoid back-surface reflections.) The dotted lines are again the values of $\mathscr{R}_{max}(f = 1)$ and $\mathscr{R}_{min}(f = 1)$ for a transition layer of zero thickness. In this case, however, $\mathscr{R}_{max} = \mathscr{R}_-$ and $\mathscr{R}_{min} = \mathscr{R}_+$ because n_1 is not between n_0 and n_2. The overall reflected intensity signal with quartz is much lower than with silicon, but the fractional modulation [defined as $(\mathscr{R}_+ - \mathscr{R}_-)/(\mathscr{R}_+ + \mathscr{R}_-)$] and sensitivity of $\mathscr{R}_{max}(f)$ and $\mathscr{R}_{min}(f)$ to the presence of a transition layer are significantly enhanced.

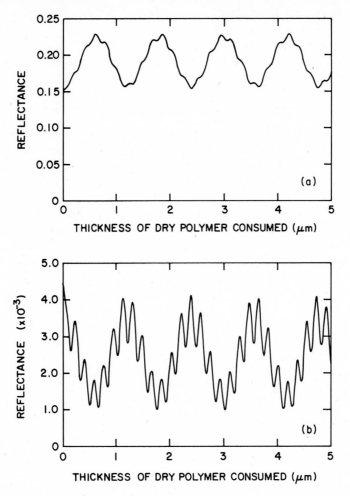

Figure   4.9:   The effect of substrate index on reflectance and sensitivity to the transition layer calculated for a 5-μm-thick PMMA film absorbing methanol from solution. The transition layer is characterized by an f value of 0.68 (L_t = 100 nm). The equilibrium swelling is taken to be 20%. Trace (a) is for a silicon substrate; trace (b) is for a quartz substrate.

Figure 4.9 gives the simulated reflected intensities during constant-rate sorption for quartz and silicon substrates [calculated with the aid of Eq. (29)]. The penetrant is methanol (n= 1.3288, Ref. 17) and the schematic concentration profile is given in Fig. 4.4b. The index of the solvent-saturated layer, n_1, is calculated to be 1.4623 from Eq. (14) for an equilibrium swelling of 20% (18). The transition layer is taken to have an f value of 0.68, corresponding to a thickness of 100 nm for a linear profile. As for the case of dissolution, the trace (a) in Fig. 4.9 for a silicon substrate gives a larger signal than that of trace (b) for the quartz substrate, but the quartz substrate is relatively more sensitive to the high-frequency oscillatory component arising from reflections from the solvent front (the boundary zone between media 1 and 2) as it moves through the film.

The reason for the enhanced sensitivity to the transition layer with quartz substrates is that the total reflectivity with a silicon substrate is dominated by reflection from the Si/polymer interface, i.e.,

$$| \, r_{polymer,Si} \, | \; >> \; | \, r_{polymer,quartz} \, | \; \sim \; | \, r_{solvent,polymer} \, |$$

Sensitivity to Thickness Changes: Interferometer III: Interferometer III consists of two partially reflecting surfaces separated by an air gap whose thickness matches that of the polymer film thickness. Figure 4.10 shows the reflectivity [calculated with Eq. (7)] as a function of optical path length difference for equal top and bottom surface reflectivities (i.e., $|r|^2 \equiv |r_{01}|^2 = |r_{12}|^2$) of 0.1 (a), 0.4 (b), and 0.95 (c). The dramatic differences between traces (a) and (c) are due to the contribution of multiple reflections, which govern the behavior of the denominator of Eq. (7). For the $|r|^2 = 0.1$ case (a), multiple reflections are relatively unimportant, so the reflectance is approximately sinusoidal about the average reflectance [i.e., the reflectance corresponds to the numerator of Eq. (7)]. As the reflectivities increase, multiple reflections become increasingly important and the dips in reflectivity become sharper. For the $|r|^2 = 0.95$ case (c), the denominator of Eq. (7) varies by three orders of magnitude, resulting in a very nonsinusoidal reflectance which is exceedingly sensitive to polymer thickness changes near reflectance minima. The width of the reflectivity dip is characterized by a quantity known as the "finesse" which is defined as the ratio of the separation of adjacent dips to the half-width of the dip. It is given by $\pi |r| / (1 - |r|^2)$ (for $|r|^2 \gtrsim 0.9$) (Ref. 2). Interferometers of design III with highly reflective surfaces are known as Fabry-Perot interferometers; their use in thermal expansion measurements will be discussed below.

4.3 APPLICATIONS

Each of several applications of laser interferometry is described in turn. The first applications involve irreversible changes such as those resulting from dissolution or drying/curing; the later applications involve reversible changes such as those resulting from thermal expansion/contraction or penetrant sorption/desorption.

4.3.1 Etch Rate Monitor (Dry Processing)

The use of interferometric techniques for monitoring the (low-pressure) etching or deposition of thin transparent films is now standard practice in the microelectronics and optical coating industries. In the mid-1960s such techniques were implemented with white light sources, with wavelength selectivity provided by an interference filter at the photodetector (19). Laser interferometers (of design II) for these applications were introduced in the mid-1970s (20, 21) and are now commercially available as complete packages (including laser, detector, etc.). With the substitution of a vacuum chamber for the penetrant-filled beaker, the experimental arrangement is analogous to that shown in Fig. 4.5b.

Common applications include polymer etch rate measurements and end-point detection during "dry processing" such as plasma etching or reactive ion etching

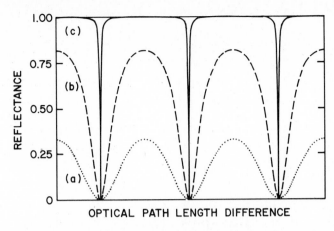

Figure 4.10:   Reflectivities as a function of optical path length difference for interferometer III. Traces are for equal top and bottom surface reflectivities of (a) 0.1, (b) 0.4, and (c) 0.95.

(22). End-point detection is used to determine when an etching process has been completed. Accurate end-point detection is increasingly critical as film and feature dimensions decrease in size, especially in cases where etching past the end-point can damage underlying structures. Typically the end-point is detected as an abrupt change in the interference pattern as a function of time, as shown in the reflectivity trace in Fig. 4.6. As stated earlier, etch rate measurements are performed by measuring the number of fringes evolved per unit time.

Two potentially complicating effects can exist during dry etching. The first is surface roughness, which can develop during etching if the polymer is doped with a nonerodible material, or if nonerodible electrode materials are sputtered onto the polymer to form "micro-masks" which shield the underlying polymer from the plasma. Under these circumstances, both the overall and fractional modulated reflected intensity will fall in time. In the worst case, the modulated component is no longer detectable after a few fringes of etching.

The second effect is due to sample heating by the plasma. We first observed this effect in etch rate measurements (in O_2/CF_4 rf plasma) on 25-μm-thick polyimide films on silicon substrates (23). Upon turning off the plasma after etching a few fringes, the film's apparent thickness continued to decrease (by as much as half a fringe or so) over a time period of tens of seconds. This behavior is illustrated in Fig. 4.11; it results from thermal contraction. The apparent etch rate of thick samples will thus be lower than the true etch rate if the sample is heating up. (Thermal expansion effects are less important with thinner samples because the effects are proportional to thickness.) This is rarely a problem for etch rate measurements since data taken before the sample has reached its equilibrium temperature are usually ignored. It can be a problem, however, if one tries to accurately etch a small thickness (of the order of one fringe) from a thick sample whose temperature is changing. Interferometric measurements of temperature are discussed in Section 4.3.5.

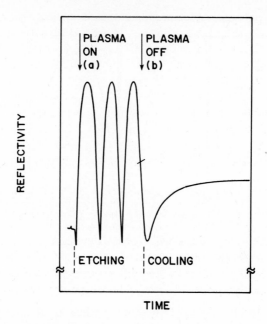

Figure 4.11: Reflected intensity of a photoresist sample (of 5.1-μm initial thickness) on a silicon wafer: (a) Dry etching followed by (b) Thermal contraction after plasma is turned off.

4.3.2 Dissolution Phenomena

The use of laser interferometry for studies of polymer dissolution started with simple etch rate measurements and has been extended to the study of transition layer phenomena by Krasicky and coworkers (6,10-12). Related optical studies of dissolution include a CARIS-related one by Konnerth and Dill (24) and two based on ellipsometry (25,26). Such studies are needed, for example, to characterize and optimize the performance of photoresist developers used in the microelectronics industry. A good developer for a positive (negative) resist will very selectively etch the exposed (unexposed) resist. Furthermore, an understanding of the mechanisms of dissolution is needed for materials and process optimization.

Early dissolution rate studies using laser interferometry in the configuration of Fig. 4.5b include one by Meyerhofer (27) on the photosolubility of diazoquinone resists and one by Cooper et al. (28) on the MEK dissolution rates of PMMA films as a function of PMMA molecular weight.

Later, Krasicky et al. (10) focused on the detection and measurement of a transition layer at the solvent/polymer interface. They measured the dissolution rate and amplitude reduction factor f [using a version of Eq. (28)] for a number of polymers as a function of polymer molecular weight, solvent, and temperature. The expression given by Eq. (25) for a linear concentration profile was used to determine the transition layer thickness. A simulation of their data has been presented in Fig. 4.8a. It was found that the transition layer thickness depends on the polymer identity and increases with polymer molecular weight. (The transition layer results from the inability of polymers to rapidly disentangle themselves from the surface of the glassy, unswollen polymer film.)   For the PMMA/MEK system, the transition layer was observable for molecular weight MW > 30 K. The transition layer thickness was not directly relatable to the rate of dissolution, since an order of magnitude change in the dissolution rate gave the same transition layer thickness.

In a more detailed study (11), Krasicky et al. calculated the transition layer parameters f and ϕ_t for different models of the concentration profile in the transition layer and compared them to f and ϕ_t values measured for the PMMA/MEK system, with PMMA molecular weights ranging from 14 to 1400 K. They found the values of the transition layer thickness L_t to be relatively insensitive to the model used for the profile.

Recently another group, Limm et al. (29) used laser interferometry (with an optical fiber) in combination with fluorescence quenching to study the dissolution of 0.8-μm-thick films of PMMA (MW ~ 411 K) on silicon or quartz substrates in a 1:1 mixture of 2-butanone and 2-propanol. They used a matrix method to calculate the value of the amplitude reduction factor $f(L_t)$ [approximately given by Eq. (25)] as a function of the transition layer thickness L_t for a cosine concentration profile. A steady-state transition layer thickness of 90 nm was then inferred from the measured value of f, in good agreement with the results of Ref. 10.

4.3.3 Monitor for Drying/Curing

Laser interferometry can be used to monitor the processing of solution-cast polymer films from drying to curing. In situ measurements of drying/curing can be useful for a number of reasons. For example, determination of solvent evaporation rates from wet films of polymer solution enables comparisons of speed and completeness of solvent removal for different heating schedules. Also, the degree of polymer film shrinkage during baking can be correlated to the completeness and reproducibility of cure.

Saenger and Tong (8) monitored a freshly spun 18.3% (by weight) solution of polyamic acid [based on the pyromellitic dianhydride (PMDA)-oxydianiline (ODA)] in n-methyl-2-pyrrolidone (NMP) as it was converted to a cured polyimide film of 6 μm in thickness. The experimental arrangement is shown in Fig. 4.5a. The samples were on silicon substrates; the refractive index of NMP is 1.4684. Their reflectivity data for the solvent removal step, a 10-min hot plate "soft" bake in air at 85°C is shown in Fig. 4.12. As the film approaches dryness, the evaporation rate slows down, and the amplitude of the fringes increases. We speculate that the low initial amplitude may be due to a diffuse solvent/air interface caused by the rapidly evaporating solvent. After the 85°C soft bake, the remaining 10-μm film was cured with 30-min "hard" bakes at 150°C (resulting film thickness 7.3 μm) and 230°C (resulting film thickness 5.7 μm). A comparison of the film thicknesses (profilometer measured) with the number of fringes evolved indicated a shortage of about 10% in the number of fringes. The discrepancy was attributed to refractive index increases during curing and verified by independent index measurements after each bake step. Negligible shrinkage (< 0.1 μm, profilometer measured) resulted from the remaining 30-min bakes at 300 and 400°C in dry N_2.

In a related application, Brekner and Feger (30) used laser interferometry in their study of the complexation of the solvent NMP with a PMDA-ODA based polyamic

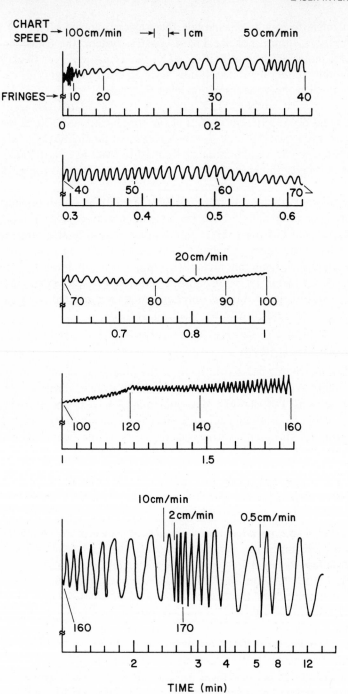

Figure 4.12: Reflectivity (with interferometer II as shown in Fig. 4.5a) of a solution of polyamic acid in NMP as a function of time during a 10-min 85 °C hot plate bake.

acid with a molecular weight of 20-28 K. Films of a polyamic acid/NMP solution (17.5 wt %) were spun onto silicon wafers and the thickness change due to loss of "free" solvent was recorded with a laser interferometer (1 fringe = $\lambda/2n_{NMP}$) under conditions ranging from a room temperature vacuum bake to a hot plate bake in air at 100°C. The thickness of the remaining film was then measured with a profilometer. A comparison of this thickness with the initial thickness (given by the sum of the "postevaporation" thickness and the measured thickness loss of NMP) allowed an estimate of the remaining NMP content per polyamic acid molecule. Their results indicated a stable $(NMP)_4$/polyamic acid molecule complex for the room temperature vacuum dried sample, and a more stable $(NMP)_2$/polyamic acid molecule complex for bakes in air above room temperature but below 85°C. Further verification of their conclusions was provided by a thermogravimetric analysis of the complexed films.

4.3.4 Structural Stability

Structural stability is another polymer property amenable to investigation with laser interferometry. This application nicely bridges the somewhat arbitrary distinction made at the beginning of this section between reversible and irreversible changes in polymer films.

Basically, a polymer film is not structurally stable if its thickness or refractive index is irreversibly changed by thermal cycling. This can be detected (with interferometer II) by a discrepancy between the number of fringes evolved during heating and the number evolved during subsequent cooling. If they are not equal, the film has changed. Conversely, a film that shows the same number of fringes during heating and cooling can be considered, at least in some context, to be structurally stable.

For example, 125 μm-thick Kapton (polyimide) films in interferometer III were thermally cycled on a hot plate in air (31). After repeated cycling 20 − 230°C and/or overnight baking at 230°C, the Kapton films finally gave the same number of fringes for the heating and cooling parts of the thermal cycle. The initial discrepancies were primarily attributable to film shrinkage [expected from the Du Pont literature (9) to be about 1%] arising from the relaxation of internal stresses placed in the film during manufacture (9). A smaller component of the discrepancy may be attributable to an annealing out of the surface topography which consists of gentle hills and valleys. As these disappear, so do the small air gaps at the polymer/reflector interfaces, with the end result that the reflectors move closer together. In the materials we have studied, our criteria for structural stability are typically met after a few thermal cycles.

4.3.5 Thermal Expansion Coefficient/Thermometry

In principle, laser interferometry can be used to measure the thermal expansion coefficient, α, in the film thickness direction by measuring the thickness change $\Delta L = \alpha L \, \Delta T$ induced by a known temperature change ΔT [see Eqs. (16) and (17)]. Conversely, if the calibration factor, ΔT/fringe (given by Eq. (15)), is known, an unknown ΔT can be determined from an interferometric measurement of the number of evolved fringes.

Laser interferometric techniques for temperature measurement (using interferometer II) are noncontact, and especially well suited to the measurements of sample temperatures during plasma processing. Bond et al. (32) first used such a technique to measure the temperatures of glass substrates a few mm thick during plasma etching. The polymer film temperature reached during plasma etching can be determined with the aid of Eq. (15) by counting the number of fringes evolved after the plasma is turned off. The datum cited earlier for a polyimide film of thickness 25 μm (i.e., half a fringe of thermal contraction) with an $\alpha + \beta$ value of $\sim 80 \times 10^{-6}/°C$, indicates a temperature rise during etching of $\sim 50°C$. We have also used laser interferometry to determine the temperature of the hot plate used in our TEC measurements. The "sensor" was a 2.5-cm diameter, 0.24-cm thick sapphire disk with a ΔT/fringe of $\sim 5°C$. In the applications just mentioned, the temperature changes were monotonic. Schemes are available (23) to extend these techniques to situations where the temperature changes are *not* monotonic, that is, where it is necessary to know the direction of the temperature change as well as its magnitude.

The TEC measurements can be done on films bonded to a substrate (the bottom reflector of interferometer III), or on films sandwiched between the two reflectors of interferometer III. Measurements of α for films bonded to low TEC substrates such as Si or quartz are especially interesting due to the possibility that the hindered in-plane expansion will be transferred to the thickness direction. Also, polymers synthesized for their low in-plane TECs (e.g., certain polyimides, Ref. 33) may have proportionately greater expansion in the thickness direction.

Previously, interferometers of design III (34) and IV (35) have been used to measure the TECs of quartz or salt materials with thicknesses of the order of 1-10 cm at high temperatures. As mentioned earlier, interferometer III is preferred over II due to the temperature dependence of the refractive index of the polymer. Our experiences (31) with TEC measurements have been with interferometer III used with 50-125 μm thick Kapton polyimide films and 26 μm thick PMDA-ODA type polyimide films (cured to 400°C) bonded to silicon substrates.

With reference to Fig. 4.5c, the construction details for interferometer III are as follows: The top reflector is a 2.5-cm diameter, 0.64-cm thick quartz beamsplitter coated (with a multilayer dielectric) to give 40% reflectance at normal incidence. The other side of the beamsplitter is antireflection coated. Four spring-loaded screws (two not shown) allow tilt adjustment of the top reflector, which must be parallel to the bottom reflector, as well as in good contact with the polymer film's top surface. The bottom reflector is either a thick (2.5-mm) silicon wafer (substrate for the bonded film) with a reflectivity of 34% or an Al coated glass mirror with a reflectivity of 91% (17). Thick substrates are required if "bending-beam effects" (36) induced by film/substrate TEC mismatches are to be negligible. The ~ 6 mm $\times \sim 3$ mm hole in the polymer was made by excimer laser ablation (37), although any process leaving a clean hole would be acceptable.

Figure 4.13 shows the reflectivity traces for (a) a 125 μm thick Kapton film on an Al reflector and (b) a 0.24-cm thick sapphire "thermometer" during thermal cycling on a hot plate between ~ 20 and 260°C. Repeated thermal cycling (not shown) was needed before reproducible results were obtained (i.e., fringes up = fringes down). We found an α value of $\sim 110 \times 10^{-6}/°C$ around 35°C (measured at a cooling rate

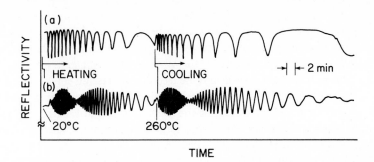

Figure   4.13:   Thermal expansion data: Reflectivity of (a) interferometer III containing a 5 mil thick film of Kapton polyimide and (b) a sapphire "thermometer" (as interferometer II) during thermal cycling (20-260°C.) A fringe in (a) corresponds to a thickness change of $\sim$ 3200 Å; a fringe in (b) corresponds to a temperature change of $\sim$ 5°C.

of 2°C/ min) and a value of $\sim$ 180 $\times$ 10^{-6}/°C around 240°C (measured at a heating rate of 2°C/ min). These values are about a factor of 3 higher than the literature values (9) of the Kapton thermal expansion coefficient in the in-plane direction. Similarly high TEC results were found for the 26-μm films on silicon wafers: $\sim$ 100 $\times$ 10^{-6}/°C around 90 $\pm$ 30°C and $\sim$ 250 $\times$ 10^{-6}/°C around 200°C.

The major difficulty in using interferometric techniques for the measurement of thin film TEC's stems from the fact that the resolution is, at best, about half a fringe, which corresponds to a thickness change ΔL of $\sim$ 160 nm [see Eq. (17)]. With ΔL = αL ΔT and thin films (small L), ΔT must be rather large to generate one fringe of expansion. (For a 25-μm-thick film and α = 100 $\times$ 10^{-6}/°C, ΔT must be $\sim$ 60°C .) Any value of α obtained from a measurement based on a complete fringe must therefore be an average over a fairly large temperature range. The strong temperature dependence of α (rapidly increasing with increasing temperature) makes it difficult to estimate the value of α at room temperature.

More sensitive interferometric schemes can be implemented at the cost of additional complexity. [For a brief review, see Touloukian et al. (38) and references contained therein.] One possibility is a relatively simple modification of interferometer III. The two reflectors are made highly reflecting ($\gtrsim$95%) so that the reflectivity will depend very sensitively on temperature where the reflectance is near a minimum (see Fig. 4.10c). The initial optical path length difference must then be adjusted (independently of polymer thickness and temperature) to be near a minimum at the temperature for which a measurement of α is desired. This can be done by etching "phase shifting" pads of different thicknesses (ranging from 0 to 300 nm in steps of several tens of nm) into the top reflector before deposition of the (multilayer dielectric) reflective coating. The initial reflectance would then be adjusted by selecting the appropriate pad. Good arguments, however, could be made for measuring thin film TECs with other methods, for example, a capacitive measurement of the air gap in a structure such as interferometer III.

4.3.6 Sorption Phenomena

Laser interferometry has been used to study the transport of low molecular weight, strong swelling agents (e.g., penetrants or solvents) in thin ($< 10\ \mu$m) supported polymer films. This topic is relevant to the microelectronics industry in that polymer films appearing in a finished device may be exposed to casting solvents (e.g., during the spin coating of a polymer solution) or stripping solvents (e.g. during photoresist removal). A thorough understanding of the mechanisms and kinetics of solvent diffusion in thin polymer films can have a significant impact on the development of new processing strategies in microfabrication.

There is also a theoretical interest in the mechanisms of penetrant transport in glassy polymers. In these systems, transport is frequently "non-Fickian" due to the effect of local, swelling-induced deformation on the solvent diffusion dynamics; typically, integral sorption of strong swelling agents is accompanied by a sharp solvent front separating a highly swollen layer from a glassy, almost dry polymer core. The range of possible behaviors can be classified in terms of two known limiting cases (39): Fickian-like (or "Pseudo-Fickian") transport where the boundary position changes with the square root of time, and "Case II" transport where the boundary moves at a constant velocity.

Using an interferometer similar to the one used for dissolution studies, Tong et al. (14) determined the thickness increase as a function of time of thin (4-8 μm) films of a fully cured (PMDA-ODA) polyimide (on silicon wafers) immersed in NMP as a function of NMP temperature (22-120°C). Also studied were 4-μm-thick films in dimethyl sulfoxide (DMSO) at 22°C. Their reflectivity data are shown in Fig. 4.14 for the case of a 4.1-μm film in NMP at 65°C. They found an equilibrium swelling of $\sim$ 20% with NMP and $\sim$ 30% with DMSO. The time scale for reaching equilibrium (at room temperature) was much faster for DMSO ($\sim$ 5 h) than for NMP ($\sim$ 2-3 days). The time scale for NMP sorption sharply decreased with increasing temperature, with an activation energy for the effective diffusion constant of $\sim$ 56 kJ/mol. Their analysis indicated that the transport of NMP was best described as anomalous, that is, intermediate between diffusion controlled and Case II transport. For DMSO the transport behavior was clearly Case II.

The "optical signature" for a sharp boundary layer moving between the penetrated, swollen layer and the dry, glassy core consists of two higher frequency oscillatory components superimposed on the slower frequency component arising from the entire film's overall thickness increase. The high frequency components arise from the interference effects of the additional reflection from the boundary as it moves through the dry film. Such optical signatures have been seen by Krasicky and coworkers (6,12) in dissolution studies involving a gel layer (e.g., PMMA in a mixture of methanol and ethyl acetate, or polystyrene in cyclohexane), but they were not present in the Tong et al. data for the polyimide/NMP (see Fig. 4.14) or /DMSO systems on silicon substrates. Recent calculations (40) similar to those shown in Fig. 4.9 suggest that the additional oscillatory component would not be detectable for boundary layers thicker than $\sim$ 150 nm. Preliminary experimental results from our lab (40) indicate that the optical signature of a "sharp" front is detectable in the PMMA/methanol system on quartz substrates, although not on silicon substrates, in accordance with the calculations of Fig. 4.9.

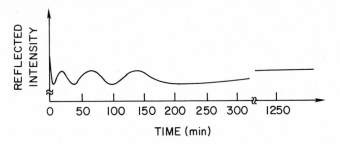

Figure 4.14: Sorption data (from Ref. 14): Reflectivity of a 4.1-μm-thick polyimide film (with interferometer II as shown in Fig. 4.5b) immersed in NMP at 65°C.

4.4 REFERENCES

1. W. A. Pliskin and S. J. Zanin, "Film Thickness and Composition," in **Handbook of Thin Film Technology**, L. I. Maissel and R. Glang, McGraw-Hill, New York (1983).

2. E. Hecht and A. Zajac, *Optics*, Addison-Wesley, Reading, MA (1975).

3. D. Rover, E. Dieulesaint, and Y. Martin, "Improved Version of A Polarized Beam Heterodyne Interferometer," *IEEE Ultrason. Symp. Proceedings*, 432 (1985).

4. J. K. Thomson, H. K. Wickramasinghe, and E. A. Ash, "A Fabry-Perot Acoustic Vibration Detector- Application to Acoustic Holography," *J. Phys.*, D6, 677 (1973).

5. O. S. Heavens, **Optical Properties of Thin Solid Films**, Dover, New York (1965).

6. P. D. Krasicky, R. J. Greole, and F. Rodriguez, "A Laser Interferometer for Monitoring Thin Film Processes: Application to Polymer Dissolution," *Chem. Eng. Comm.*, 54, 279 (1987).

7. A. W. Crook, "The Reflection and Transmission of Light by Any System of Parallel Isotropic Films," *J. Opt. Soc. Am.*, 38, 954 (1948).

8. K. L. Saenger and H. M. Tong, "Laser Interferometric Measurement of Polymer Thin Film Thickness Changes During Processing," *J. Appl. Polym. Sci.*, 33, 1777 (1987).

9. **Kapton Polyimide Film: Summary of Properties**, Du Pont Company, Electronics Department, High Performance Film Division, Wilmington, DE (1987).

10. P. D. Krasicky, R. J. Groele, J. A. Jubinsky, F. Rodriguez, Y. M. N. Namaste, and S. K. Obendorf, "Studies of Dissolution Phenomena in Microlithography," *Polym. Eng. Sci.*, 27, 282 (1987).

11. P. D. Krasicky, R. J. Groele, and F. Rodriguez, "Measuring and Modelling the Transition Layer During the Dissolution of Glassy Polymer Films," *J. Appl. Polym. Sci.*, 35, 641 (1988).

12. F. Rodriguez, P. D. Krasicky, and R. J. Groele, "Dissolution Rate Measurements," *Solid State Tech.*, 28, 125 (1985).

13. R. Jacobsson, "Light Reflection from Films of Continuously Varying Refractive Index," *Progr. Optics*, 5, 247 (1966).

14. H. M. Tong, K. L. Saenger, and C. J. Durning, "A Study of Solvent Diffusion in Thin Polyimide Films Using Laser Interferometry," *J. Polym. Sci., Polym. Phys. Ed.*, 27, 689 (1989)

15. **Optics Guide 4**, Melles Griot, Irvine, CA (1988).

16. D. Dijkkamp, A. S. Gozdz, T. Venkatesan, and X. D. Wu, "Evidence for the Thermal Nature of Laser-Induced Polymer Ablation," *Phys. Rev. Lett.*, 58, 2142 (1987).

17. R. C. Weast, Ed., **Handbook of Chemistry and Physics**, CRC Press, Boca Raton (1987).

18. N. L. Thomas and A. H. Windle, "Diffusion Mechanics of the System PMMA-Methanol," *Polym.*, 22, 627 (1981).

19. R. Glang, "Vacuum Evaporation," *Handbook of Thin Film Technology*, L. I. Maissel and R. Glang, Eds., McGraw-Hill, New York (1983).

20. J. A. Bondur, W. R. Case, and H. A. Clark, "Interferometric Method for *In Situ* Plasma Etch Rate Monitoring," *ECS Abstract 303*, 78-1 (1978).

21. H. A. Clark and J. A. Bondur, "Applications of Interferometric Plasma Etch Rate Monitoring," *ECS Abstract 304*, 78-1 (1978).

22. B. Chapman, **Glow Discharge Processes**, Wiley, New York (1980).

23. G. Appleby-Hougham, B. Robinson, K. L. Saenger, and C. P. Sun, "Nonintrusive Thermometry for Transparent Thin Films by Laser Interferometric Measurement of Thermal Expansion (LIMOTEX) Using Single or Dual Beams," *IBM Tech. Disc. Bull.*, 30, 239 (1987).

24. K. L. Konnerth and F. H. Dill, "*In Situ* Measurement of Dielectric Thickness During Etching or Developing Processes," *IEEE Trans. Electron Dev.*, ED-22, 452 (1975).

25. W. W. Flack, J. S. Papanu, D. W. Hess, D. S. Soong, and A. T. Bell, "*In Situ* Measurement of Resist Dissolution With a Psi-Meter," *J. Electrochem. Soc.*, 131, 2200 (1984).

26. J. Manjkow, J. S. Papanu, D. W. Hess, D. S. Soane (Soong), and A.T. Bell, "Influence of Processing and Molecular Parameters on the Dissolution Rate of Poly-(Methyl Methacrylate) Thin Films," *J. Electrochem. Soc.*, 134, 2003 (1987).

27. D. Meyerhoffer, "Photosolubility of Diazoquinone Resists," *IEEE Trans. Electron Dev.*, ED-27 921 (1980).

28. W. J. Cooper, P. D. Krasicky, and F. Rodriguez, "Effects of Molecular Weight and Plasticization on Dissolution Rates of Thin Polymer Films," *Polymer*, 26, 1069 (1985).

29. W. Limm, G. P. Dimnik, D. Stanton, M. A. Winnik, and B. A. Smith, "Solvent Penetration and Photoresist Dissolution: A Fluorescence Quenching and Interferometry Study," *J. Appl. Polym. Sci.*, 35, 2099 (1988).

30. M.-J. Brekner and C. Feger, "Curing Studies of a Polyimide Precursor. II. Polyamic Acid," *J. Polym. Sci.*, A-1, 25, 2479 (1987).

31.   K. L. Saenger and H. M. Tong, unpublished results.

32.   R. A. Bond, S. Dzioba, and H. M. Naguib, "Temperature Measurements of Glass Substrates During Plasma Etching," *J. Vac. Sci. Technol.*, 18, 335 (1981).

33.   S. Numata, S. Oohara, K. Fujisaki, J. Imaizumi, and N. Kinjo, "Thermal Expansion Behavior of Various Aromatic Polyimides," *J. Appl. Polym. Sci.*, 31, 101 (1986).

34.   E. G. Wolff and S. A. Eselun, "Thermal Expansion of A Fused Quartz Tube in A Dimensional Stability Test Facility," *Rev. Sci. Instrum.*, 50, 502 (1979).

35.   T. S. Aurora, S. M. Day, V. King, and D. O. Pederson, "High-temperature Laser Interferometer for Thermal Expansion and Optical-length Measurements," *Rev. Sci. Instrum.*, 55, 149 (1984).

36.   H. M. Tong and K. L. Saenger, "Bending - Beam Characterization of Thin Polymer Films," this volume.

37.   J. T. C. Yeh, "Laser Ablation of Polymers," *J. Vac. Sci. Technol.*, A4, 653 (1986).

38.   Y. S. Touloukian, R. K. Kirby, R. E. Taylor, and T. Y. R. Lee, *Thermal Expansion - Nonmetallic Solids*, Thermophysical Properties of Matter, Vol. 13, IFI/Plenum, New York (1977).

39.   T. Alfrey, E. F. Gurnee, and W. G. Lloyd, "Diffusion in Glassy Polymers," *J. Polym. Sci.*, C12, 249 (1966).

40.   C. J. Durning, K. L. Saenger, and H. M. Tong, "A Study of Case II Transport by Laser Interferometry," *47th ANTEC Conference Proceedings*, p. 400, Society of Plastics Engineers, Brookfield, CT (1989).

5

PIEZOELECTRIC RESONATORS: CONSIDERATIONS FOR USE WITH MECHANICALLY LOSSY MEDIA

K. K. Kanazawa

Almaden Research Center
IBM Research Division
San Jose, California

Following the first description by Sauerbrey (1) in 1958 of the sensitivity of the resonant frequency of a quartz resonator to the mass of an overlying thin film, the quartz microbalance has become a workhorse in the measurement of the rate and thickness of deposited films. The theoretical relation between the resonant frequency of the quartz to the mass density of the film has been developed for materials characterized by an elastic shear modulus having no viscous component. Although there have been a number of other applications for this versatile device (2,3), the majority of its uses has been for the measurement of changes in the mass of low-loss elastic films. In the period between 1958 and 1980, the relation between the frequency of the resonator and mass of the overlayer was described with increasing precision, making possible the use of the resonator over an increasing range of thickness deposits. These studies were concerned primarily with loss-free, elastic films. There were a few attempts (4,5) to explore the use of the resonator with lossy viscous liquid films, but these studies have not seen wide application to date. One of the interesting uses of the microbalance in the lossy regime was the study (6) of the superfluid behavior of He^4. Further activity in the area of lossy overlayers was discouraged by the common belief that the losses induced by the viscous nature of an overlayer would be large enough to prevent oscillation of the resonator (7,8).

This situation changed dramatically early in the 1980s with the demonstration of oscillation of quartz resonators in liquid environments (9-12). This empirical proof of the operability of the resonator in liquid has sparked a resurgence of activity in new applications involving lossy overlayers. The resonator is now being used not only with liquids (13), but also with polymer fluids (14), with polymer films (15), and with polymer films in liquids (16,17). Both experimental and theoretical understanding of the device operation in these new applications are required for its effective, quantitative use. There is an ongoing effort to continually improve these understandings.

In the following, the experimental requirements for operation are first considered, discussing not only the use of the resonator with lossy media, but the interfacing of the resonator with electrochemical instrumentation as well. In the final section, the state of understanding of the relation between the changes in the observable behavior of the resonator with the parameters of the overlayer is discussed.

5.1 GENERAL

The basic principle in the use of the quartz resonator for gravimetric applications is fairly easy to describe. The crystal takes the form of a flat platelet (generally circular) having electrodes on opposing faces as illustrated in Fig. 5.1. The platelet must assume a specific crystalline orientation to generate the desired shear waves. The AT cut of crystalline quartz is frequently used for this purpose. An excellent review (18) has recently appeared which discusses many of the growth and device theories for quartz resonators. The behavior of this resonator is studied electrically, with the crystal being connected to external instrumentation by way of the electrodes. Electrical stimulation generates acoustic shear waves in the crystal which can interfere constructively at certain frequencies to give rise to its resonances. The property of interest is generally this resonant frequency and for that purpose, the crystal acts as the frequency determining element of an electronic oscillator. The frequency of the output signal from the oscillator is observable and is easily measured using commercially available counting circuitry. This frequency changes as a film is deposited onto one of the surfaces, reflecting the acoustic properties of the overlayer. The changes in this frequency as the resonator is loaded with an overlayer can be related to the changes in deposited mass by way of a model. Recently, electronic impedance analyzers make possible the detailed study of the resonant spectrum, permitting the measurement of more than one observable and the determination of more than one overlayer property. However, this technique is still in its infancy and will not be discussed further.

5.2 EXPERIMENTAL CONSIDERATIONS

5.2.1 The Electrical Admittance, Bode Representation

An understanding of the behavior of the electrical admittance (the inverse of the impedance) of the resonator is extremely valuable to gain an understanding of the some of the important empirical constraints imposed by its properties. It has long been known that at any given resonant frequency, the equivalent circuit for the resonator can be represented (19) as in Fig. 5.2. It consists of one arm which contains the capacitor C_P representing the simple dielectric behavior of the quartz platelet in parallel with an arm consisting of the series combination of an inductor L, a series capacitance C_S, and a resistor R. Because C_P can be described in terms of the appropriate dielectric constant of quartz ε_{22}, that arm has been called the "static" arm, insensitive to the dynamics of the motion. The behavior of the shear waves and its changes with loading are represented by the series arm.

The behavior of the admittance as a function of frequency can be plotted as shown in Fig. 5.3, where the upper half of the figure shows the magnitude of the admittance and the lower half of the figure shows its phase. The magnitude has extremal values at two frequencies and the phase passes through zero near these two frequencies. The relative change of phase with frequency is seen to be very large near the zero phase regions. Since the short-term stability of crystal oscillators improves with

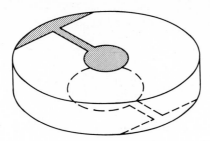

Figure 5.1:   The physical configuration of an AT-cut quartz resonator prepared for gravimetric studies is shown. Thin metal electrodes have been vacuum deposited on the opposing surfaces.

increasing phase sensitivity (20), the oscillator should be operated close to zero phase for high sensitivity measurements.

5.2.2 The Complex Admittance Representation

There is a different representation of this same information which is considerably more illuminating.   This is the admittance plot in which the real part of the admittance called the conductance G is plotted along the abscissa and the imaginary part called the susceptance B is plotted along the ordinate. The frequency interval for this plot has been centered around the frequency at which the series arm resonates, and is illustrated in Fig. 5.4. The plot traces a circle in a clockwise manner as the frequency is increased. The center of the circle is located at the coordinates $(1/2R,\ \omega_S C_P)$ with the radius of the circle being $1/2R$. For the purposes of representing the admittance plot of a resonance, the ordinate of the center is taken to be $\omega_S C_P$ instead of the actual value of the susceptance ωC_P because the range of frequencies covered by the plot is so small compared to ω_S. The ordinate axis is a tangent line to the circle. In use, the crystal is connected to external instrumentation so that the capacitor C_P not only represents the dielectric capacitance of the quartz, but also the input capacitance of the instrumentation as well as capacitance of the cabling coupling the crystal to the instrument.

Shown in Fig. 5.4 are three important frequencies. The rightmost resonance labeled by f_S is the frequency at which the real part of the admittance (the conductance), is a maximum. This occurs precisely at the series resonance of the right-hand arm and represents the dynamics of the shear waves. The two additional critical frequencies are defined as those frequencies at which the phase of the admittance vanishes. The lower of these two resonances lies very close to f_S as shown in the figure and is labeled f_L. The magnitude of the admittance at f_L is seen to be relatively large. In the parlance of circuit design, this critical frequency is called the "series resonant" frequency. The higher of the two frequencies where the phase of the admittance vanishes is labeled f_H and the magnitude of the admittance is seen to be small. Thus, this critical frequency is associated with the "parallel resonant" frequency. The practical importance of these two zero phase frequencies is that an ideal oscillator circuit using this crystal will oscillate at one of these two frequencies,

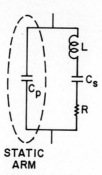

STATIC
ARM

Figure 5.2: This is the classical equivalent circuit representation for the electrical behavior of the quartz resonator. This representation is valid in the neighborhood of one of the various harmonic resonances.

depending on the particular design of the oscillator. This is discussed in considerable detail by Matthys (21). Although oscillators operating near f_H are used in certain applications (22), most of the circuits used for gravimetric purposes oscillate near f_L and for the sake of definiteness, this case will be assumed. Typically, an unloaded crystal which has a low-frequency resonance at 5 MHz would have its high-frequency resonance at a frequency 16 KHz higher if the wiring and instrumental capacitances are negligible.

The difference between f_S and f_L is usually quite small. If the stray capacitance connecting the crystal is negligible compared to its intrinsic capacitance (typically 4 pF), then f_L is only about 1.4 Hz higher in frequency than f_S. This difference will increase, of course, with the addition of cabling capacitance and instrumental capacitance. Since the oscillation frequency is sensitive to these capacitances, it is important to keep these values fixed. It is also important to keep the magnitude of these capacitances as small as practical. Since the vertical position of the center of the circle increases with increasing C_P, it is clear that the frequency difference between f_S and f_L increases as the capacitance increases. Not only does that imply a greater difference between the true frequency of series resonance and the oscillation frequency, it also causes the oscillator to function at points of decreasing phase sensitivity, compromising its stability.

5.2.3 Influences of Loss

From this same figure, the deleterious effect of increasing losses on the crystal resonance can be discussed. When the resonator is interfaced to a medium having appreciable viscosity, the losses are represented in the equivalent circuit by an increased resistance R. Since the dielectric capacitance is not changed, then the ordinate of the circle center is not changed but the diameter of the circle is reduced and the abscissa coordinate of the center is moved to the left to keep the circle tangent to the ordinate. Note that with sufficient losses, the diameter of the circle may be reduced to such an extent that the circle no longer intersects the abscissa. In that case, oscillations can no longer be supported, and the output of the oscillator goes to zero. The amount of

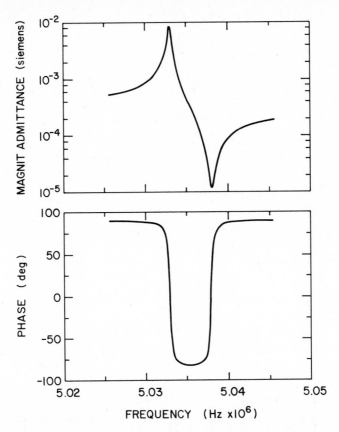

Figure 5.3:   The frequency dependence of the magnitude of the admittance is illustrated in the upper portion of the figure. The corresponding phase of the admittance is shown below.

loss that can be tolerated decreases with increasing C_P, giving another reason to minimize the stray capacitances. Even in the case where oscillations are sustained, the phase sensitivity is drastically reduced not only because of the decreased Q of the resonator, but also because of the further separation of f_L from f_S.

These considerations have considerable implications for the implementation of the quartz resonator loaded with a lossy medium. To keep the wiring capacitance and its changes small it is important to locate the oscillator circuitry as close to the crystal as possible. The use of coaxial cabling should be kept to a minimum. In the case where the crystal is being interfaced to a liquid, it is preferable that only one face contact the liquid. Shunting capacitance and resistance (in the case of a conducting liquid) are thereby kept to a minimum. It is useful to use electronic circuitry where one of the electrodes is at a true ground potential, and to interface that grounded electrode to the liquid. This will further minimize the shunting of the crystal with stray impedances. These concepts are illustrated in the mounting assembly shown in Fig. 5.5. It shows the mounting for a quartz crystal used in a plating monitor, where the mount is immersed into a plating bath. The electronic circuit board containing the

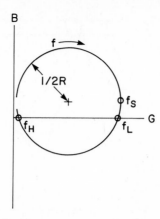

Figure 5.4:   This shows an alternate method of viewing the frequency dependence of the admittance. Here the real part of the admittance, G, is plotted along the abscissa and the imaginary part, B, is plotted along the ordinate. The key frequencies are circled.

oscillator is mounted in the same housing with the crystal. Fixed wiring is used to connect the crystal to the oscillator. Only one face of the crystal is exposed to the solution, an O-ring serving as a seal. The exposed electrode is at the ground potential of the circuitry.

The grounding of the exposed electrode is also useful to minimize any changes in the shunt capacitance resulting from the motion of nearby metallic parts. But this is not the only requirement on oscillator circuitry. The mechanical loss of viscous films is reflected as an increased resistance of the impedance at resonance, and in order to maintain oscillations, the open loop gain of the oscillator circuit must be increased. When the crystal is located in a controlled environment along with its oscillator circuit, the power leads and the cabling to the frequency measuring instrumentation must be led through that environment. To minimize the cabling required for these purposes, a circuit has been used which combines the function of the power leads and the rf cabling so that only a single coaxial cable is required. An example of such a circuit is shown in Fig. 5.6. Included in the figure are the details of the output stage of the power supply providing the DC. The output impedance of power sources are typically inductive, its magnitude rising to rather high values in the rf regime. A battery driven unity gain buffer using the OPA 111 operational amplifier is used to provide a stable DC source having a low output impedance such that the impedance terminating the connecting cable remains sensibly close to the 50-ohm value, even at radio frequencies. The radio frequency from the oscillator is coupled to the frequency counting circuitry through a 1500-pF capacitance. The oscillator pictured is used for lower-frequency applications with crystals resonating up to 7 MHz. It uses the general-purpose transistors, the 2N3904 and 2N3906. The steady-state dissipation is only 35 mW, minimizing thermal stability problems. It has a grounded electrode arrangement as required, and has sufficient gain to permit sustained oscillations when the crystal is loaded with common solvents.

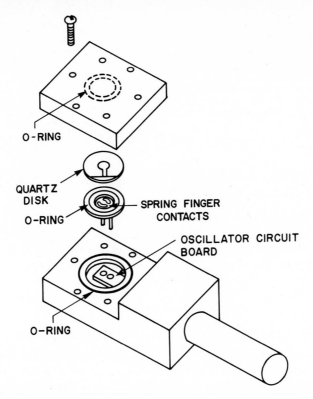

Figure 5.5:   This illustrates a crystal mount assembly used for plating applications. It incorporates many of the design considerations discussed in the text.

5.2.4 The Electrochemical Interface

It is not uncommon to find applications of lossy quartz resonators where they are coated with polymer films which are infused with or interfaced to a liquid. This occurs in studies such as polymer solvation (23) and polymer dissolution (16). There are other applications where conducting polymer films (24) are placed into a conducting liquid such that the interfacial electrical properties can be controlled. External electrochemical instrumentation is used to control the interface. These instruments control the potential or current between a conducting electrode and the immersing liquid. In these situations, the feature of the grounded electrode in the oscillator has an additional benefit.   If the electrochemical instrumentation is designed in such a way that the interfacial electrode is grounded, then it can be used with the electronic oscillator, sharing only a common ground. In this fashion the two circuits are uncoupled, the rf signals of the oscillator remaining in the quartz, and the low-frequency electrochemical signals in the liquid.   The so-called Wenking potentiostat is an electrochemical instrument designed to control the potential between an electrode and the immersing electrolyte and features the grounded electrode. A block diagram for such a potentiostat is shown in Fig. 5.7. The electrochemical cell is shown diagrammatically enclosed with the dotted lines. The

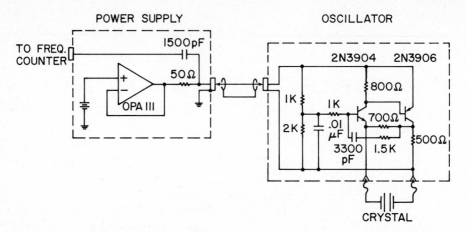

Figure 5.6: This schematic diagram shows details of the oscillator and power supply output stage. It is configured so that only a single coaxial cable is required as a connecting link between the oscillator and external instrumentation.

crystal is shown at the lower part of the cell, with the grounded electrode extending into the cell and the rf electrode outside the cell. The grounded quartz electrode inside the cell is the electrode of electrochemical interest, and is called the working electrode. In order to maintain its potential at some value relative to the liquid electrolyte, a special electrode which maintains a constant potential with respect to the liquid is placed in the cell near the working electrode. The circuitry is arranged to draw negligible current from the reference electrode so that it is operated near equilibrium. A third electrode called the counter electrode serves to permit cell current to be drawn through the cell. The operational amplifier to which the reference electrode is connected generally has a high input impedance and serves to maintain the potential of the reference electrode equal to the input voltage e_{in}. The current required to establish that potential passes through one of the current sensing resistors selected by the sensitivity switch. The instrumentation amplifier (AD 522) produces an output voltage e_1 which is proportional to the cell current.

5.3 THEORETICAL CONSIDERATIONS

5.3.1 Elastic Overlayers

The initial work of Sauerbrey (1) described the linear relation between the decrease in resonant frequency and the mass density of an overlaying elastic film. This relationship was found to be valid up to changes in frequency of 2% and can be written in the form

$$\Delta f = -\frac{2f_0^2 \rho \varepsilon}{\sqrt{\rho_Q \mu_Q}} \tag{1}$$

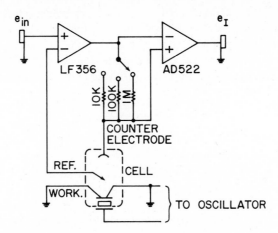

Figure   5.7:   This block diagram is illustrative of a Wenking-style potentiostat used in conjunction with the oscillator for electrochemical studies.

Here Δf is the change in frequency, f_0 is the unloaded resonant frequency, ρ the density of the film, ε its thickness, ρ_Q is the quartz density, and μ_Q is the elastic shear modulus for quartz. The product $\rho\varepsilon$ is the mass density of the overlying film. The density of quartz is $\rho_q = 2.649 \times 10^3$ kg/m³ and the elastic shear modulus for the quartz is taken to be that of its "piezoelectrically stiffened" modulus $\mu_Q = 2.924 \times 10^{10}$ kg/m².s.   The acoustic properties of the overlayer are not required. For thicker films, the acoustic delay of the shear wave in the overlayer is no longer negligible.

Later studies (25) have treated the quartz/overlayer as a compound resonator consisting of two layers of materials having different acoustic properties. These studies have treated the resonance as resulting from the entrapped shear waves, without regard to the influence of the electrical forcing function. In this sense, these provide a purely mechanical description of the resonant frequencies. For thick elastic films, this relationship was described by Lu and Lewis (26):

$$\tan\left(\pi\frac{\Delta f}{f_0}\right) = -\sqrt{\frac{\rho\mu}{\rho_Q\mu_Q}}\,\tan\left(\frac{2\pi f\rho\varepsilon}{\sqrt{\rho\mu}}\right) \tag{2}$$

Here μ is the elastic shear modulus of the overlayer. The relationship now includes the acoustic behavior of the film, and is routinely used with great success in analyzing the deposition of elastic films.

5.3.2 Viscous Overlayers

Crane and Fischer (4) recognized the importance of loss in the film and used an electrical transmission line equivalent to represent the acoustical behavior of the quartz and a viscous overlayer to model this behavior. Unfortunately, their analysis showed that the change in frequency goes to zero in the elastic limit where the viscosity of the film is vanishingly small instead of reducing to the Lu and Lewis result. A different approach was taken by Glassford (5) in his study of polymeric fluid deposits. He treated the loss using a perturbation technique, obtaining agreement with some experimental results using silicone oil. The relationship of the change in frequency resulting from an immersion of one electrode into a solvent was described by Kanazawa (12). The change in frequency resulting from this immersion is given by

$$\Delta f = -\frac{f_0^{3/2}}{\sqrt{\pi \rho_Q \mu_Q}} \sqrt{\rho \eta} \tag{3}$$

where η is the viscosity of the liquid.

Recently, an analysis due to Benes (27) has appeared using an equivalent circuit to represent not only the acoustic behavior of the quartz and the overlayer, but also to specifically include the influence of electrical forces. The resonances of this system are described in terms of the frequency behavior of the electrical impedance of the crystal. The behavior of the impedance of a quartz crystal loaded with thick metallic layers was treated. The analysis not only gave the changes in the frequency of the fundamental but that of the various harmonics as well. Although it remains to be demonstrated, Benes has pointed out that this treatment could be extended to include the effects of viscous losses by using complex values for the shear modulus of the overlayer. The description of the impedance is sufficiently complex that those results are not included here.

A mechanical model (28) describing the resonant frequencies of a viscoelastically loaded resonator has been used as a basis for a study of stress effects. This mechanical model can be criticized for its lack of rigor in the neglect of electrical forcing effects, but is based on the same physical picture employed in the earlier studies of the elastic compound resonator, which also viewed the resonances from a purely mechanical point of view. The results from this mechanical study can be written

$$\tan\left(\pi \frac{\Delta f}{f_0}\right) = -\frac{1}{\sqrt{\rho_Q \mu_Q}} \operatorname{Re}\left[\sqrt{\rho \tilde{\mu}} \tan\left(\omega \sqrt{\frac{\rho}{\tilde{\mu}}} \varepsilon\right)\right] \tag{4}$$

The complex shear modulus $\tilde{\mu}$ represents the ac behavior of a viscoelastic medium and is related to the shear modulus μ and the viscosity η by $\tilde{\mu} = \mu + i\omega\eta$. In the limit of a nearly elastic film in which $(\omega\eta/\mu)^2$ can be neglected with respect to unity, this reduces to the Lu and Lewis expression. This shows that Eq. (2) can be used to describe the mass loading on the quartz resonator even in the case where small losses are present.

5.3.3 Sources of Frequency Change

When the resonator, or resonator with film is in contact with a liquid, the interpretation of the frequency shift must be made with some care. Roughness of the surface will entrap fluid and give rise to an additional frequency decrease resulting from the mass of the entrapped fluid (29). It has been useful to describe this frequency shift using Eq. (1), where ρ is now the fluid density and ε is a mean depth of entrapped fluid. The effect of the roughness can be minimized by using a crystal whose surface has been chemically polished to a smooth finish. Such crystals are relatively efficient for operation in the harmonic modes since the surface scattering of the shear waves is minimized and are sometimes called "overtone polished" crystals. In the general case of a relatively low loss-film deposited on a quartz resonator immersed in a fluid, the frequency shift has three components, one due to the mass density of the film as expressed in Eq. (2), another due to the shift of a simple Newtonian fluid as expressed in Eq. (3), and finally, a shift due to entrapped fluid, as described in the foregoing.

The high sensitivity of the quartz resonator can be utilized even when it is coupled to a lossy medium by using care in the details of its implementation and in the interpretation of the data. Particular care must be exercised with shunting capacitances. Cross talk with other instruments, such as electrochemical analyzers, must be avoided. Possible extraneous sources of frequency shifts, such as surface roughness, need to be considered. The appropriate model coupling the overlayer parameters with observed frequency shifts must be used.

In the case of extremely lossy materials, where oscillations can no longer be reasonably sustained, it is still possible to obtain data describing the resonances by the use of impedance analyzers. This will require more detailed models for the interpretation of such data and provides an exciting area for continuing studies.

5.4 APPLICATIONS

5.4.1 Polymer Dissolution

An excellent example of the use of the quartz resonator with polymer films immersed in liquid is in the measurement of the kinetics of the dissolution process (16). In this study, 5-MHz quartz crystals were coated with a commercial photoresist which then were given varying degrees of exposure. Following this, these films were immersed in a developer. The sensitivity of the frequency to film thickness amounted to about 7 kHz/μm. Upon immersion, an initial frequency decrease was observed of the order of 2 kHz. This decrease was due primarily to the solvent viscosity as described by Eq. (3). The surface roughness was unknown and its contribution could not be estimated. However, both of these contributions to the frequency shift were relatively small and could be presumed constant. They were treated as a constant offset.

As the film dissolved, the change in its thickness was calculated using Eq. (1), knowing the film density. In this case, these polymer overlayers of approximately ten microns thickness could be treated as thin, elastic films. Of course, much thicker films would require the use of Eq. (2), or perhaps even Eq. (4), if the viscosity of those thick films were to become important. Interesting results were obtained with films exposed using monochromatic radiation where the amount of exposure varied periodically with distance from the substrate, an effect arising from optical interference. The subsequent dissolution showed periodic fluctuations whose separations were consistent with the known index of refraction.

5.4.2 Redox of Conducting Polymers

In an application coupled with electrochemistry (24), thin (4000 Å) films of the conducting polymer polypyrrole were grown onto gold electrodes on the quartz from a $AgClO_4$ /acetonitrile solution containing the pyrrole monomer. After growth, the films were rinsed, dried, and immersed in a 1 M $LiClO_4$ solution in tetrahydrofuran. The potential at the interface was stepped through intervals (ΔV) in a quasistatic manner and the change in frequency Δf and the change in charge ΔQ were recorded. A common mechanism for the oxidation and reduction of conducting polymer films is the diffusion of anions into the film during oxidation and their diffusion back out into solution during reduction. This would predict a mass increase during oxidation and a mass decrease during the reduction. A study of $(\Delta Q/\Delta V)$ during a redox cycle showed the collection of negative charge during oxidation in two steps and the subsequent loss of that charge during reduction, also in two steps. The mass changes were calculated using Eq. (1), where the product $\rho\varepsilon$ is interpreted as the mass change per unit area on the film. As in the previous example, the changes in frequency due to the solvent and surface roughness were treated as a constant offset.

Surprisingly, in the first oxidative step, the film mass decreased, while the mass increased in the second oxidative step. During reduction, the mass was observed to increase in two steps. This net mass increase during reduction indicates the net uptake of the Li^+ cations. During the first oxidative step, these cations are believed to sweep back out of the film, accounting for the initial mass decrease. The second oxidative step, in which the mass is observed to increase, is believed due to the incorporation of the perchlorate anions. This ability to simultaneously measure the charge and mass changes have been extremely valuable and informative in electrochemical applications of the resonator.

5.4.3 The Future

Emerging techniques for extracting additional information concerning the nature of the quartz resonances promise vigorous activity in the near future. Increased theoretical understanding of the viscoelastically loaded resonator such as the work of Benes (27) is vital for the quantitative interpretation of the data. Studies such as the clotting of blood (15) in which the real part of the shear wave impedance was determined indicate one possible direction. Another important direction is indicated by the study of stress effects arising from polymer films (28) since these can give rise to important frequency changes, particularly in temperature dependent studies. Many other directions for study will doubtless emerge with the reawakening of interest in the versatile bulk wave quartz resonator.

5.5 REFERENCES

1. G. Sauerbrey, "Verwendung von Schwingquarzen zur Waegung duenner Schichten und zur Mikrowaegung," *Z. F. Physik*, 155, 205 (1959).

2. J. Hlavay and G. G. Guilbault, "Applications of the Piezoelectric Crystal Detector in Analytical Chemistry," *Anal. Chem.*, 49, 1890 (1977).

3. C. S. Lu and A. W. Czanderna, Eds., **Applications of Piezoelectric Quartz Crystal Microbalances**, Elsevier, Amsterdam (1984).

4. R. A. Crane and G. Fischer, "Analysis of a Quartz Crystal Microbalance with Coatings of Finite Viscosity," *J. Phys. D: Appl. Phys*, 12, 2019 (1979).

5. A. P. M. Glassford, "Response of a Quartz Crystal Microbalance to a Liquid Deposit," *J. Vac. Sci. Technol.*, 15, 1836 (1978).

6. M. Chester, L. C. Yang, and J. B. Stephens, "Quartz Microbalance Studies of an Adsorbed Helium Film," *Phys. Rev. Lett.*, 29, 211 (1972).

7. W. W. Schulz and W. H. King, Jr., "A Universal Mass Detector for Liquid Chromatography," *J. Chromatogr. Sci.*, 11, 343 (1973).

8. T. Nomura and O. Hattori, "Determination of Micromolar Concentrations of Cyanide in Solution with a Piezoelectric Resonator," *Anal. Chim. Acta*, 115, 323 (1980).

9. T. Nomura and A. Minemura, "Behavior of a Piezoelectric Quartz Crystal in an Aqeous Solution and the Application to the Determination of Minute Amount of Cyanide," *Nippon Kagaku Kaishi*, 1980, 1621 (1980).

10. P. L. Konash and G. J. Bastiaans, "Piezoelectric Crystals as Detectors in Liquid Chromatography," *Anal. Chem.*, 52, 1929 (1980).

11. S. Bruckenstein and M. Shay, "An In-Situ Weighing Study of the Mechanism for the Formation of the Adsorbed Oxygen Monolayer at a Gold Electrode," *J. Electroanal. Chem.*, 188, 131 (1985).

12. K. K. Kanazawa and J. G. Gordon II, "The Oscillation Frequency of a Quartz Resonator in Contact with a Liquid," *Anal. Chim. Acta*, 175, 99 (1985).

13. H. Muramatsu, E. Tamiya, and I. Karube, "Computation of Equivalent Circuit Parameters of Quartz Crystals in Contact with Liquids and Study of Liquid Properties," *Anal. Chem.*, 60, 2142 (1988).

14. K. K. Kanazawa and C. E. Reed, "Studies of Liquids in Shear Using a Quartz Resonator," *Proc. 41st Freq. Control Symp.*, 350 (1987).

15. T. Funck and F. Eggers, "Clotting of Blood at a Gold Surface Probed by MHz Shear Quartz Resonator," *Naturwissensch.*, 69, 499 (1982).

16. W. D. Hinsberg, C. G. Willson, and K. K. Kanazawa, "Measurement of Thin-Film Dissolution Kinetics Using a Quartz Crystal Microbalance," *J. Electrochem. Soc.*, 137, 1451 (1986).

17. H. Muramatsu, J. M. Dicks, E. Tamiya, and I. Karube, "Piezoelectric Crystal Biosensor Modified with Protein A for Determination of Immunoglobins,", *Anal. Chem.*, 59, 2760 (1987).

18. J. C. Brice, "Crystals for Quartz Resonators", *Rev. Mod. Phys.*, 57, 105 (1985).

19. K. S. van Dyke, "The Electric Network Equivalent of a Piezo-electric Resonator," *Phys. Rev.*, 25, 895 (1925).

20. See, for example, A. Benjaminson, *Res. Dev. Tech. Rept.*, SLCET-TR-85-0445-F.

21. R. J. Mathys, *Crystal Oscillator Circuits*, Wiley, New York (1983).

22. See, for example, M. Benje, M. Eiermann, U. Pettermann and K. G. Weil, "An Improved Quartz Microbalance. Applications to the Electrocrystallization and Dissolution of Nickel," *Ber. Bunsenges. Phys. Chem.*, 90, 435 (1986).

23. M. H. Ho, G. G. Guilbault, and B. Rietz, "Continuous Detection of Toluene in Ambient Air with a Coated Piezoelectric Crystal," *Anal. Chem.*, 52, 1489 (1980).

24. J. G. Miller and D. I. Bolef, "Sensitivity Enhancement by the Use of Acoustic Resonatoris in CW Ultrasonic Spectroscopy," *J. Appl. Phys.*, 39, 4589 (1968).

25.   V. Mecea and R. V. Bucur, "The Mechanism of the Interaction of Thin Films with Resonating Quartz Crystal Substrates: The Energy Transfer Model," *Thin Solid Films*, 60, 73 (1979).

26.   C. S. Lu and O. Lewis, "Investigation of Film-thickness Determination by Oscillating Quartz Resonators with Large Mass Load," *J. Appl. Phys.*, 43, 4385 (1972).

27.   E. Benes, "Improved Quartz Crystal Microbalance Technique," *J. Appl. Phys.*, 56, 608 (1984).

28.   M. C. Chu, H. M. Tong, and K. Kanazawa, unpublished work.

29.   R. Schumacher, G. Borges, and K. K. Kanazawa, "The Quartz Microbalance: A Sensitive Tool to Probe Surface Reconstructions on Gold Electrodes in Liquid," *Surf. Sci.*, 163, L621 (1985).

6
ION BEAM ANALYSIS OF THIN POLYMER FILMS

P. F. Green and B. L. Doyle

Sandia National Laboratories
Albuquerque, New Mexico

The use of ion beams for the modification, development, and characterization of materials has had a major impact in areas of technological importance such as the microelectronics and biomedical technologies. This usage is bifurcated: beams with energies less than 1 MeV are usually used to modify materials, while beams with energies in excess of 1 MeV are typically used for materials analysis.

The most commonly recognized beam modification treatment is the implantation doping of semiconductors (1). Because of its high degree of control and reproducibility, the implantation process is routinely employed in the microelectronics industry. Ion implantation, however, is not restricted to electronic applications, as evidenced by several emerging technologies which use implantation to modify the near-surface regions of other types of solids. For example, the implantation of ions in various metals has recently been shown to improve the materials' resistance to wear, oxidation, and corrosion (2). This discovery has led to a myriad of applications, ranging from the development of improved artificial knee and hip joint replacements for biomedical applications (3) to the fabrication of wear resistant bearings used in the space shuttle (4).

A high-energy (MeV) ion, unlike a neutron, loses energy in a well-defined way that is characteristic of its energy and of the elemental composition of the material it traverses. In contrast to an electron, it undergoes very little lateral scattering as it travels through materials. For this reason, MeV ions are useful for determining the concentration versus depth distribution of elements in materials. The depth resolution of MeV ions in materials varies from a few tens to a few hundred angstroms, depending on the type of ion used for the incident beam, the beam energy, and the target composition. High-energy ions can probe depths in the near-surface regions ranging from 0.01 to over 10 μm. They offer distinct advantages over the use of techniques such as secondary ion mass spectroscopy (SIMS) (5) and X-ray photoelectron spectroscopy (XPS) (6) in the case of depth profiling materials. This is because the use of SIMS or XPS requires the sputtering of the target material surface, which is a highly destructive process and may alter the composition of the material being probed. On the other hand, ion beam analysis is considered to be nondestructive.

Generally, ion beam analysis experiments used for materials characterization fall into three main classes, depending on the nature of the interaction between the beam and the target nuclei. In typical experiments, the ions are produced by an ion source in an accelerator (e.g., Van de Graaff); these ions are then accelerated to MeV energies, and travel through an evacuated beam pipe toward a target. In Rutherford backscattering spectrometry (RBS) (7), a small percentage of the incident ions, upon approaching sufficiently close to a target nuclei, are backscattered. A projectile that is backscattered from a nucleus in a target has a final energy that is dependent upon the nucleus from which it backscattered and the depth in the target at which the interaction occurred. The backscattered particles are detected by a silicon surface barrier detector. A detected particle creates a current pulse in the detector, the height of which is proportional to its energy. A plot of the number of pulses versus the height of each pulse is displayed on a multichannel analyzer. This plot provides the backscattering spectrum of number of particles detected with a given energy E_d as a function of E_d, which can readily be converted to a plot of concentration versus depth as shown later. A nuclear reaction analysis (NRA) (8) experiment involves the detection of the products of a single nuclear reaction between an incident ion and a target nucleus, as opposed to simple elastic scattering as with RBS. The nuclear reaction of an incident ion with target nucleus results in the emission of characteristic subatomic particles whose energy is indicative of the depth in the material at which the reaction occurred. One exception to this is the detection of γ-ray products since the γ-ray energy does not change with depth. The third set of experiments involves the kinematic recoil of target nuclei by the incident ions. The energy of these recoiled particles is characteristic of the depth at which they recoiled and their masses. The number of these particles that recoil from a given depth x_0 is a measure of their concentration at x_0. One also gets from this experiment a spectrum of the number of detected particles versus energy, E_d. This measurement technique is called elastic recoil detection (ERD) (9-15) or forward recoil spectrometry (FRES) (16,17).

In general the use of one technique may be favored over another. RBS is often used to profile elements of medium atomic mass. When the mass of the element that is being detected is lighter than the matrix in which it resides, the sensitivity of RBS is reduced because of the high background. ERD is a much better alternative in this case because it is well suited for the analysis of light mass elements such as hydrogen. NRA is used when one chooses to profile a specific isotope and is, in general, not restricted to the detection of particles of any given mass range. The advantage of RBS/ERD over NRA is that they can simultaneously profile different elements. On the other hand, there exist some NRA techniques which have a better depth resolution than ERD or RBS.

Recently a new technique, which combines ERD and RBS into a coincidence measurement over very large detector solid angles in comparison to that used in RBS or ERD, was developed by Chu and Wu (18). This technique, known as scattering recoil coincidence spectrometry (SRCS), works best for profiling light elements such as hydrogen or deuterium in self-supporting foils. Because of the large solid angles used, the typical dose requirement to obtain a good spectrum is reduced by a factor of 10^{-3} over that required in typical RBS and ERD experiments. This is useful for samples which are prone to radiation damage. One drawback of SRCS is that its depth resolution is poor in comparison to RBS or ERD.

The use of ion beams to analyze polymers is a new area (15,16). Both RBS and ERD have been used extensively to study diffusion in polymer/solvent (19-21) and polymer/polymer systems. Polymer/polymer (16,17, 22-39) diffusion is an important area because it controls a number of important physical processes such as the heat-sealing and friction-welding of plastics (41). It is also believed to be important for adhesion at polymer/polymer interfaces, and in the kinetics of phase separation and crystallization in polymer systems (42).

The translational diffusion coefficients in polymers are in general very low, 10^{-8}-10^{-18} cm^2/s. For this reason, techniques with excellent depth resolution are important if measurements are to be done in reasonable time scales. A number of different techniques have been used to study polymer/polymer diffusion. These include holographic grating techniques (forced Rayleigh scattering, (FRS) (43) and fluorescence redistribution after pattern photobleaching (FRAPP) (44), small angle neutron scattering (SANS) measurements (45), radio tracer methods (46), NMR techniques (47), electron microprobe energy dispersive X-ray composition analysis (48), and infrared microdensitometry (49, 50). RBS and ERD (FRES) are particularly well suited for diffusion measurements because of their excellent depth resolution and exceptionally good sensitivity (ppm). They also provide concentration versus depth information of various species of interest. Most of the techniques mentioned above, for example, FRS and FRAPP, provide indirect information about the motion of the molecules. The electron microprobe energy dispersive X-ray composition analysis and infrared microdensitometry techniques do provide concentration versus depth information. However, their depth resolution is poor (microns) in comparison to RBS and ERD. A discussion of various techniques used to study polymer/polymer diffusion may be found elsewhere (40).

The intent of this chapter is to give the reader a detailed description of how one analyzes data from an RBS or ERD experiment. A series of examples are provided to show how useful the techniques can be for the study of polymers. The major portion of this chapter will be devoted to a detailed discussion of the ERD (FRES) technique. To a lesser extent, a discussion of RBS will be presented. Applications of these techniques will be restricted to those from the area of polymer/polymer diffusion. One application will be concerned with the use of ERD to investigate the validity of fundamental theories describing the translational dynamics of polymer molecules. A second will discuss how one may learn about the thermodynamics of polymer mixtures. The applications of ion beam analysis are indeed limitless, since there exists innumerable cases where one may choose to depth profile elements.

6.1 FUNDAMENTALS OF ELASTIC RECOIL DETECTION

6.1.1 Thin Target Approximation

Shown in Fig. 6.1 is a schematic of the ERD scattering experiment. A monoenergetic beam of ions of energy E_0 (MeV) impinges on a target at an angle θ_1 with respect to the target normal. The incident ions undergo a number of kinematic collisions with the nuclei in the target which result in the recoil of some of these target nuclei at angles θ . An energy sensitive detector is placed at a fixed angle in

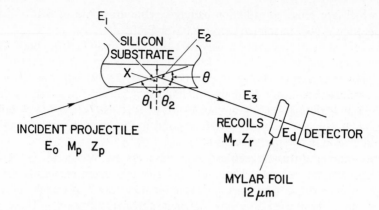

Figure 6.1: Schematic of the ERD scattering experiment; the "thin target approximation". Collision occurs at a depth x beneath the surface.

order to detect these nuclei which have energies that are characteristic of their masses and the depth from which they recoiled. A nucleus that recoils from the surface has an energy E_2 which is related to E_0 by

$$E_2 = KE_0 \tag{1}$$

where

$$K = \left(\frac{4M_p M_r}{(M_p + M_r)^2} \right) \cos^2 \theta \tag{2}$$

is the kinematic factor for elastic scattering; M_p is the mass of a projectile and M_r is the mass of a recoil. It is clear that through knowledge of E_0, M_p, θ, and measurement of E_2 that ERD can be used to identify light elements at the surface of solids by determining their masses, M_r. For the situation in which a particle recoils from a depth x beneath the surface (Fig. 6.1), the energy with which the particle leaves the sample is E_3

$$E_3 = KE_0 - [S]x \tag{3}$$

E_2 is the energy of the recoil after the collision. [S] is defined in terms of the stopping powers S_1 of the projectile as it traverses the sample and S_2 of the recoiling nucleus as it travels out of the sample. The stopping powers are a direct measure of the energy loss rates of the projectile and recoils as they travel through the sample. If S_1 and S_2 (see Section 6.5) are assumed to be approximately constant with respect to energy, then

$$[S] = \frac{KS_1}{\cos \theta_1} + \frac{S_2}{\cos \theta_2} \tag{4}$$

In this experiment, it is customary to place a foil (see range foil in Fig. 6.1) in front of the detector to stop any of the forward scattered projectiles from being detected. Therefore, the detector "sees" only the less massive, but more penetrating, recoils. The recoils lose an amount of energy δE upon passing through the foil, so the energy with which they are detected is given by

$$E_d = E_3 - \delta E \tag{5}$$

From this measurement, a spectrum of particle yield (i.e., the number of particles detected with energy E_d whose energy lies between $E_d + \Delta E/2$ and $E_d - \Delta E/2$, where ΔE is the width of an energy channel in the multichannel analyzer) versus E_d is obtained. One now has a definite relationship between the energy and depth scale, E_d versus x. The concentration of nuclei, $N(x)$ at a given depth x may be expressed in terms of the scattering cross section $\sigma (E_1, \theta)$, which is a measure of the probability that a scattering event will occur. The recoil yield, $Y_r(E_d)$ (11), is then

$$Y_r(E_d) = \frac{\Phi_p N(x)\sigma(E_1, \theta)\Omega\delta E_d}{\cos \theta_1 \dfrac{dE_d}{dx}} \tag{6}$$

In the above equation, Φ_p is the flux of incident particles and Ω is the solid angle subtended by the detector. The parameter dE_d/dx is the effective stopping cross section of the recoils in the range foil, and is given by (11)

$$\frac{dE_d}{dx} = R[S] \tag{7}$$

Here, $R = S(E_d)/S(E_3)$, which is the ratio of the stopping powers of the recoil in the range foil at energies E_3 and E_d defined at the surfaces. In some cases, the scattering process may be described by the Rutherford formula which, if defined in terms of the laboratory coordinates, is given by

$$\sigma(E_1, \theta) = \left[\frac{Z_p Z_r e^2 (M_p + M_r)}{2M_r E_1} \right]^2 \left[\frac{1}{(\cos \theta)^3} \right] \tag{8}$$

Z_p and Z_r are the atomic numbers of the projectile and recoils, respectively, and e is the charge of an electron. In cases where the scattering cross section is not well known, another procedure may be used. This is discussed in the section where the application to the study of polymers is presented.

6.1.2 The Thick Target Approximation

The approach given above for analyzing the data is somewhat simplified since the stopping powers [S] are, in general, not constant and are a strong function of energy (depth). Furthermore, the energy loss rates in the foil are also not constant. A detailed analysis of the scattering process, discussed below, indicates how these effects may be accounted for (14). Essentially, the sample is divided into a series of sublayers which are sufficiently thin (see Section 6.5) that S_1 and S_2 can be considered to be constant. The analysis of each layer is then performed separately.

Shown in Fig. 6.2 is another schematic of the scattering experiment where the sample is divided into a series of sublayers, each of known thickness. The monoenergetic particles of energy E_0 are directed toward the sample at an angle of θ_1 with respect to the target normal. The target nuclei that recoil at angles of θ_2 traverse the range foil of thickness $x^{(0)}$ before being detected with energy E_d by a silicon surface barrier detector.

By considering in detail the energy loss process, the depth at which the collision occurred may be related to the experimental observable E_d. In addition, the recoil yield may be related to the concentration of recoils through knowledge of the energy loss rates (effective stopping powers) and scattering cross sections.

We will begin by considering the simple case where the sample is composed of two layers and subsequently generalize it to n sublayers (14). The first layer of this sample is of thickness x_1 and the other of x_2. The energy of an incident projectile at the depth x_1 is E'_0. This depth of penetration may be expressed in terms of the stopping powers $S_p^{(1)}$ of the projectile p in layer 1 of the sample and the energies E_0 and E_0' (14)

$$x^{(1)} = c_1 \int_{E'_0}^{E_0} \frac{dE}{S_p^{(1)}(E)}$$

(9)

In this equation $c_1 = \cos \theta_1$. The following terminology will be used in all subsequent discussions. The superscript i in the symbol denoting the stopping power, $S_j^{(i)}$, or the layer thickness $x^{(i)}$, refers to the layer and the subscript j in $S_j^{(i)}$ refers to either the incident projectile, p, or a recoil r. We can express the depth $x^{(2)}$ in a manner similar to that in Eq. (9)

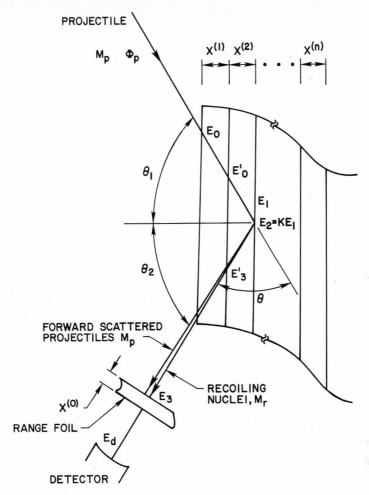

Figure   6.2:   Schematic of the ERD experiment where the target is divided into a series of sublayers; the "thick target approximation."

$$x^{(2)} = c_1 \int_{E_1}^{E'_0} \frac{dE}{S_p^{(2)}(E)} \tag{10}$$

where E_1 is the energy of the projectile at the interface $x^{(2)}$. If a nucleus of mass M_r recoils from the interface at depth $x^{(2)}$ with energy $E_2 = KE_1$, then $x^{(2)}$ may be defined in terms of the stopping powers of the recoil, r, in layer 2 and the energies E_2 and E'_3 defined at both interfaces

$$x^{(2)} = c_2 \int_{E'_3}^{E_2} \frac{dE}{S_r^{(2)}(E)} \tag{11}$$

where $c_2 = \cos\theta_2$. At the inner interface of layer 1, the recoil has energy E'_3 and outside the sample it has energy E_3 (see Fig. 6.2). The depth x_1 may then be expressed in terms of the energies E_3 and E'_3 of the recoil at those positions

$$x^{(1)} = c_2 \int_{E_3}^{E'_3} \frac{dE}{S_r^{(1)}(E)} \tag{12}$$

The thickness of the range foil may also be expressed in terms of the energy E_3 that the recoil possesses before it enters the range foil and E_d, that which it possesses after it traverses the range foil

$$x^{(0)} = \int_{E_d}^{E_3} \frac{dE}{S_r^{(0)}(E)} \tag{13}$$

The yield, $Y_r^{(2)}(E_d)$, of particles which originate at depth x_2 and which have energy E_d, which lies between $E_d - \Delta E_d/2$ and $E_d + \Delta E_d/2$, may be expressed in terms of $N_r^{(2)}(x^{(2)})$, the number density of ions that recoil from depth $x^{(2)}$ at the interface of the layer

$$Y_r^{(2)}(E_d) = \frac{\Phi_p N_r^{(2)}(x^{(2)})\sigma_r(E_1, \theta)\Omega\delta E_d}{c_1 \dfrac{dE_d}{dx^{(2)}}} \tag{14}$$

In Eq. (14), $dE_d/dx^{(2)}$ is the effective stopping power in layer 2 and it encompasses the effects of the stopping powers of the projectile and the recoil particles in the sample and in the range foil. An expression for $dE_d/dx^{(2)}$ may be obtained as follows. It may be shown from Eqs. (10) and (11) that the effective stopping powers in layer 2 may be expressed as (14)

$$\frac{dE'_3}{dx^{(2)}} = \frac{S_r^{(2)}(E'_3)}{S_r^{(2)}(E_2)} [S]_{p,r}^{(2)} \tag{15}$$

In this equation,

$$[S]_{p,r}^{(2)} = \frac{KS_p^{(2)}(E_1)}{c_1} + \frac{S_r^{(2)}(E_2)}{c_2} \tag{16}$$

Note that this is equivalent to Eq. (4), except here it is defined for layer 2 where $S_p^{(2)}$ and $S_r^{(2)}$ are constant. The term $S_r^{(2)}(E'_3)/S_r^{(2)}(E_2)$ on the right-hand side of Eq. (15) is the ratio of the stopping powers of the recoil in the second layer at energies E'_3 and E_2 defined at both interfaces. It has been shown (14) that upon calculating expressions for dE'_3 and dE_3 using Eqs. (12) and (13) that

$$\frac{dE_d}{dx^{(2)}} = \frac{S_r^{(2)}(E'_3)}{S_r^{(2)}(E_2)} \cdot \frac{S_r^{(1)}(E_3)}{S_r^{(1)}(E'_3)} \cdot \frac{S_r^{(0)}(E_d)}{S_r^{(0)}(E_3)} \cdot [S]_{p,r}^{(2)} \tag{17}$$

The term $S_r^{(2)}(E_3)/S_r^{(2)}(E_2)$ is the ratio of the stopping powers in layer 2 defined at the interfaces where the recoil has energies E_3 and E'_3. The second ratio, $S_r^{(1)}(E_3)/S_r^{(1)}(E'_3)$, is defined accordingly for layer 1 and the third term is associated with the range foil.

The above procedure may be generalized to include scattering from the nth interface. The effective stopping power for recoils from the nth interface may be shown to be (14)

$$\frac{dE_d}{dx^{(n)}} = \prod_n [S]_{p,r}^{(n)} \tag{18}$$

where

$$\prod_n = \frac{S_r^{(n)}(E_3^{(n)})}{S_r^{(n)}(E_2)} \cdot \frac{S_r^{(n-1)}(E_3^{(n-1)})}{S_r^{(n-1)}(E_3^{(n)})} \cdots \frac{S_r^{(0)}(E_d)}{S_r^{(0)}(E_3^{(1)})} \tag{19}$$

Each term in this product represents the ratio of the stopping powers of the recoil energies at the interfaces of a given layer. The layer is denoted by the index n. The yield of recoils from the nth interface is

$$Y_r^{(n)}(E_d) = \frac{\Phi_p N_r^{(n)}(x^{(n)})\sigma_r(E_1, \theta)\Omega\delta E_d}{c_1 \dfrac{dE_d}{dx^{(n)}}} \tag{20}$$

Equations (18), (19), and (20) provide a general formalism for the analysis of materials using ERD (FRES). Note that the expression for the yield for a particle that recoils from depth $x^{(1)}$ in the first layer is

$$Y_r^{(1)}(E_d) = \frac{\Phi_p N_r^{(1)}(x^{(1)})\,\sigma_r(E_1, \theta)\Omega\delta E_d}{c_1 \dfrac{dE_d}{dx^{(1)}}} \tag{21}$$

An expression for the effective stopping power for ions that recoil from the same depth is

$$\frac{dE_d}{dx^{(1)}} = \frac{S_r^{(1)}(E_3)}{S_r^{(1)}(E_2)} \cdot \frac{S_r^{(0)}(E_d)}{S_r^{(0)}(E_3)} [S]_{p,r}^{(1)} \tag{22}$$

The first term in the product on the right-hand side is the ratio of the stopping powers of the recoil at energies E_2 and E_3 defined at the interfaces of layer 1, and the second term is associated with the range foil. In each sublayer the stopping power is assumed to be constant. It is important to note that in the limit where $S_r^{(1)}$ is constant, that Eq. (22) is identical to Eq. (7) that describes the "thin film approximation" since $S_r^{(1)}(E_3)/S_r^{(1)}(E_3) = 1$. In order to do an analysis, one must in general solve Eqs. (18) and (19) numerically to find $dE_d/dx^{(n)}$. In order to find the energy of the beam at the interface of each layer, the following recursion relation may be used

$$E_1 = E_0^{(n)} = E_0^{(n-1)} - x^{(n)}S_p^{(n)}(E_0^{(n-1)}) \tag{23}$$

The stopping powers $[S]_{p,r}^{(n)}$ are determined at each interface after the energy is calculated using Eq. (23). This enables the subsequent determination of $dE_d/dx^{(n)}$. Therefore, at each interface, one calculates E_1, E_2, $[S]_{p,r}^{(n)}$, dE_d/dx, and $\sigma(E_1, q)$ (the Rutherford scattering formula is not always applicable).

6.1.3 Optimizing Experimental Conditions

In conducting these experiments, one has control over a number of parameters such as the type of beam used, the incident beam energy as well as the incident and exit angles, θ_1 and θ_2. As discussed later, one does have some degree of control over the thickness of the range foil. The choice of these parameters will influence the sensitivity, the probing depth and the depth resolution of this technique.

Sensitivity and Probing Depth: There is a limitation on the value of θ, the recoil angle, that can be used in the experiment. From Eq. (1), the energy, E_2, with which a nucleus recoils after a collision decreases monotonically to zero as θ increases from 0 to 90° These recoils lose an appreciable amount of energy upon passing through the range foil. Therefore, only particles that recoil at angles smaller than some critical angle will have enough energy to be detected. This angle is dictated by the material used for the range foil, its thickness, the projectile, and the energy of the projectile. For example, $K = 0.495$ for a proton which recoils at $\theta = 30°$ as a result of a collision with a helium particle. For an incident energy $E_0 = 3.0$ MeV of the helium ion, the recoil energy of the proton is $E_2 = 1.485$ MeV. However, approximately 0.4 MeV of its energy is lost as it passes through the mylar foil of thickness 12 microns. The practical range of θ is therefore limited.

There is a further restriction on the selection of θ. By decreasing θ_1 and keeping θ_2 fixed, θ could in principle be decreased. However, the probing depth is decreased and the analysis depth increased. It may also be noted that decreasing θ_2 and keeping θ_1 fixed, hence decreasing θ, would have the same effect on the recoil particle trajectory. In many cases, this will affect the depth resolution, as discussed in the following section. It may appear that one could in principle improve the situation by using higher incident energies. However, the increase in E_0 has to be compensated by using a thicker range foil in order to prevent any of the forward scattered ions from reaching the detector. One of the major limitations in the use of the range foil is that it adversely affects the depth resolution due to energy straggling of the recoils and foil nonuniformities. These effects are discussed later.

The sensitivity of this technique is determined primarily by the scattering cross section. In many instances, the cross section is Rutherford [Eq. (6)]. This is true for silicon ions, of energy less than 28 MeV, which recoil protons or deuterons. The Rutherford cross section decreases with decreasing θ. This reduced cross section is an additional restriction on the use of small values of θ. A notable exception to Rutherford scattering cross section occurs when helium ions are used to recoil protons or deuterons. A plot of the scattering cross section as a function of energy at different angles is shown for deuterium in Fig. 6.3a, and for protons in Fig. 6.3b (13). The angular dependence of the proton recoil cross section is weak over a wide energy range, unlike the cross section for the recoil of a deuteron. The depth resolution, which is discussed below, is a somewhat more important problem.

Depth Resolution: The depth resolution of the ERD technique is affected by a number of variables such as the particle used to obtain the depth profiles, thickness nonuniformities in the range foil, the surface roughness of the sample, the angular spread of the beam, and straggling of the beam in the material. These factors are discussed in detail below.

The depth resolution, (δx), is defined in terms of the effective stopping powers dE_d/dx such that (11)

$$(\delta x) = \frac{\delta E_d}{dE_d/dx} \qquad (24)$$

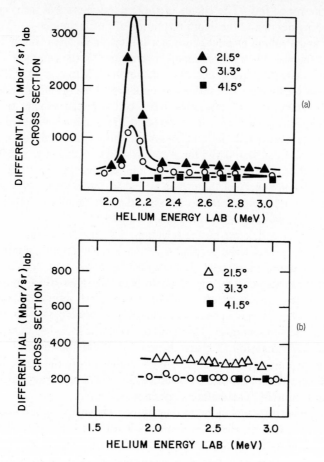

Figure 6.3: (a) A plot of the scattering cross section for the recoil of protons by MeV helium ions as a function of the incident energy at different scattering angles. (b) A plot of the scattering cross section for the recoil of deuterons by MeV helium ions as a function of incident energy at different scattering angles (From Ref. 13.)

This suggests that all the uncertainties that affect E_d also affect the depth resolution. Below is a detailed analysis of the factors that contribute to (δx). A general expression for $(\delta x)^2$ is

$$(\delta x)^2 = (\delta E_d)_r^2 + (\delta E_d)_s^2 + [S_r^{(0)}(E_d)\,\delta x^{(0)}]^2 + \left[\frac{S_r^{(0)}(E_d)}{S_r^{(3)}(E_3)}\right]^2 (\delta E_3)^2 \quad (25)$$

where the contributions are added in quadrature (14)

The first term on the right-hand side of this expression is due to the detector resolution. This is independent of the experimental conditions. The second is due to energy straggling of the recoils in the range foil. An energetic ion that travels through a material loses energy through collisions with electrons. Since this is a quantized process, it is subject to statistical fluctuations. Consequently, a series of particles that initially have identical energy E'' will, upon traversing a distance ΔX, have an energy distribution $\Delta E''$. This is energy straggling. The third term is due to nonuniformities in the thickness of the range foil. Effects due to straggling in the range foil and to nonuniformities are usually the most important limitation to the depth resolution. The final term results from the uncertainties of the energies, E_3, of the recoils just before they enter the range foil. The uncertainties in E_3 are due to nonuniformities in the sample surface, and straggling of the recoil from its point of origin and the sample surface. These are usually trivial contributions. The most important contribution is the uncertainty of E_2 which is described by the following expression

$$(\delta E_2)^2 \;=\; (E_1 \, \delta K)^2 \,+\, (K \, \delta E_1)^2 \tag{26}$$

The first term in this expression is due to the kinematic spreading of the recoiled beam. This uncertainty in E_2 contributes significantly to the depth resolution and steps can be taken to minimize it. A rectangular acceptance slit may be placed in front of the detector at a distance r from the beam spot on the sample. The height, h, of the slit is related to its width, w, and to r such that

$$h \;=\; 2(rw \tan \theta)^{1/2} \tag{27}$$

This condition leads to a minimum broadening of

$$\delta K \;=\; 3K \tan \theta \, \delta\theta \tag{28}$$

The slit should be positioned vertically with its center in the plane of scattering. The uncertainties in E_1 make only minor corrections to the depth resolution. They arise from straggling of the projectile before a collision and from path length variations.

Many of the contributions discussed above can be minimized in order to improve the depth resolution. Some are more important than others. In performing ERD experiments, some general guidelines may be followed. The single most important limitation to the depth resolution is the presence of the range foil. This is because the recoils lose an appreciable amount of their energy traveling through it. Producing very uniform films is, therefore, essential because of the resulting improvement in beam energy definition. The condition dictated by Eq. (27) which minimizes the kinematic spread should be followed. The samples that are produced should be very uniform which is usually not a problem for polymer films. A highly collimated beam should also be used. Finally, an improvement may be realized by using a heavier ion to obtain depth profiles. Figure 6.4 shows a plot of the depth resolution as a function of mass of the accelerating particle (up to $^{14}N_7$) for the recoil of hydrogen in silicon. It is clear that the use of a heavier accelerating particle results in improvements in

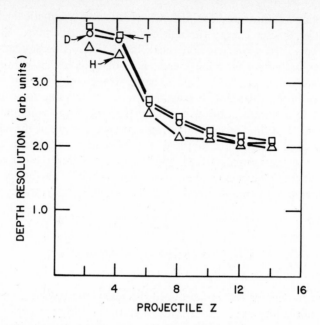

Figure 6.4: The variation of the depth resolution of hydrogen and its isotopes that are recoiled by incident ions of different masses.

(δx). One should be cautioned, however, that there are drawbacks associated with beam damage from the use of heavier ions.

6.2 FUNDAMENTALS OF RUTHERFORD BACKSCATTERING SPECTROMETRY

For reasons discussed below, the effective use of RBS (7) is limited to the detection of ions that are heavier than the ions that constitute the analysis beam. Shown in Fig. 6.5 is a schematic of the experiment. A monoenergetic beam of ions (typically helium) of energy E_0, which is produced by a Van de Graaff accelerator, is directed toward the sample surface. A small fraction of these ions get sufficiently close to the target nuclei that they undergo elastic collisions. A portion of these particles get backscattered into a silicon surface barrier detector. Recall that in ERD, it is the energy of a recoiled particle that is detected. The energy with which the incident ions get backscattered from the surface is given by

$$E_d = kE_0 \tag{29}$$

where the kinematic factor for elastic scattering is

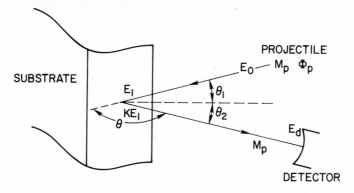

Figure   6.5:   Schematic of the RBS scattering experiment.

$$k = \left| \frac{M_p \cos\theta + (M_t^2 - M_t^2 \sin^2\theta)^{1/2}}{M_p + M_t} \right|^2 \tag{30}$$

The parameter θ is still defined as the scattering angle. Note the differences in the kinematic factors for scattering between ERD and RBS. In a limiting case where an annular detector is set at $180°$ then Eq. (30) becomes

$$k = \left(\frac{M_p - M_t}{M_p + M_t} \right)^2 \tag{31}$$

If one assumes that the energy loss rates of the inward and outward paths of the projectile are constant, which may be true of thin targets, then the energy of the particle at depth x is

$$E_1 = E_0 - \frac{x}{\cos\theta_1} \frac{dE}{dx_{in}} \tag{32}$$

and the energy of the backscattered particle outside the sample is

$$E_d = KE_1 - \frac{x}{\cos\theta_2} \frac{dE}{dx_{out}} \tag{33}$$

The energy of the detected particle can therefore be written as

$$E_d = KE_0 - [S]x \tag{34}$$

where the energy loss factor is

$$[S] = \left(\frac{K}{\cos \theta_1} \frac{dE}{dx_{in}} + \frac{1}{\cos \theta_2} \frac{dE}{dx_{out}} \right) \tag{35}$$

Note that Eq. (34) is analogous to Eq. (2). The aim is now to relate the spectrum yield, $Y_p(E_d)$, to the number of atoms per unit area at a depth x in the sample. The spectrum height may be written as follows

$$Y_p(E_d) = \sigma(E_1, \theta_1) \Omega \Phi_p \frac{N(x)}{\cos \theta_1} \tag{36}$$

where E_1 is the energy of the incident projectile just before the collision at depth x, Ω is the solid angle subtended by the detector, and $N(x)$ is the number of ions per unit area. The scattering cross section $\sigma(E_1, \theta_1)$ may be defined in terms of a differential scattering cross section

$$\frac{d\sigma}{d\Omega} = \left(\frac{Z_1 Z_2 e^2}{2E \sin^2\theta} \right)^2 \frac{(\cos \theta_1 + \{1 - [(M_p/M_t) \sin \theta_1]\}^{1/2})^2}{\{1 - [(M_p/M_t) \sin \theta_1]^2\}^{1/2}} \tag{37}$$

where e is the charge of an electron and Z_t and Z_p are the atomic numbers of the target nuclei and projectile, respectively. It is clear that the relative scattering probabilities are strong functions of target atomic number, Z_t. For example, using helium ions, the relative probability for scattering off oxygen versus uranium is 1:149. Heavier elements give a much more significant contribution to the spectrum. However, the mass resolution is poorer at higher masses. In general, RBS is ideally suited for medium atomic number elements.

The above analysis is the equivalent of the "thin target approximation" discussed in Section 6.1. For thicker films, the procedure outlined in Section 6.1.2 may be used.

It may be pointed out that there now exists a computer code "RUMP" written by Doolittle to analyze RBS data (51,52). This program allows the reader to input specific experimental parameters (beam energy, projectile mass, detector resolution, target composition, etc.), and then generate a simulated spectrum for comparison with the experimental data. A nonlinear least-squares analysis is then done to obtain the best fit. The routine is sophisticated enough to include effects such as detector resolution and straggling.

6.3 POLYMER/POLYMER DIFFUSION (THEORETICAL BACKGROUND)

6.3.1 Tracer Diffusion

The current understanding of the translational diffusion of a single chain comprised of N monomer segments in a highly entangled melt of chains, each comprised of P monomer segments, is based on the concept of reptation which was introduced by deGennes in 1971. In the melt, the diffusion of this N-mer chain is restricted to a tube-like region formed as a result of its intersections with the neighboring P-mer chains. Shown in Fig. 6.6a is a schematic of a labeled chain in a melt; Figure 6.6b shows a schematic of the labeled chain restricted to the "tube." The dynamical modes of this chain in its "tube" are assumed to be described by the Rouse model (53); therefore, the diffusion coefficient of the chain along an average trajectory defined by the "tube" is denoted by (54-56)

$$D_t = \frac{k_B T}{N \zeta} \tag{38}$$

where ζ is the monomeric friction coefficient, k_B is the Boltzmann constant, and T is the temperature; $N = M/M_0$. Here M_0 is the molecular weight of a monomer in the chain and M is the total molecular weight of the chain. The ends of the chain that emerge from the "tube" choose random directions in space in which to move. As the chain diffuses, portions of a new "tube" are created ahead, while portions of the old are abandoned. On long time scales defined by $\tau_d \sim \zeta N^3$, the chain loses complete memory of the old "tube." The center of mass diffusion coefficient of a chain diffusing by reptation alone is given by (54-58)

$$D^* = D_{REP} = D_0 M^{-2} \tag{39}$$

where

$$D_0 = \frac{4}{15} \frac{M_0 M_e k_B T}{\zeta} \tag{40}$$

Here M_e is the molecular weight between entanglement. Both D^* and the zero shear rate viscosity, η_0, are related through τ_d, the longest relaxation time in the melt. Graessley has shown that one may easily express D^* in terms of η_0 (59):

$$D^* = \frac{G_N^0}{135} M_e^2 L \frac{M_c}{\eta_0(M_c)M^2} \tag{41}$$

(a)

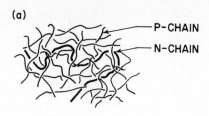

P-CHAIN

N-CHAIN

(b)

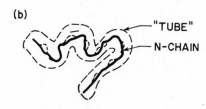

"TUBE"

N-CHAIN

Figure 6.6: (a) Picture of a labeled chain comprised of N monomer segments in a melt of chains, each comprised of P-monomer segments. (b) The N-mer chain is confined to a tubelike region.

In this equation $G_N^0 = \rho RT/M_c$ is the plateau modulus where ρ is the density of the polymer. M_c is the critical molecular weight for viscous flow and $L = <R^2>/M$ where $<R^2>$ is the root mean square end-to-end vector. Equation (41) allows one to calculate D^* based on knowledge of the viscoelastic parameters of the polymer. It establishes a definite link between diffusion and viscoelasticity.

The temperature dependence of the diffusion coefficient may be obtained by noting that Eq. (41) suggests that D^*/T can be expressed in terms of η_0 as follows

$$\log \frac{D^*}{T} = C(M) - \log \eta_0 \qquad (42)$$

$C(M)$ is a strong function of the molecular weight. This function does, however, vary slowly with temperature; it may change by a few percent over 100 degrees, whereas η_0 and D^* will vary by a few orders of magnitude in most polymers over this temperature range. It is well known that the temperature dependence of η_0 is accurately described by the Vogel-Fulcher equation, or equivalently by the WLF (Williams-Landel-Ferry) equation (60-62)

$$\log \eta_0 = A + \frac{B}{T - T_0} \qquad (43)$$

where T_0 and B are Vogel constants. The temperature dependence of D^*/T is easily shown to be (25,63)

$$\log \left(\frac{\dfrac{D^*}{T}}{\dfrac{D^*_{ref}}{T_{ref}}} \right) = \frac{B}{T_{ref} - T_0} - \frac{B}{T - T_0} \qquad (44)$$

In the above equation, T_{ref} is a reference temperature at which D^*_{ref} is determined.

When the P-mer chains are sufficiently short, the surroundings of the N-mer chain become altered on time scales comparable to τ_d. Consequently, the "tube" may undergo a series of displacements. The foregoing description of the motion of the N-mer chain occurring in a fixed "tube" is, therefore, inadequate. Theories (64-68) have shown that under these conditions, the D^* of the N-mer chain may be modified by the addition of a term, D_{CR}, which describes the effects of the environment on its motion; $D^* = D_{REP} + D_{CR}$. One of a number of theories which were designed to describe the effect of the environment on the N-mer chain is due to Graessley (65). He showed that the contribution of constraint release should be described by

$$D_{CR} = \alpha_{CR} D_0 M_e^2 M^{-1} P^{-3} \qquad (45)$$

where $\alpha_{CR} = (48/25)(12/\pi^2)^{z-1}$ depends on z, the number of "suitably situated" constraints per M_e. The tracer diffusion coefficient therefore becomes

$$D^* = D_0 M^{-2} (1 + \alpha_{CR} D_0 M_e^2 M P^{-3}) \qquad (46)$$

Below is a discussion on mutual diffusion in polymer mixtures.

6.3.2 Mutual Diffusion

The driving force for tracer diffusion is entropic in origin . The mutual diffusion coefficient, $D(\phi)$, in a mixture of two polymers A and B is highly composition dependent, and is influenced by enthalpic effects. For this reason it is possible, based on measurement of $D(\phi)$, to obtain information about the thermodynamics of a polymer mixture. In particular, information about the Flory-Huggins (58,69) interaction parameter, χ, which measures the strength of the interactions between the A and B segments may be learned. If the A/B interactions strongly favor mixing, $\chi < 0$, then $D(\phi)$ is enhanced over the case where $\chi = 0$. On the other hand, if $\chi > 0$, but the mixture is still within the stable regime defined by $\chi < \chi_s$ where (58)

$$\chi_s(\phi) = \{[N_A \phi]^{-1} + [N_B(1 - \phi)]^{-1}\} /2 \qquad (47)$$

then $D(\phi)$ is observed to undergo the well-known "thermodynamic slowing down." Here it is retarded in comparison to the case where $\chi = 0$. In the above equation, N_A

is the number of segments that comprise an A chain and N_B is the corresponding number in a B chain; ϕ is the volume fraction of A chains and $1 - \phi$ is that of the B chains.

The free-energy change per segment that arises from mixing A and B segments may be approximated by the mean field Flory-Huggins expression (58,69)

$$\frac{\Delta F_{mix}}{k_B T} = \frac{1}{N_A} \ln \phi + \frac{1}{N_B} \ln(1 - \phi) + \phi(1 - \phi)\chi \qquad (48)$$

The first two terms in this equation represent the combinatorial entropy of mixing. The third represents contributions from the enthalpic or noncombinatorial entropy of mixing. Since the combinatorial entropy of mixing varies as N^{-1}, the third term, which includes χ, is primarily responsible for the phase equilibrium behavior of polymer mixtures. Within the mean field approximation, the compositional dependence of the mutual diffusion coefficient $D(\phi)$ may be expressed as (70-74)

$$D(\phi) = 2\phi(1 - \phi)D_T[\chi_s(\phi) - \chi] \qquad (49)$$

The first term, $\phi(1 - \phi)D_T\chi_s$, in this expression describes the situation in which the driving force for interdiffusion is controlled by the combinatorial entropy of mixing. The second term is the correction to $\phi(1 - \phi)D_T\chi_s$ in the presence of enthalpic and non-combinatorial entropy of mixing contributions to interdiffusion. The parameter D_T is the Onsager transport coefficient. The precise form of D_T for polymer/polymer diffusion was, until recently, uncertain (75). It was demonstrated by a series of FRES experiments (36,37) that

$$D_T = D^*_A N_A(1 - \phi) + D^*_B N_B\phi \qquad (50)$$

In this equation, D^*_A is the tracer diffusion coefficient of species A into species B, and D^*_B is defined accordingly. In the section that follows, a discussion of the application of Helium ERD (FRES) to the study of diffusion in polymers is presented.

6.4 HELIUM ERD DEPTH PROFILING OF POLYMERS (GENERAL)

The vast majority of ERD experiments that have been performed on polymers have been in the area of diffusion. In typical experiments, a thin film of a deuterated polymer of thickness approximately 15 nm is placed in contact with a thick film (4 μm) of normal (undeuterated) polymer. The insert in Fig. 6.7 shows a schematic of the sample. The layer of deuterated polymer is then allowed to diffuse into the unlabeled polymer host. ERD is then used to obtain the concentration versus depth profile of the diffusant. In the actual experiment helium ions, produced by an ion source and accelerated, impinge on the sample at an angle of $\theta_1 = 15°$. The exit angle of the recoils are $\theta_2 = 15°$ and $\theta = 30°$. Note that the incident and exit angles are chosen to be equal. The energy of the incident helium ions is chosen to be

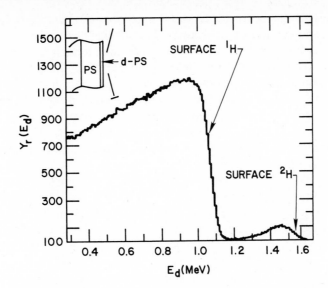

Figure   6.7:   An ERD profile, $Y(E_d)$ versus E_d, of D-PS chains which diffused into a host of PS. The deuterium profile, the surface of which is identified as "surface ^2H," is seen at higher energies. The hydrogen profile, whose surface is denoted by "surface ^1H," is seen at lower energies.

$E_0 = 3.0$ MeV. The slit distance $r = 5.08$ cm, a slit height of $h = 1.69$ cm, and a slit width of $w = 0.16$ cm were chosen for experiments conducted at Sandia National Laboratories.   Shown in Fig. 6.7 is a typical $Y_r(E_d)$ versus E_d spectrum for deuterated polystyrene, D-PS, chains of which diffused into a high molecular weight polystyrene, PS, matrix. The surface peak of the hydrogen (^1H) is at 1.1 MeV, and that of deuterium (^2H) is at 1.5 MeV. Both profiles are well separated in energy and, therefore, easily distinguished. The surface peaks of the ^1H profile and the ^2H profile are not sharp, and this is due to the finite resolution of the technique. As discussed earlier, the most important limitation of the depth resolution is the energy straggling of the recoils through the mylar foil, the foil thickness nonuniformities, and the detector resolution. The depth resolution here is 40 keV which is 80 nm. The depth resolution, δx, of the ERD experiment may be determined by obtaining an ERD profile of a thin polymer film on a substrate; if the thickness of the film is small in comparison to δx, then the profile will have a line shape that can be described by a Gaussian function. The full width at half-maximum is δx. Typically, polymer films of $\sim$ 15 nm are easily spun on a silicon substrate and its ERD profile is then obtained. A discussion concerning an empirical determination of δx may be found in Ref. 17. Reference 7 may be consulted for further details. Below is a description of how one converts the $Y(E_d)$ versus E_d profile to one of concentration versus depth.

Before each experiment is conducted, the Y_r versus E_d profile of a silicon nitride standard possessing a known concentration of hydrogen is obtained. This calibration run is done to ensure that the experimental parameters remain constant for each subsequent experiment. The channel number (CN_0) of the front edge of the hydrogen profile for this sample is recorded; this ensures that the detector is in the correct position from one experiment to another. The channel number of the back edge of this profile is then located in order to ensure that the tilt angle of $\theta_1 = 75°$ is constant in each experiment. Finally, the total number of counts (number of particles detected per microcoulomb of incident particle charge) in each channel in the multichannel analyzer used to collect the data must be checked for constancy from one experiment to another. This is done by comparing the spectrum for each subsequent run. Information about the experimental parameters such as E_0, θ_1, θ_2, and θ is recorded. Information about the target such as the target density and composition is also recorded for subsequent use.

An energy, E_d, versus channel number scale is then computed as follows. The channel number of the hydrogen surface peak is recorded. Note that a polymer standard could be used to locate the surface peak of deuterium or of hydrogen. The choice is dictated by the identity of the isotope for which the depth scale is being computed. In most polymer diffusion experiments, one is concerned with the deuterium profile. The location (i.e., channel number CN_0) of the D surface peak is therefore recorded in such cases. The kinematic factor, K, for the collision is calculated. This is followed by a calculation of the energy loss of the surface recoils in the range foil (12-μm mylar foil is used here). This energy loss is calculated by dividing the foil into 10 sublayers in each of which the stopping cross sections are considered constant. At each interface the energy loss is calculated. At this point, one now has a correlation between the energy at the surface and CN_0. One, therefore, knows the E_d /channel ratio.

The energy versus depth scale of the target is then calculated as follows. The target is divided into a series of sublayers each of thickness 1000 Å. At the interface of each sublayer, the program calculates the energy, E_1, of the projectile just before a collision and E_2, the energy of the projectile just after the collision, and then E_d, the energy with which a particle is detected. An effective stopping power dE_d/dx is calculated numerically using the information from a table of stopping powers (76,77). A table of E_1, E_2, E_d, dE_d /dx, x, and channel number (CN) is then compiled. One then, therefore, has obtained the energy versus depth scale.

The yield per channel of the spectrum is affected by the scattering cross section, the concentration of nuclei, $N(x)$, and the effective stopping cross sections, dE_d/dx. The effect of the changing effective stopping cross sections is corrected by scaling the yield at each channel CN by a factor $S(CN)/S(CN_0)$ where S is the stopping power.

In order to determine the quantity of diffusant (D-PS into a PS host) as a function of depth, the normal procedure is to use the Rutherford scattering cross sections. While this is true for many experiments, it is not true for the recoil of protons or deuterons by an alpha particle. This apparent difficulty is circumvented by the use of a standard for which the concentration of deuterium is known. The most common approach is to use a thick film of purely deuterated PS. The profile is obtained and it is used to divide the profile of the D-PS present in the PS host. Figure 6.8 shows the profile of volume fraction as a function of depth of D-PS chains of molecular weight M = 200,000 which was allowed to diffuse into a high molecular weight PS host of molecular weight P = 900,000 for 115 min at 163°C.

The procedure to obtain profiles of PS in a D-PS matrix is similar. There is a limitation on the thickness of the D-PS layer. Note that its surface peak is positioned

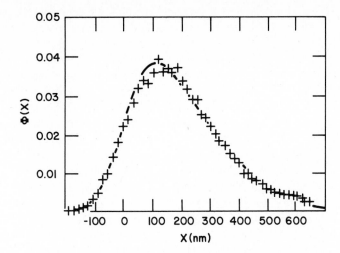

Figure   6.8:   A profile of volume fraction, Φ (x), of D-PS as a function of depth, x, into the PS host. The D-PS chains are of molecular weight M = 2 x 10^5 and the PS chains are of molecular weight P = 9 x 10^5. The chains were allowed to diffuse for 115 min at 163°C; the tracer diffusion coefficient extracted from this profile is D* = 4.2 x 10^{-14} cm^2/s.

at a higher energy than that of PS, and if it is very thick then deuterons that recoil from deep in the sample will produce a spectrum that overlaps with the ^{1}H peak. The procedure outlined above to obtain the volume fraction versus depth profile should be followed. The only difference is that the surface peak of hydrogen should be used, since a depth scale for recoiled protons is needed. It may be pointed out, however, that the depth scale for hydrogen is very close to that for deuterons.

6.5 MEASUREMENT OF TRACER AND INTERDIFFUSION IN POLYMERS

In this section, examples are presented concerning the use of RBS and ERD study tracer and mutual diffusion in polymers. It is shown here that the diffusion coefficients determined by these techniques are in excellent agreement with those determined by other techniques. Some of the examples presented below concern the use of RBS and ERD to investigate fundamental theories of polymer/polymer diffusion. We also show how one may learn about the thermodynamics of polymer mixtures using these techniques.

6.5.1 RBS Studies of Diffusion in Polymers

The experiment that is about to be described is a new version of the Kirkendall effect in metallic alloys.   Here, RBS is used to determine the displacement of gold "markers" initially placed at the interface of two polymer films, one of molecular weight M and the other of P, where M_c < M << P. The marker displacement was observed to vary as $t^{1/2}$. The tracer diffusion coefficient, D* , of the M-chains was determined from the marker displacement and was shown to vary as M^{-2}.

A silicon wafer was coated with a layer of ultrahigh molecular weight polystyrene of molecular weight P = 2.0 x 10⁶ by drawing the wafer from a viscous solution of the polymer at a constant velocity. A discontinuous layer of gold [particles were approximately 50 Å in diameter (22)] was evaporated onto the polymer surface under a vacuum of 10⁻⁵ torr. Transmission electron microscopy measurements show that the gold is distributed on the surface as a series of islands of mean dimensions of 1.8 nm with a mean spacing of 2.5 nm. One may consult Ref. 22 for the details. A second film of molecular weight M << P was prepared on a glass slide in a manner similar to that described above. This film, of thickness 1 μm, was floated off the glass slide unto a bath of distilled water from which it was picked up onto the surface of the gold decorated PS layer. A schematic of the sample is shown in Fig. 6.9. The depth of the "markers" below the front surface was determined using RBS. A helium beam of energy 2.4 MeV was used for the analysis.

In an RBS spectrum, gold at the surface would register a peak at 2.12 MeV [k_{Au} = 0.9225; see Eq. (32)] and that below the surface would register a peak at lower energies dictated by Eq. (35), assuming the thin film approximation is valid. The energy loss factor could be calculated based on the density and the stopping cross sections of the individual elements in PS using Bragg's rule and Zeigler's (76) or Brice's (77) tables. However, it can be determined empirically by measuring the shift of the Au peak by films of known thickness. This latter procedure was used and $[S]^{-1}$ was found to be approximately constant, 2.2 nm/keV, or 11 nm/channel on a multichannel analyzer , over the depth range of interest.

The RBS spectra for PS chains of M = 33,000 diffusing into a PS matrix of P = 2.0 x 10⁶ at 174°C is shown in Fig. 6.10. Figure 6.10a is the spectrum that was obtained before the diffusion process. The peak whose surface appears just below channel 110 (E = 0.454 MeV) is due to the helium ions which scatter off carbon. The sharp peak at channel number 290 (E = 1.436) is due to scattering of the ions off Au at the interface. There is no peak from hydrogen since these masses are too light to backscatter the more massive helium ions. Parts (b) and (c) show spectra for situations in which the films were allowed to diffuse for 0.5 and 1.3 h, respectively. The results in this figure clearly show that the gold "markers" approach the surface (direction of the faster diffusing species) as the M- and P-chains interdiffuse. During these experiments, the beam is moved to a different region each time an analysis is done after subsequent anneals. At no time during any of the experiments did we observe a broadening of the Au peak, the peaks remained sharp. Figure 6.11 shows the behavior of the marker displacement as a function of time. These data clearly show that $\Delta X \sim t^{1/2}$.

An in-depth analysis of diffusion under these experimental conditions has shown that the marker displacement is

$$\Delta X = C(D^* t)^{1/2} \tag{51}$$

where $C = \phi'(0) \, \alpha \, [1 - \alpha \, \phi(0)]$ is a function of the ratio $\alpha = M/P$, the composition at the interface, $\phi(0)$ and $\phi'(0)$. The analysis considers the fact that since the M- and P-chains have different diffusion coefficients, then there should be a net flux of matter flowing in the direction of the slower diffusing species (P-chains). This

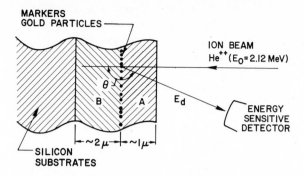

Figure   6.9:   Schematic of the RBS "marker" experiment.

would be compensated by a bulk or "vacancy" flow in order to avoid a buildup in osmotic pressure. The result of this is a movement of the "markers" toward the side of the faster diffusing species (M-chains). The value of C is computed by numerically solving the diffusion equation. The details of this may be found in Ref. 22. Using the calculated values of C and the experimentally determined values of $\Delta X/t^{1/2}$, the tracer diffusion coefficients of the M-chains diffusing into two different hosts P = 2.0 x 10^6 and P = 1.8 x 10^6 were computed. The results are plotted in Fig. 6.12, where D* for the diffusion of the M-chains into the P = 2.0 x 10^6 host is represented by the circles and the triangles represent diffusion into the P = 2.0 x 10^6 matrix. Notice that both sets of data superimpose on the same line. This is expected since the matrix is of sufficiently high molecular weight that it is not expected to contribute to the motion of the M-chain. The experimental results may be described by the equation

$$D^* \;=\; 0.008 \; M^{-2} \tag{52}$$

In the section that follows, these experimentally determined values of D* are compared with those obtained using other techniques and with theoretical predictions made using Eq. (41).

There have been a number of other applications of RBS to the study of diffusion in polymer/solvent systems. The reader is encouraged to see Refs. 19-21 where these are discussed in detail.

6.5.2 ERD Studies of Diffusion

Tracer Diffusion: The sample which consists of a thin layer of D-PS (~15 nm) on top of a thick layer (~2 μm) of PS is supported by a polished silicon wafer. A schematic is shown in the insert in Fig. 6.7. The thick layer of PS was produced by drawing a polished silicon wafer at a constant velocity from a viscous solution of PS. The thin layer was produced by spinning a dilute solution of D-PS on a glass slide from which it was floated onto a bath of distilled water to be picked up onto the surface of the PS coated layer.

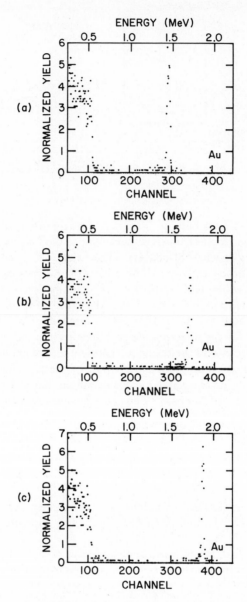

Figure 6.10: RBS profiles of yield versus energy for PS/PS interdiffusion; (a) The marker at its initial position before interdiffusion. (b) The position of the marker after interdiffusion for 30 min at 174°C. (c) The marker position after annealing for 78 min at 174°C.

Determination of D*: Since the thin layer of D-PS becomes dilute ($\phi < 0.07$) soon after the diffusion commences, the process can be described by a solution to the diffusion equation assuming the diffusion coefficient is a tracer diffusion coefficient.

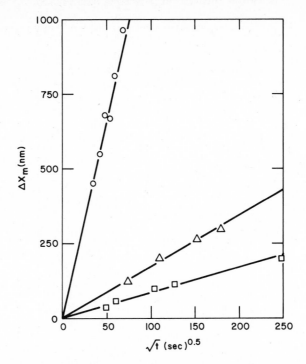

Figure 6.11: The marker displacement, Δx_m, is shown here as a function of $t^{1/2}$ for the diffusion of PS chains of M = 3.3 x 10^4 (O), M = 2.55 x 10^5, (Δ) and M = 3.9 x 10^5 ($\square$) into a host of PS chains of P = 2 x 10^7.

The solution for a thin film of thickness H that is allowed to diffuse into a semi-infinite region is given by

$$\phi(x) = \frac{1}{2} \left\{ \mathrm{erf} \left[\frac{H + x}{(4D^*t)^{1/2}} \right] + \mathrm{erf} \left[\frac{H - x}{(4D^*t)^{1/2}} \right] \right\} \qquad (53)$$

The line drawn through the data in Fig. 6.8 is a fit obtained by convoluting the above equation with the instrumental resolution function which is a Gaussian with a full width at half maximum of 80 nm. The D^* was varied to give the best fit to the data. The D^* that was determined was 4.2 x 10^{-14} cm^2 /s.

Molecular Weight Dependence of D^*: In what follows, we discuss the molecular weight dependence of the dynamics of D-PS chains in PS hosts. The results of a FRES analysis of the M dependence of the diffusion of D-PS into PS at 171°C are represented by the triangles in Fig. 6.13. The squares represent the diffusion of PS into D-PS. Both sets of data are identical. For a comparison, the results of the Rutherford backscattering (RBS) marker experiment of the diffusion of PS into PS are represented by the filled circles. Independent measurements of the diffusion of

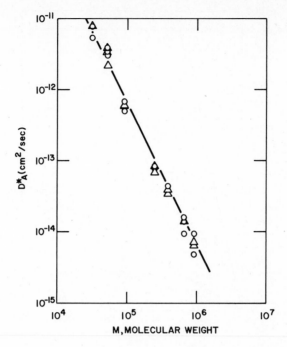

Figure 6.12: A plot of the tracer diffusion of PS chains of molecular weight M which diffused into PS hosts of $P = 2.0 \times 10^7$ (O) and $P = 2.0 \times 10^6$ (Δ) at $174°C$; $D^* \sim M^{-2}$.

fluorescent labeled PS into PS by a holographic grating technique (49) are represented by the open circles. It is clear, however, that an independent comparison of studies performed using different techniques yields results that are in excellent agreement with each other. All the data may adequately be described by the equation

$$D^* = 0.005 \, M^{-2} \tag{54}$$

which is in agreement with the reptation prediction. The prefactor in Eq. (56) was calculated using Eq. (41) to be (16,17,22)

$$D^* = 0.003 \, M^{-2} \tag{55}$$

This is in excellent agreement with the measured results.

   Temperature Dependence: In what follows, we discuss the temperature dependence of diffusion and of the zero shear rate viscosity of the polymer. Recall that the ratio D^*/T and η_0^{-1} should exhibit nearly the same temperature dependence. The filled triangles in Fig. 6.14 represent the ratio D^*/T as a function of $(T - T_0)^{-1}$ for a D-PS chain of $M = 1.1 \times 10^5$ which diffused into a PS host. The open triangles represent data for the diffusion of D-PS of $M = 4.3 \times 10^5$ into a PS host that was nor-

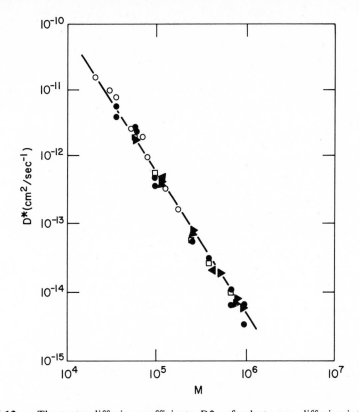

Figure   6.13:   The tracer diffusion coefficients, D* , of polystyrene, diffusing into polystyrene hosts using different techniques at 171°C is plotted as a function of M. The open circles represent the diffusion of fluorescent-labeled PS into PS using a holographic grating technique (Ref. 49). The filled circles represent the data obtained from the RBS marker experiment; the host in this case is the P = 2.0 x 10[7] host; these data were shifted using the temperature shift factor for PS (see text). The open squares represent the diffusion of PS into D-PS hosts at 171°C. The filled left triangles are from the diffusion of D-PS into PS at 171°C, and the filled right triangles are from the same experiment conducted at 174°C. This latter set of data was shifted using the temperature shift factor for PS.

malized by a factor of $(110/430)^2$ so that it could be superimposed on the other set of D-PS data. The line drawn through the data was computed with Eq. (44) using constants B = 710 and T_0 = 49°C. These constants are the identical ones used to fit the η_0 data of PS by Graessley and Roovers (78). A reference temperature of T_{ref} = 170°C was used.

The temperature dependent data of polymethylmethacrylate (PMMA) are shown in Fig. 6.15 in which D*/T (O) is plotted as a function of $(T - T_0)^{-1}$. The line drawn through the data was computed using Eq. 44. The constants B = 1118 and T_0 = 35°C were obtained from fits to the viscosity data of PMMA (60). These data clearly demonstrate that one should be able to predict the temperature dependence of D* (T)/T from the temperature dependence of zero shear rate viscosity.

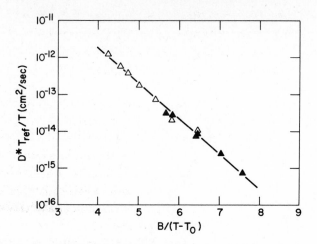

Figure 6.14:   The temperature dependence of the diffusion of D-PS into PS, D^* T/T_{ref} versus $B/(T - T_0)$, is shown here. The open triangles represent the diffusion of D-PS of M = 4.3×10^5, and the filled ones that of D-PS of M = 1.1×10^5 into a high molecular weight PS host.

Constraint Release: A series of experiments were performed (23,24) where the tracer diffusion coefficient, D^*, of a chain of molecular weight M diffused into a series of melts, each of molecular weight P.   Figure 6.16 shows a plot of D^* as a function of P for different values of M. The lines drawn through the data were computed using Eq. (46) in which $\alpha_{CR} = 11$ and $M_e = 18,000$.

It is clear from these measurements that at large values of P, D^* is independent of P which one should anticipate based on earlier discussions in Section 6.3.1. In this regime, D^* is described by reptation of the M-chain in a "fixed" environment; D^* varies as M^{-2}, independent of P. Below a characteristic molecular weight P', D^* increases appreciably with decreasing P. At P', the contribution of constraint release to the diffusion process becomes comparable to the reptation contribution. As P decreases the contribution due to constraint release becomes increasingly important.

Mutual Diffusion: There is great interest in the thermodynamics of isotopic polymer mixtures (80-90). Recent small angle neutron scattering (SANS) measurements strongly suggest that the phase equilibrium properties of a polymer system are affected by isotopic substitution. These systems are characterized by a slightly positive Flory-Huggins interaction parameter, χ, and exhibit an upper critical solution temperature (UCST) (29,30,80-83, 85, 87). The general method for obtaining χ from the SANS measurements is to fit the experimental scattering curves to deGennes' random phase approximation (RPA) of the static structure factor using χ as an adjustable parameter (58). In this section, we discuss a method by which one may determine the χ parameter from measurement of the mutual diffusion coefficient (29,30,36,37), $D(\phi)$, where ϕ is the composition of the deuterated polymer in the mixture, in this system. The UCST is determined from the temperature depend-

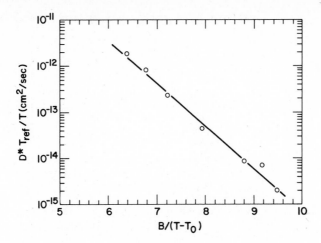

Figure 6.15:   The temperature dependence of D-PMMA; $D^* T / T_{ref}$ versus $B/(T - T_0)$, into PMMA.

ence of $D(\phi)$. Comparisons are made with the results of the SANS measurements and with theoretical predictions of χ (81-83,87).

In systems for which the value of the χ parameter is positive but the mixture is still in the range of single phase stability, defined by $\chi < \chi_s$, $D(\phi)$ should experience a minimum or a "thermodynamic slowing down" in the vicinity of the critical composition ϕ_c where

$$\phi_c = \frac{N_H^{1/2}}{N_H^{1/2} + N_D^{1/2}} \tag{56}$$

In the above equations, N_D and N_H are the number of monomer segments that comprise the deuterated and undeuterated polymer chains, respectively. As χ approaches χ_s , or equivalently, as the temperature approaches the UCST, the system experiences large fluctuations in composition; therefore, the extent of the "critical slowing down" should be enhanced. In cases where $\chi > c_s$, the mixture is unstable and undergoes phase separation. "Thermodynamic slowing down" effects should not, however, be observable when $\chi = 0$.

We now discuss the sample preparation procedure. High molecular weight mixtures of deuterated and undeuterated polystyrenes ($N_D = 104$, $N_H = 8.8 \times 10^3$) were used in this study. Diffusion couples were produced in which on either side of the interface, is a blend of D-PS and H-PS , one of composition ϕ_0 and the other of $\phi_0 + \Delta\phi$; $\Delta\phi$ is chosen to be about 10%. The samples were prepared by coating a polished silicon wafer with ~1-μm film of a blend of a given composition using the solution casting procedure described earlier. A second film (~0.45 μm) was produced separately by spinning a solution of a composition that differed from the

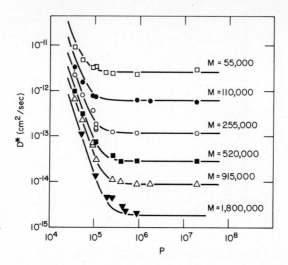

Figure 6.16: Constraint release of D-PS chains of molecular weight M which diffuse into PS hosts of molecular weight P. The plot here is one of D* of the M-chains as a function of P. These data are at 174°C.

previous film by ~ 10% on a glass slide. The film was then floated onto a bath of distilled water from which it was transferred onto the surface of the coated silicon wafer. Figure 6.17a shows the FRES profile of a film of average composition ϕ_{ave} = 0.55 of D-PS; $\Delta\phi$ = 0.09. The line drawn through the data represents the instrumental broadening which has a full width at half-maximum of 100 nm. The resolution of this system is fixed; this line is not a fit to the data. The initially steplike concentration profile in this sample was allowed to broaden by interdiffusion at elevated temperatures. Shown in Fig. 6.17b is an FRES profile of the sample that was allowed to diffuse at 174°C for 15.5 h. Since the composition of the couple changes by only 10%, a single mutual diffusion coefficient should control the interdiffusion process. The mutual diffusion coefficient was then extracted by fitting it with a solution to the diffusion equation

$$\Phi(x) \ = \ \overline{\phi} \ - \ \frac{1}{2} \Delta\Phi \, \mathrm{erf} \left[\ \frac{x}{(4D^*t)^{1/2}} \ \right] \tag{57}$$

which was convoluted with the instrumental resolution function. In the above equation, t is the diffusion time. The mutual diffusion coefficient extracted from the data at this composition is D(0.55) = 2.0 x 10⁻¹⁵ cm²/s. Equation (53) is equally applicable and it yielded the same values of D(ϕ). Mutual diffusion coefficients were extracted at different temperatures using couples of varying compositions.

D(ϕ) is plotted as a function of the average blend composition, ϕ_{ave}, in Fig. 6.18 at temperatures of 166, 174, 190, and 205°C. The data at each temperature exhibit a pronounced minimum or "critical slowing down" in the vicinity of ϕ_{crit} = 0.5. The

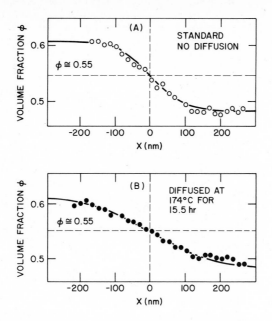

Figure 6.17:   Volume fraction versus depth for the diffusion of two layers, one of average composition $\Phi = 50\%$ and the other of $\Phi = 60\%$; (a) prior to interdiffusion, and (b) after interdiffusion for 15.5 h at 174°C. The diffusion coefficient extracted from this data is 2.0 x 10^{-15} cm^2/s.

magnitude of this effect increases with decreasing temperature (χ approaches χ_s), which one would anticipate if the system exhibits a UCST. Note that the correlation length, ξ, of the concentration fluctuations increases,

$$\xi^{-2} \sim [\phi\,(1 - \phi)(\chi_s(\phi) - \chi)] \tag{58}$$

which suggests that D(ϕ) should decrease. Equation (49) may be expressed in terms of the correlation length D(ϕ) $\sim \xi^{-2}$, which shows that when the correlation length of the concentration fluctuations in the system increases, χ approaches χ_s, and D(ϕ) decreases.

Since D(ϕ) is sensitive to small changes in χ, especially near the stability limit, knowledge of D(ϕ) is a useful way to determine χ. The lines drawn through each set of data are computed using Eqs. (49) and (50) in which χ is the only adjustable parameter. Its value is adjusted to yield the best fit to the data in the middle of the concentration regime. The data of the FRES experiments may be fit by the following equation:

$$\chi = \frac{0.22\,(+0.06)}{T} - 3.2\,(+1.2) \times 10^{-4} \tag{59}$$

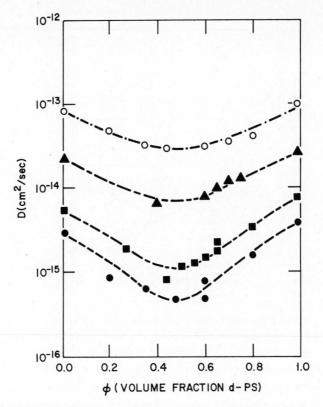

Figure 6.18: The interdiffusion coefficient as a function of the average composition of the various diffusion couples. Data were taken at 166°C (filled circle), 174°C (filled square), 190°C (filled triangle), and 205°C (open circle).

A plot of this data is shown in Fig. 6.19. The stability limit of this mixture was calculated, using Eq. (47), to be $\chi_s(\phi_{crit}) = 2.1 \times 10^{-4}$. In the terminology used here, the D-PS species is the A species and H-PS is the B species. Equation (59) predicts the UCST to be about 130°C. The SANS data of Bates and Wignall (81) may be expressed as

$$\chi = \frac{0.02\,(+0.01)}{T} - 2.9\,(+0.4) \times 10^{-4} \qquad (60)$$

The agreement between both sets of experimental data is excellent. One should easily anticipate these effects for the following reason. The combinatorial entropy of mixing varies as $1/N$ which approaches zero in the limit of high N, suggesting that, within the framework of the Flory-Huggins free energy, the enthalpic contributions, embodied by χ, dominate the free energy. Since $\chi > 0$, completely random mixing is not favored.

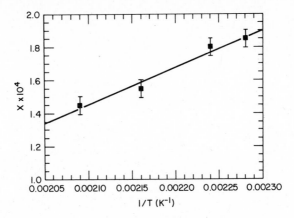

Figure   6.19:   Plot of Flory-Huggins interaction parameter versus $1/T$.

There have recently been works (83,86,87) which suggest that χ is composition dependent in the composition range of ϕ close to 1 and to 0, but relatively independent of composition in the middle of the composition regime. It turns out that this technique of determining χ is not very sensitive near the composition extremes.

In conclusion, these results clearly demonstrate that when $\chi < 0$, the mutual diffusion coefficient is retarded in comparison to the situation for which $\chi = 0$.

Mutual Diffusion in the Polystyrene-Poly(2,6-Dimethyl 1,4-Phenyleneoxide) System ($\chi < 0$): The first set of diffusion experiments to determine χ in a polymer blend using the procedure described above were performed on this system. The sample preparation procedure was essentially the same. Equation (53) was used to extract the mutual diffusion coefficient. One essential difference between this and the previous experiment is that, since the glass transition temperatures of PS and poly(2,6-dimethyl 1,4-phenyleneoxide) system (PXE) differ by $112°$, the temperature at which each diffusion couple was annealed was chosen such that $T - T_g = 66°$. The reference T_g was chosen to be that of the average composition of the blend. The main purpose of the discussion below is to demonstrate that in cases where $\chi < 0$, $D(\phi)$ is enhanced over the case where $\chi = 0$.

Figure 6.20 shows a plot of the results of measurement of $D(\phi)$ as well as D^* in this blend. The circles and squares represent the D^*'s of the diffusion of D-PS (M = 255,000) and D-PXE (M = 35,000) chains, respectively, into the blend of varying compositions. These are highly composition dependent and vary a few orders of magnitude. This behavior is attributed to variations in the monomeric friction coefficient, ξ. The diamond-shaped symbols are the $D(\phi)$ values measured at $T - T_g = 66°$. The solid line drawn through these data was computed with Eqs. (50) and (51) using the measured D^*'s and $\chi = 0.112 - 62/T$ (36,37) which were determined using the procedures outlined above. One may compare this with the broken line which was computed using $\chi = 0$. These experiments show conclusively that in situations where $\chi < 0$, the mutual diffusion coefficient is enhanced over the case where $\chi = 0$.

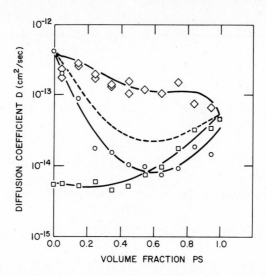

Figure 6.20: Tracer and interdiffusion in the PXE/PS system. The diamond-shaped figures represent the interdiffusion in the PXE/PS blends of varying composition; $\chi < 0$. The broken line represents interdiffusion if $\chi = 0$. The circles represent the tracer diffusion of D-PS chains into PXE/PS blends of varying volume fractions and the squares that of D-PXE into the same blends (from Ref. 70).

6.6 CONCLUDING REMARKS

We have shown how elastic recoil detection (ERD) and Rutherford backscattering spectrometry (RBS) may be used to determine the composition versus depth profile of various elements in materials. This presentation included a description of how one may construct the geometry of both scattering experiments in order to obtain optimal results as well as a detailed description of how one may analyze data obtained from both experiments. A number of applications were chosen from the field of polymer/polymer diffusion since most of the work of ion beam analysis has been concentrated in this area. It was shown how these techniques have been used to investigate the validity of fundamental theories describing the translational dynamics of polymer molecules, and how they may be used to gain information on the thermodynamics of polymer mixtures. These techniques have had a major impact on the area of polymer/solvent diffusion. This was not discussed here in detail, but a number of important references have been provided. Another field where the use of ion beams is now of current interest is that of ion implantation of polymers (90-93), which was not discussed here. It is our hope that this chapter will serve as a useful reference for someone intending to analyze polymers using ion beam analysis.

6.7 ACKNOWLEDGMENTS

The work described here benefited from important collaborations particularly with Prof. E. J. Kramer and Prof. J. W. Mayer of Cornell University. Other useful collaborations occurred with Prof. P. J. Mills, Dr. R. J. Composto, Dr. C. J. Palmstrom and Dr. P. S. Peercy. This work was performed at Sandia National Laboratories and was supported by the U.S. Department of Energy under contract number DE-AC04-76DP00789.

6.8 REFERENCES AND NOTES

1. J. W. Mayer, L. Eriksson and J. A. Davies, **Ion Implantation in Semiconductors**, Academic Press, New York (1970).

2. G. K. Hubler, "Surface Alloying by Ion Beams," 287, in **Surface Alloying Using Ion, Electron, and Laser Beams**, L. E. Rehn, S. T. Picraux, and H. Wiedersich, Eds., American Society of Metals, Metals Park, Ohio (1987); C. R. Clayton, "Chemical Effects of Ion Implantation: Oxidation, Corrosion and Catalysis," 325, ibid.

3. P. Sioshansi, "Surface Modification of Industrial Components by Ion Implantation," *Mat. Sci. Eng.*, 90, 373 (1987).

4. L. Ng and Y. Naevheim, unpublished work presented at the *Ball and Bearing Technical Symposium*, Orlando, FL (1987).

5. A. Benninghoven, Ed., **Secondary Ion Mass Spectroscopy SIMSII-V**, Springer Verlag, New York (1979-1985).

6. D. Briggs, Ed., **Handbook of X-ray and Ultraviolet Plotoelectron Spectroscopy**, Heyden and Son, Ltd., London (1978).

7. W. K. Chu, J. W. Mayer, and M. A. Nicolet, **Backscattering Spectrometry**, Academic Press, New York (1978).

8. G. Amsel, J. P. Nadai, E. d'Artemare, D. David, E. Girard, and J. Moulin, "Microanalysis by the Direct Observation of Nuclear Reactions Using a 2 MeV Van Der Graff Accelerator," *Nuclear Instruments Methods*, 92, 481 (1971); R. A. Langley, S. T. Picraux, and F. L. Vook, "Depth Distribution Profiling of Deuterium with ^{3}He," *J. Nuclear Mat.*, 53, 257 (1974); C. Cardinal, C. Brassard, J. Chabbal, L. Deschenes, J. P Labrie, J. L'Eucyler, and M. Lernoux, "A Method of Utilizing Lithium-Induced Nuclear Reactions in the Determination of Carbon Nitrogen and Oxygen in Thin Films," *Appl. Phys. Lett.*, 26, 543 (1975).

9. J. L'Eucyler, C. Brassard, C. Cardinal, J. Chabbal, L. Deschenes, J. P. Labrie, B. Terrault, J. G. Martel, and St. Jacues, "An Accurate and Sensitive Method for the Determination of the Depth Distribution of Light Elements in Heavy Materials," *J. Appl. Phys.*, 47, 381 (1976).

10. B. L. Doyle and P. S. Peercy, "Techniques for Profiling H with 2.5 MeV Van Der Graff Accelerators," *Appl. Phys. Lett.*, 34, 811 (1979).

11. B.L. Doyle and P.S. Peercy, "Hydrogen Depth Profiling Using Elastic Recoil Detection," *Internal Sandia National Laboratories Report*, SAND79-02106 (1979).

12. A. Turos and O. Meyer, "Depth Profiling of Hydrogen by the Recoiling of Protons," *Nuclear Instr. Meth. Phys. Res.*, B4, 92 (1984).

13. S. Nagata, S. Yamaguchi, Y. Fujino, Y. Hori, N. Sugiyama, and K. Kamada, "Depth Resolution and Recoil Cross Section for Analyzing Hydrogen in Solids

Using Elastic Recoil Detection with a ^{4}He Beam," *Nuclear Instr. Meth. Phys. Res.*, B6, 533 (1985).

14. B. L. Doyle and D. K. Brice, "The Analysis of Elastic Recoil Detection Data," *Nuclear Instr. Meth. Phys. Res.*, B35, 301 (1988).

15. G. W. Arnold, B. L. Doyle, and B.C. Bunker, "Use of a Tandem Accelerator for Light Element Analysis," 50, in *Proc. Three-Day In-Depth Review of the Nuclear Accelerator Impact in the Interdisciplinary Field*, P. Mazzoldi and G. Moschini, Eds., Laboratori Nazionali di Legnaro, Padua, Italy (1984).

16. P. F. Green, P. J. Mills, and E. J. Kramer, "Diffusion Studies of Polymer Melts by Ion Beam Depth Profiling of Hydrogen," *Polymer*, 27, 1063 (1986).

17. P. J. Mills, P. F. Green, C. J. Palmstrom, J. W. Mayer, and E. J. Kramer, "Analysis of Diffusion in Polymers by Forward Recoil Spectrometry," *Appl. Phys. Lett.*, 45, 958 (1984)

18. W. K. Chu, "Scattering Recoil Coincidence Spectrometry," *Nuclear Instr. Meth. Phys. Res.*, B35, 518 (1988).

19. R. L. Lasky, Ph.D. Thesis, *Rutherford Backscattering Analysis of Case II Diffusion in Polystyrene*, Cornell University (1987).

20. P. J. Mills, C. J. Palmstrom, and J. W. Mayer, "Concentration Profiles of Non-Fickian Diffusants in Glassy Polymers Using Rutherford Backscattering Spectrometry," *J. Mater. Sci.*, 21, 1479 (1986).

21. J. F. Romanelli, J. W. Mayer, E.J. Kramer, and T. P. Russell, "Rutherford Backscattering Spectrometry Studies of Iodine Diffusion in Polystyrene," *J. Polym. Sci., Polym. Phys. Ed.*, 24, 263 (1986).

22. P. F. Green, C. J. Palmstrom, J. W. Mayer, and E. J. Kramer, "Marker Displacement Measurements of Polymer-Polymer Interdiffusion," *Macromolecules*, 18, 501 (1985).

23. P. F. Green, P. J. Mills, C. J. Palmstrom, J. W. Mayer, and E. J. Kramer, "The Limits of Reptation in Polymer Melts," *Phys. Rev. Lett.*, 53, 2145 (1984).

24. P. F. Green and E. J. Kramer, "Matrix Effects on the Diffusion of Long Polymer Chains," *Macromolecules*, 19, 1108 (1986).

25. P. F. Green and E. J. Kramer, "Temperature Dependence of Tracer Diffusion Coefficients in Polystyrene," *J. Mat. Res.*, 1, 202 (1986).

26. P. J. Mills, P. F. Green, C. J. Palmstrom, J. W. Mayer, and E. J. Kramer, "Polydispersity Effects on Diffusion in Polymers: Concentration Profiles of d-Polystyrene Measured by Forward Recoil Spectrometry," *J. Polym. Sci., Polym. Phys. Ed.*, 24, 1 (1986).

27. H. R. Brown, A. C. M. Yang, T. P. Russell, and W. Volksen, "Diffusion and Self Adhesion of the Polyimide PMDA-ODA," *Polymer*, 29, 1807 (1989).

28. P. F. Green and B. L. Doyle, "Silicon Elastic Recoil Detection Studies of Polymer-Polymer Diffusion: Advantages and Disadvantages," *Nuclear Instr. Meth. Phys. Res.*, B18, 64 (1986).

29. P. F. Green and B. L. Doyle, "Isotope Effects on Interdiffusion in Blends of Normal and Deuterated Polymers," *Phys. Rev. Lett.*, 57, 2407 (1986).

30. P. F. Green and B. L. Doyle, "Thermodynamic Slowing Down of Interdiffusion in Isotopic Polymer Mixtures," *Macromolecules*, 20, 2471 (1987).

31. P. F. Green, T. P. Russell, M. Granville, and R. Jerome, "Diffusion of Homopolymers into Non-equilibrium Diblock Copolymer Structures: 1. Molecular Weight Dependence," *Macromolecules*, 21, 3266 (1988).

32. P. F. Green, T. P. Russell, M. Granville, and R. Jerome, "Temperatures Dependence of Tracer Diffusion of Homopolymers into Non-equilibrium Diblock Copolymer Structures," *Macromolecules*, 22, 908 (1989).

33. P. F. Green, "Diffusion in Block Copolymers and Isotopic Polymer Mixtures," in **New Trends in the Physics and Physical Chemistry of Polymers**, S. Lee, Ed., to be published (1989).

34. S. F. Tead and E. J. Kramer, "Polymer Diffusion in Melt Blends of Low and High Molecular Weight Polystyrene," *Macromolecules*, 21, 1513 (1988).

35. S. F. Tead, W. E. Vanderline, A. L. Ruoff, and E. J. Kramer, "Diffusion of Reactive Ion Beam Etched Polymers," *Appl. Phys. Lett.* (1988).

36. R. J. Composto, Ph.D. Thesis, *Diffusion in Polymer Blends*, Cornell University (1987).

37. R. J. Composto, J. W. Mayer, E. J. Kramer, and D. M. White, "Fast Mutual Diffusion in Polymer Blends," *Phys. Rev. Lett.*, 57, 1312 (1986).

38. R. J. Composto, E. J. Kramer, and D. M. White, "Fast Macromolecules Control Diffusion in Polymer Blends," *Nature*, 323, 1980 (1987).

39. P. J. Mills, J. W. Mayer, E. J. Kramer, G. Hadziioannou, P. Lutz, C. Strazielle, P. Remp, and A. J. Kovacs, "Diffusion of Polymer Rings in Linear Polymer Matrices," *Macromolecules*, 20, 513 (1987).

40. M. Tirrell, "Polymer Self Diffusion in Entangled Systems," *Rubber Chem. Tech.*, 57, 523 (1984).

41. S. S. Voyutskii, **Adhesion and Autohesion of High Polymers**, Wiley-Interscience, New York (1963).

42. P. G. deGennes, "Dynamics of Fluctuations and Spinodal Decomposition in Polymer Blends," *J. Chem. Phys.*, 72, 4756 (1980).

43. L. Leger, H. Hervet, and F. Rondalez, "Reptation in Entangled Polymer Solutions by Forced Rayleigh Scattering," *Macromolecules*, 14, 1732 (1981).

44. B. A. Smith, E. T. Samulski, L. P. Yu, and M. A. Winnik, "Tube Renewal Versus Reptation: Polymer Diffusion in Molten Poly(Propylene Oxide)," *Phys. Rev. Lett.*, 52, 45 (1984); S. J. Mumby, B. A. Smith, E. T. Samulski, Li-P. Yu, and M. A. Winnik, "Temperature Dependence of the Diffusion Coefficient in Poly(propylene Oxide) in the Undiluted State," *Polymer*, 27, 1826 (1986).

45. C. R. Bartels, W. W. Graesley, and B. Crist, "Measurement of Self Diffusion in Polymer Melts," *J. Polym. Sci., Polym. Lett. Ed.*, 21, 495 (1983).

46. Y. Kumagai, H. Wantanabe, K. Miyaasaki, and T. Hata, "Diffusion Measurement of Tritium Labelled Polystyrene in the Polymer Bulk," *J. Chem. Eng., Japan*, 12, 1 (1979).

47. G. Fleischer, "Self Diffusion in Melts of Polystyrene and Polyethylene in the Bulk by Pulsed Field Gradient NMR," *Polym. Bull.*, 9, 152 (1983); "Chain

Length Dependence of Self Diffusion in Melts of Polystyrene and Polyethylene," *Colloid Polym. Sci.*, 265, 89 (1987).

48. P. T. Gilmore, R. Falabella, and R. L. Lawrence, "Polymer/Polymer Diffusion. 2. Efects of Temperature and Molecular Weight on Macromoleculer Diffusion in Blends of Poly(vinylchloride) and Poly(ε-caprolactone)," *Macromolecules*, 13, 880 (1980).

49. M. Antonietti, J. Coutandin, and H. Sillescu, "Chain Length and Temperature Dependence of the Self Diffusion Coefficients in Polystyrene," *Macromol. Chem., Rapid Commun.*, 5, 525 (1984).

50. J. Klein and C. Briscoe, "Diffusion of Long Chain Molecules Through Bulk Polyethylene," *Proc. Soc. London A*, 365, 53 (1979).

51. L. R. Doolittle, "Algorithims for the Rapid Simulation of RBS Spectra," *Nuclear Instr. Meth. Phys. Res.*, B9, 344 (1985).

52. L. R. Doolittle, "A Simulation Algorithm for RBS Analysis," *Nuclear Instr. Meth. Phys. Res.*, B15, 227 (1986).

53. P. E. Rouse, "A Theory of Linear Viscoelastic Properties of Dilute Solutions of Coiling Polymers," *J. Chem. Phys.*, 21, 1272 (1953).

54. M. Doi and S. F. Edwards, "Dynamics of Concentrated Polymer Systems," *J. Chem. Soc., Faraday Trans. II*, 10, 1789 (1978).

55. M. Doi and S. F. Edwards, **The Theory of Polymer Dynamics**, Oxford University Press, Oxford, United Kingdoms (1986).

56. W. W. Graessley, *Roy. Soc. Chem. Faraday Div., Faraday Symp.*, 18, 1 (1983).

57. P. G. deGennes, "Reptation of A Polymer Chain in the Presence of Fixed Obstacles," *J. Chem. Phys.*, 55, 572 (1971).

58. P. G. deGennes, **Scaling Concepts in Polymer Physics**, Cornell University Press, Ithaca, NY (1979).

59. W. W. Graessley, "Some Phenomenological Consequences of the Doi Edwards Theory of Viscoelasticity," *J. Polym. Sci., Polym. Phys. Ed.*, 18, 27 (1980).

60. G. C. Berry and T. G. Fox, "The Viscosity of Some Polymers and Their Concentrated Solutions," *Adv. Polym. Sci.*, 5, 261 (1968).

61. J. D. Ferry, **Viscoelastic Properties of Polymers**, Wiley, New York (1980).

62. T. G. Fox and V.R. Allen, "The Dependence of the Zero Shear Melt Viscosity and the Related Friction Coefficient and Critical Chain Length on Measurable Characteristics of Chain Polymers," *J. Chem. Phys.*, 41, 344 (1964).

63. N. Nemoto, M. R. Landry, I. Noh, and H. Yu, "Temperature Dependence of the Self Diffusion Coefficient of Polyisoprene in the Bulk State," *Polym. Comm.*, 25, 141 (1984).

64. W. Hess, "Generalized Theory for Entangled Polymer Liquids," *Macromolecules*, 21, 2587 (1988).

65. W. W. Graessley, "Entangled Branched and Network Polymer Systems: Molecular Theories," *Adv. Polym. Sci.*, 47, 67 (1982).

66. H. Wantanabe and M. Tirrell, "Reptation with Configuration-Dependent Constraint Release in the Dynamics of Flexible Polymers," *Macromolecules*, 22, 927 (1989).

67. M. Daoud and P. G. deGennes, "Some Remarks on the Dynamics of Polymer Melts," *J. Polym. Sci., Polym. Phys. Ed.*, 17, 1971 (1979).

68. J. Klein, "Dynamics of Entangled Linear, Branched and Cyclic Polymers," *Macromolecules*, 19, 105 (1986).

69. P. J. Flory, **Principles of Polymer Chemistry**, Cornell University Press, Ithaca, NY (1953).

70. E. J. Kramer, P. F. Green, and C. J. Palmstrom, "Interdiffusion and Marker Movements in Concentrated Polymer-Polymer Diffusion Couples," *Polymer*, 25, 473 (1984).

71. H. Sillescu, "The Relation of Interdiffusion and Self Diffusion in Polymer Mixtures," *Makromol. Chem., Rapid Comm.*, 5, 519 (1984).

72. H. Sillescu, "Relations of Interdiffusion and Tracer Diffusion Coefficients in Polymer Blends," *Makromol. Chem., Rapid Comm.*, 8, 393 (1987).

73. G. Foley and C. Cohen, "Diffusion in Polymer-Polymer Blends," *J. Polym. Sci., Polym. Phys. Ed.*, 25, 2027 (1987).

74. F. Brochard, J. Jouffroy, and P. Levinston, *Macromolecules*, 17, 2925 (1984).

75. There is a second prediction for D(f) (Ref. 74) which differs from Eq. 51 such that $D^{-1}_T = (D^*_{d-PS}N_D)^{-1}(1 - \phi) + (D^*_{h-PS}N_H)^{-1}\phi$. Since we have chosen $N_H = N_D$ both predictions yield the same result. References 22, 37, and 38 provide strong evidence showing Eq. (51) is correct.

76. J. F. Zeigler, **The Stopping Powers and Ranges of Ions in Materials**, Vol. 4, Pergamon Press, New York (1977); H. H. Anderson and J. F. Zeigler, ibid., Vol. 3.

77. D. K. Brice, "Three Parameter Formula for the Electronic Stopping Cross Section at Non-Relativistic Velocities," *Phys. Rev. A*, 6, 1791 (1972).

78. J. Roovers and W. W. Graesley, "Melt Rheology of 4-Arm and 6-Arm Polystyrenes," *Macromolecules*, 12, 959 (1979).

79. D. J. Plazek, V. Tan, and V. M. O'Rourke, "The Creep Behavior of Ideally Actic and Commercial Polymethylmethacrylate," *Rheol. Acta*, 13, 367 (1974).

80. F. S. Bates, G. D. Wignall, and W. C. Koehler, "Critical Behavior of Binary Liquid Mixtures of Deuterated and Protonated Polymers," *Phys. Rev. Lett.*, 55, 2425 (1985).

81. F. S. Bates and G. D. Wignall, "Nonideal Mixing in Binary Blends of Perdeuterated and Protonated Polystyrenes," *Macromolecules*, 19, 932 (1986).

82. F. S. Bates and G. D. Wignall, "Isotope Induced Quantum Phase Transitions in the Liquid State," *Phys. Rev. Lett.*, 57, 1425 (1986).

83. F. S. Bates, M. Muthukumar, G. D. Wignall, and L. J. Fetters, "Thermodynamics of Isotopic Polymer Mixtures: Significance of Local Structure and Symmetry," submitted to *J. Chem. Phys.*

84. H. Yang, R. S. Stein, C. Han, B. J. Bauer, and E. J. Kramer, "Compatibility of Hydrogenated and Deuterated Polystyrene," *Polym. Comm.*, 27, 132 (1986).

85. A. Lapp, C. Picot, and H. Benoit, "Determination of the Flory Interaction Parameter in Miscible Polymer Blends by Measurement of the Apparent Radius of Gyration," *Macromolecules*, 18, 2437 (1985).

86. K. S. Schweizer and J. G. Curro, "Microscopic Theory of the Structure, Thermodynamics and Apparent χ Parameter in Polymer Blends," *Phys. Rev. Lett.*, 60, 809 (1988).

87. J. G. Curro and K. S. Schweizer, "Theory of the χ Parameter of Polymer Blends: Effective of Attractive Interactions," *J. Chem. Phys.*, 88, 7242 (1988).

88. F. S. Bates, "Thermodynamics of Isotopic Polymer Mixtures: Poly(vinyl ethylene) and Poly(ethyl ethylene)," *Macromolecules*, 20, 2221 (1987).

89. R. A. L. Jones, J. Klein, and A. M. Donald, "Mutual Diffusion in a Miscible Polymer Blend," *Nature (London)*, 321, 161 (1986).

90. M. S. Dresselhaus, B. Wasserman, and G. E. Wnek, 413, in **Ion Implantation and ion Beam Processing of Materials**, G. K. Hubler, O. W. Holland, C. R. Clayton, and C. W. White, Eds., Materials Research Society, Vol. 27, North-Holland, New York (1984).

91. G. E. Wenek, B. Wsserman, M. S. Dresselhaus, S. E. Tunneyand, and J. K. Stille, "Ion Implantation of a Polyquinoline," *J. Polym. Sci., Polym. Lett. Ed.*, 23, 609 (1985).

92. B. Wasserman, M. S. Dresselhaus, M. Wolf, G. E. Wnek, and J. D. Woodhouse, "Transient Transport Measurements of Ion Implanted Polymers," *J. Appl. Phys.*, 60, 668 (1986).

93. T. Venkatesen, S. R. Forrest, M. L. Kaplan, P. H. Schmitt, C. A. Murray, W. L. Brown, L. Rupp, and H. Schonhorn, "Conductive Regions of Highly Resistive Organic Films," *Struct. Chem. Anal. Ion Beam*, 56(10), 2778 (1984).

7
FLUORESCENCE REDISTRIBUTION AFTER PATTERN PHOTOBLEACHING

Barton A. Smith

Almaden Research Center
IBM Research Division
San Jose, California

In the 1970s, a new technique, fluorescence recovery after photobleaching, was developed for the measurement of lateral diffusion in living cell membranes and model membrane systems. The first use of photobleaching to measure tracer diffusion in membranes was reported by Poo and Cone of Johns Hopkins University in 1974. They made use of the intrinsic optical absorbance of rhodopsin to follow its diffusion in rod outer segment membranes (1). The use of fluorescence rather than absorbance allowed the method to be extended to molecules which are in low concentration and lack intrinsic visibility (2). The molecules were labeled with fluorescent dyes which could be bleached by the intense light of a focused laser beam. The diffusion was measured by observing the return of fluorescence in the dark spot as intact molecules diffused into the bleached region. The method was well suited to the study of membranes due to the high sensitivity with which fluorescence can be detected, and to the well-developed chemistry for fluorescent labeling of biological molecules.

At the same time, a number of transient optical holographic grating techniques were being developed for the measurement of transport properties in solids and liquids. In 1971, Eichler et al. of the I. Physikalisches Institut der Technischen Universität Berlin, reported the first detection of a transient grating (3). They used a ruby laser to produce a thermal refractive index grating in dye-containing methanol. In 1972, they reported additional measurements, including the time decay of the grating due to thermal diffusion (4).

In 1973, Pohl et al. of the IBM Zurich Research Laboratory reported the first application of the transient grating method to study transport properties. They measured the thermal diffusivity in NaF as a function of temperature, and called this technique *forced Rayleigh scattering* (5). The grating was created by a CO_2 laser and detected by its diffraction of a HeNe laser beam. The time scale was 100 μs.

In 1978, Salcedo et al. at Stanford University reported the measurement of singlet electronic excitation transport in a molecular crystal (6). The time scale of their measurement was 1 ns. Also in 1978, the first reports were published of the measurement of mass transport by a transient diffraction grating. Hervet et al. of the Collège de France reported the translational diffusion coefficient of the photochromic dye methyl red in an aligned liquid crystal, as a function of direction (7). The time scale was 0.1 s.

In 1977, Smith and McConnell at Stanford University combined the earlier methods by bleaching fluorescent molecules in a pattern of stripes. They called the new technique "fluorescence redistribution after pattern photobleaching" (FRAPP). It was first used to measure the diffusion coefficient of dye-labeled phospholipids in oriented films (8). FRAPP combines the sensitivity of fluorescence measurements with the precision of the holographic methods, and is thus the method of choice for many systems.

7.1 THE METHOD

In FRAPP, an intense burst of laser light is used to irreversibly photochemically destroy (bleach) fluorescent dye molecules in the illuminated regions of the sample. FRAPP differs from forced Rayleigh scattering (FRS) in that fluorescence rather than diffracted beam intensity is used to detect the amplitude of the transient grating. It differs from fluorescence photobleaching recovery (FPR) in that a periodic pattern, rather than a single bleached spot, is created in the sample.

Patterns can be created with dimensions comparable to the wavelength of light or longer. The distance over which the diffusion takes place, and thus the time scale of the measurement, can be adjusted for convenience. The shortest diffusion time which can be studied is determined by the time required to bleach the fluorescent molecules. The longest time is limited by the physical or chemical stability of the sample, or by the patience of the investigator. The concentration gradients are created in situ, without any mechanical manipulation of the samples. Artifacts which may arise due to surface phenomena are eliminated. Diffusion is measured along one or more specified directions. Anisotropic diffusion can be thoroughly characterized.

The measurement proceeds as follows:

1. The sample is prepared with a low concentration of the fluorescent species uniformly distributed. The concentration of fluorescent molecules should be low enough that the measured fluorescence intensity is proportional to concentration and the dye does not perturb the transport properties of the sample. Under such conditions, the measurement will yield the tracer diffusion coefficient of the fluorescent species.

2. The sample is exposed to an intense light which has a periodic pattern of intensity, such as alternating bright and dark stripes. The light causes irreversible destruction of the fluorescence of the molecules by photochemical reaction. These bleached molecules are no longer observed in the measurement. Thus, a periodic pattern of fluorescent molecules is produced in the sample.

3. The pattern of fluorescence is observed to change with time, as the fluorescent molecules diffuse. The rate with which the fluorescence intensity becomes uniform across the bleached region of the sample is a measure of the diffusion coefficient of the dye molecules.

7.2 THEORY

The following is a detailed discussion of the case in which the pattern is periodic in one dimension. Such patterns can be created in many ways, such as by projection of a periodic image (8), illumination through a contact mask (9), or by the interference pattern created by intersecting laser beams (10).

We will assume that the profile of light intensity as a function of position in the sample is a sine or square wave. A similar analysis can be done for more complex patterns. The light intensity P in the sample can be described by

$$P(x) = P_0 - P_1 \sin(ax) - P_2 \sin(3ax) - P_3 \sin(5ax) - \cdots \qquad (1)$$

in which P_0 is the average intensity, and a is the spatial frequency of the pattern, defined as $a = 2\pi/p$, p being the pitch or repeat distance of the pattern. The coefficients $P_1 \ldots$ represent the amplitudes of the Fourier components of the pattern. We define the contrast of the bleaching pattern as the ratio P_1/P_0. The origin of the spatial coordinate x is arbitrarily chosen. Immediately after the photobleaching has been done, the concentration C of the fluorescent molecules as a function of position x in the sample will be given by

$$C(x,0) = C_0 + C_1 \sin(ax) + C_2 \sin(3ax) + C_3 \sin(5ax) + \cdots \qquad (2)$$

in which C_0 is the average concentration of the dye molecules at $t = 0$. For a perfect square pattern, $C_2 = 1/3C_1$, $C_3 = 1/5C_1$, and so forth. In general, $C_{n+1} < C_n$. We will refer to the ratio C_1/C_0 as the contrast of the concentration pattern. It is important to note that even if the bleaching pattern $P(x)$ is a pure single sine wave, the concentration pattern $C(x, t)$ will generally contain higher-order terms, due to the nature of the photochemical kinetics (11,12).

Assuming that the diffusion coefficient of the dye is independent of its concentration at low concentration, and since the gradients in concentration are created by the pattern in only one direction, we can use the one-dimensional diffusion equation:

$$\frac{\partial C(x, t)}{\partial t} = D \frac{\partial^2 C(x, t)}{\partial x^2} \qquad (3)$$

in which x is the spatial coordinate, t is the time, and D is the diffusion coefficient. The solution to Eq. (3) with initial conditions given by Eq. (2) for $t > 0$ is

$$C(x, t) = C_0 + e^{-Da^2t}C_1 \sin(ax) + e^{-9Da^2t}C_2 \sin(3ax) + e^{-25Da^2t}C_3 \sin(5ax) + \quad (4)$$

Equation (4) indicates that the amplitudes of the Fourier components of the concentration pattern decay exponentially, with time constants inversely proportional to the square of the spatial frequency. Thus, the higher-frequency components, which were initially smaller, decay more rapidly. In many experiments the data can be adequately represented by the first two terms of Eq. (4).

There are many ways in which the decay of the concentration pattern can be observed. The simplest is to illuminate the sample with the same pattern which was used for bleaching, but with greatly reduced intensity, and to record the total fluorescence intensity from the sample as a function of time.

$$I(t) = \int_0^{2\pi/a} P(x)C(x, t)\, dx \tag{5}$$

$$I(t) = \frac{2\pi}{a} P_0 C_0 - P_1 C_1 e^{-Da^2 t} - P_2 C_2 e^{-9Da^2 t} - \cdots \tag{6}$$

For these conditions, the observed intensity I(t) is at a minimum immediately after the bleaching reaction, and increases with time. In practice, the pattern of light on the sample cannot be a perfect square wave due to diffraction and limitations of optical resolution. Thus, the higher-frequency terms are made smaller with respect to the second term. Since they also decay more rapidly than the fundamental frequency term, they often become insignificant shortly after the bleaching burst and may be neglected in the data analysis. In that case the observed intensity can be fit to a single exponential recovery curve, which has a time constant $\tau = 1/Da^2$.

We have assumed for this derivation that the pattern is infinite in extent. It has been shown that diffusion coefficients which are accurate to within 1% can be obtained with patterns containing only nine repeat periods, and accuracy to within 5% can be obtained with only three repeat periods (13).

For reasonable precision in measurement of the time constant for the recovery, data must be collected for approximately $3 \times \tau$. The time constant is inversely proportional to the diffusion coefficient and to the square of the spatial frequency. The spatial frequency is the parameter which gives the experimenter the latitude in selecting the time scale of the experiment. Variation of this parameter allows for the validation of the assumptions leading to Eq. (6), since the measured time constant must be proportional to $1/a^2$. The spatial frequency must be known accurately for an accurate determination of D. If the pattern is formed by the projection of a mask, the optical system must be precisely aligned and its magnification be accurately calibrated. If the pattern if formed by intersecting coherent beams, the angle of intersection Θ and wavelength λ must be accurately known. Then the frequency is calculated

$$a = \frac{4\pi \sin(\Theta/2)}{\lambda} \tag{7}$$

To apply Eq. (7) directly, the angle and wavelength in the sample must be known, which requires knowledge of the refractive index of the sample. However, by using certain geometries, this requirement can be avoided. Figure 7.1 illustrates one such geometry. If the beams of light enter the sample through a window which has plane, parallel surfaces, that window can be neglected in the calculation of angles within the sample. If the orientation of the sample is adjusted so that the two beams intersect the air/sample interface at equal angles, then the effects of refraction at the interface and change of wavelength in the sample cancel. In this case the interference pattern spacing can be calculated from the external (in air) angle and wavelength. Since the angle can be measured by simple trigonometry, and the emission wavelength of lasers are very accurately known, the spatial frequency can be quite accurately determined.

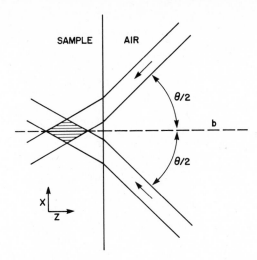

Figure 7.1:   Convenient geometry for interference pattern. The plane b which is perpendicular to the plane of incidence and bisects the angle between the beams Θ, is also perpendicular to the air/sample interface.

7.3 APPARATUS

Details will be given for the apparatus which has been used by the author to measure diffusion in polymer melts and solid films. Figure 7.2 is a schematic diagram of this apparatus. Many design considerations will be mentioned for the benefit of readers who wish to set up a FRAPP experiment. The references provide details of the many other variations of the apparatus which have been used for different applications.

7.3.1 Laser

Although measurements have been made with other light sources, such as mercury arc lamp, a visible laser is most often the best source for photobleaching experiments. Light intensities of from 10 to 500 W/cm² are typically required, which can be easily obtained from a 1-W argon ion laser at 488.0 or 514.5 nm. There are a number of dyes which can be excited and bleached at these wavelengths. When using interference patterns, it is valuable to have an intracavity etalon to increase the coherence length of the laser, thereby increasing the contrast of the projected pattern and allowing more latitude in the design of the interferometer since the path lengths do not have to be precisely equal. An optical spectrum analyzer is required for alignment of the etalon.

7.3.2 Excitation Optics

A number of devices are required between the laser and the sample, to direct and focus the beam, and control its intensity and polarization. High-quality front surface mirrors are used to direct the beam from the laser to the sample. A spatial filter may be required to reduce point-to-point variations in intensity of the beam across its

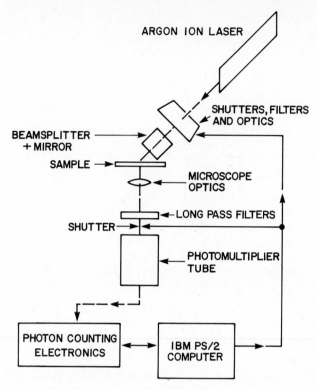

Figure 7.2: FRAPP apparatus. A PS/2 model 80 computer is used to control the timing of the experiment and collect the data. The sample is in a temperature-controlled holder with fused silica windows. An inverted fluorescence microscope is used to collect the fluorescence emission, and for visual observation of sample and beam alignment. The photomultiplier tube is chosen for low dark current, and cooled to -20°C.

profile. Some means is needed to change from the high intensity required for photobleaching to low intensity for observation. Figure 7.3 shows a device which produces coaxial high- and low-intensity beams. In order to preserve the polarization of the laser beam, the input is polarized perpendicular (E-field) to the plane of the drawing, and all of the beams lie in the same plane. The beamsplitters are coated to have 5% reflectance from one surface, and less than 1% from the other. The shutters are electromechanical, with reflective blades to prevent damage by the laser. They have 3-mm apertures, and can open or close in approximately 1 ms (Uniblitz, Vincent Associates, Rochester, NY.)

It is important that all of the lenses and beamsplitters in the excitation path be free of absorption at the excitation wavelength. Even a very small absorbance will lead to thermal lensing in the optical system during the bleaching period, which will distort both the bleaching beam and subsequent observation beam. Antireflection coatings on all optical elements are valuable both in avoiding losses in beam intensity, and in reducing unwanted reflections.

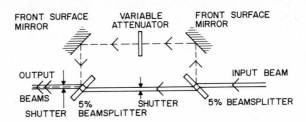

Figure 7.3:   Intensity switch. The high intensity output beam has 90% of the original laser power. The low-intensity output has between 2×10^{-3} and 1×10^{-6} of the input power. Apertures (not shown) eliminate undesired reflections.

For real image projection systems, a Ronchi ruling is a convenient mask. It consists of alternating transparent and opaque lines of equal width. Ronchi rulings can be obtained from Edmund Scientific Company. Custom masks, usually metal on transparent substrate, can be obtained from photolithography vendors which serve the microelectronics industry. The first FRAPP measurements (8) were made by projecting the image of a Ronchi ruling onto the sample through the excitation optics and objective of a standard epifluorescence microscope. This is a particularly easy system to set up, as the microscope contains most of the optical elements and filters that are required. It allows observation of the sample area to be measured by both transmitted light and fluorescence. This visual observation is very useful for alignment of the optics and selection of sample to avoid defects. It also allows the selection of certain structures within the sample, and of specific orientations of the measured diffusion direction within those structures for cases of anisotropic diffusion (14). This method is limited to samples which are thinner than the depth of field of the microscope objective. The smallest stripe period available is limited by the resolution of the microscope objective.

In order to produce the pattern by the coherent interference or holographic method, the laser beam must be divided into two beams which are then directed to intersect in the sample at an angle which is usually between 5 and 90°. In order to achieve maximum contrast in the pattern, the beams must be of equal intensity and have the same polarization. In addition, the difference in the lengths of the paths which the two beams traverse must be much less than the coherence length of the laser. A beamsplitter that has a nominal 50% reflectance for s-polarized light at 45° incidence should be used. The reflectance can be adjusted to be precisely 50% by variation of the angle of incidence. A laser power meter or other quantitative optical detector is required for this adjustment, and to provide knowledge of the light flux at the sample both for bleaching and observation. The meter should thus have a dynamic range of at least 10^6.

Automatic shutters are required to protect the photodetector from the bleaching intensity light, and to protect the sample from unnecessary exposure to the excitation. Even the low-intensity light used for observation of the pattern after bleaching will produce additional photochemistry over long periods of time. Thus, for measurements that take more than 100 s, it is usual to illuminate the sample only for brief periods of time which are long enough to make an intensity measurement. These measurement periods are separated by appropriate dark times, so that the total exposure of the sample is limited. Most photon counting systems have a provision for chopped detection, which also allows for subtraction of dark noise or other background signal during this process.

For very thin samples it is particularly important to reduce the intensity from scattering of the excitation light and background fluorescence or phosphorescence. A useful procedure is to prepare the sample on the surface of a high-quality optical substrate, and illuminate the sample film by total internal reflection (15). This eliminates passage of the excitation through additional interfaces or materials which may contribute to the background.

7.3.3 Sample

It is necessary of course for the sample to be sufficiently transparent to both the excitation and fluorescence emission light so that the pattern can be produced and the emission measured. The surface of the sample must be optically smooth, and the bulk must be free from excessive scattering. Sample holders or substrates should produce no fluorescence or phosphorescence which will interfere with the measurement. Fused silica is a good choice for substrates in most cases, and spin coating of polymer films produces samples of good optical quality.

The sample holder must often provide temperature control, and sometimes environment control for the sample. In this case, the windows of the holder must be included in the design of the excitation and observation optical systems, and are subject to the same considerations in quality and choice of materials.

7.3.4 Fluorescent Dye

Although intrinsic fluorescence can be used in a few cases (1), it is usually necessary to add a fluorescent molecule to the sample or attach one to the particle of interest. This requirement is the most serious disadvantage FRAPP compared to other techniques for the measurement of diffusion. A dye must be selected that:

1. Is soluble in the sample, or when attached to the species of interest produces a soluble product.
2. Does not perturb the dynamics of the system too much.
3. Is chemically compatible with the sample.
4. Can be excited at a wavelength at which the sample is transparent.
5. Undergoes irreversible photochemical bleaching under intense excitation.
6. Can be bleached on a time scale much shorter than the measurement time, at a dose that does no photochemical or thermal damage to the sample.

NBD chloride is a readily available compound that has been used to produce fluorescent labels for a number of polymer studies (9,10). The chloride, 4-chloro-7-nitrobenzofurazan (also known as 7-chloro-4-nitrobenz-2-oxa-1,3-diazole) is easily reacted with any primary or secondary amine to produce a fluorescent product. It can be photobleached in the presence of oxygen or hydroxyl groups. Rhodamine, fluorescein, and derivatives such as fluorescein isothiocyanate have been used in many diffusion studies, particularly those in biological or other aqueous systems (16-18).

7.3.5 Detection Optics

An objective lens with high numerical aperture (low f/number) should be used to collect the fluorescence emission from the sample. Microscope objectives are often suitable. When only the total intensity is being measured, chromatic and spherical aberration are not a problem, so relatively cheap objectives can be used. Numerical apertures of about 0.5 are usually convenient to use. Higher numerical aperture (NA) objectives have a shorter working distance which may present difficulties with temperature control of the sample when working far from room temperature.

A long pass optical filter is placed between the sample and detector to prevent scattered excitation light from reaching the detector. A photomultiplier with photon counting electronics provides the greatest sensitivity for measuring the total emission intensity.

7.3.6 Motion Isolation

The sample must remain fixed with respect to the bleaching and observation patterns for the duration of the measurement, to a precision that is small with respect to the pattern spacing. The entire apparatus must be designed to minimize vibration, mechanical motion due to thermal expansion, and thermal convection currents in the air which can perturb the light beams. Standard vibration isolation tables and rigid optical mounts that are used for holographic applications are suitable. The optics should be enclosed to minimize air currents and keep the optics clean. Room temperature that is constant to within a few degrees Celsius helps to maintain the optical alignment.

7.4 EXAMPLE

FRAPP was used to measure the diffusion coefficient of poly(propylene oxide) in the melt at room temperature, 21.6°C. The NBD dye-labeled polymer (19) was mixed in the polymer melt at a concentration of 10^{-4} M/liter, resulting in an absorbance of 0.005 at 488 nm for the 100-μm-thick sample. The sample was contained in a fixed-path length cuvette with 1-mm-thick fused silica windows. It was necessary to allow the polymer melt sample to relax for 1 h after being placed in the cuvette to prevent bulk motion of the material during the measurement. A Spectra Physics model 165 argon-ion laser with intracavity etalon had an output of 0.6 W at 488 nm. The power incident on the sample for bleaching was approximately 0.4 W, and for observation was 0.5 μW. The pattern was formed by intersecting beams, with a period of 14.1 μm. The pattern period was measured by photography of the fluorescence image and a stage micrometer through the microscope optical system. The pattern period was also calculated from the angle of intersection of the beams. The two measurements agreed to within 1%. The beams had a diameter of 250 μm in the sample.

Fluorescence intensity from the sample was observed prior to bleaching as 2.55×10^4 photon counts per second. The sample was bleached for 600 ms. Fluorescence intensity was then recorded at 1 s intervals, with a 1-s counting period for each point. Emission was collected by a 10x, NA = 0.3 microscope objective which projected a real image of the sample on the photocathode. The photodetector was an RCA C31034-02 photomultiplier in a Products for Research TE104-TS refrigerated housing.

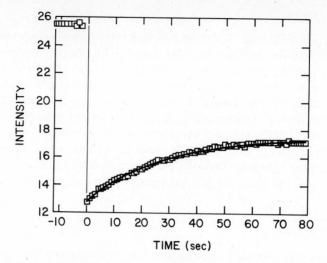

Figure 7.4:   FRAPP experimental data. The intensity is in units of 10^3 photon counts per second. The time is in seconds, with t = 0 defined as the end of the bleaching burst. Squares represent the individual intensity measurements. Data for t < 0 correspond to prebleach observations. The solid curve is the single exponential least-squares fit to the data.

Figure 7.4 shows the result of the measurement. Observed intensity for the first second after bleaching was 50% of the prebleach value. A single exponential fit to the recovery curve yielded a time constant of 27.8 s, corresponding to a diffusion coefficient D = 1.8 × 10^{-9} cm^2/s .

Since a single exponential curve provided a good fit to the data, it was not necessary to include any of the higher-order terms of Eq. (6). This is usually the case when the observation pattern can be approximated as a single sine wave. The good fit also indicates that the fluorescent molecules had a narrow distribution of diffusion coefficients.

Similar measurements were made at other pattern spacings, to confirm that the measured time constant reflected diffusive motion. Measurements were made with one beam blocked (no pattern) to rule out the possibility of recovery of the fluorescence by chemical reaction. Observation was made with no bleaching beforehand, to confirm that the intensity was low enough to avoid bleaching by the observation beam.

7.5 OTHER APPLICATIONS

A number of variations of the method have been used for other applications. Localized pattern photobleaching (20) allows simultaneous measurement of relatively slow and fast diffusion. The slowly diffusing species moves from the dark to the bright part of the pattern, while the rapidly diffusing one moves into the pattern bleached area from other parts of the sample. Scanning detection (21) provides information on both diffusion and bulk motion of material. Diffusion decreases the amplitude of the pattern, while bulk motion changes the position, or phase, of the pattern. Modulation detection allows measurements in the presence of specimen motion and photobleaching during observation (22). It adds the usual improvement of signal-to-noise ratio by increasing the frequency at which the measurement is made. Total internal reflection illumination has been used to localize the measurement to a thin surface on a thicker sample (15). Thus, molecules bound to the surface could be specifically detected in the presence of unbound molecules in the bulk.

FRAPP has been used in many studies of diffusion in model and living cell membranes. Diffusion of both proteins (23) and lipids (24) in the membranes, and proteins attached by specific binding (25) have been measured. The technique has been adapted to study the diffusion of ions in polyelectrolyte solutions (26,27), and has been used extensively to probe the polymerization dynamics and equilibria of actin (28-31). The diffusion of labeled polymer has recently been used to characterize the transitions in thermally reversible gels (32). The large number and wide range of applications for FRAPP during the 10 years after its invention demonstrate the versatility, sensitivity, and precision of this technique.

7.6 REFERENCES

1. M. Poo and R. A. Cone, "Lateral Diffusion of Rhodopsin in the Photoreceptor Membrane," *Nature*, 247, 438 (1974).

2. D. Axelrod, D. E. Koppel, J. Schlessinger, E. Elson, and W. W. Webb, "Mobility Measurements by Analysis of Fluorescence Photobleaching Recovery Kinetics," *Biophysical J.*, 16, 1055 (1976).

3. H. Eichler, G. Enterlein, J. Munschau, and H. Stahl, "Lichtinduzierte, Thermische Phasengitter in Absorbierenden Flssigkeiten," *Z. Angewandte Phys.*, 31, 1 (1971).

4. H. Eichler, G. Enterlein, P. Glozbach, J. Munschau, and H. Stahl, "Power Requirements and Resolution of Real-time Holograms in Saturable Absorbers and Absorbing Liquids," *Appl. Opt.*, 11, 372 (1972).

5. D. W. Pohl, S. E. Schwarz, and V. Irniger, "Forced Rayleigh Scattering," *Phys. Rev. Lett.*, 31, 32 (1973).

6. J. R. Salcedo, A. E. Siegman, D. D. Dlott, and M. D. Fayer, "Dynamics of Energy Transport in Molecular Crystals: the Picosecond Transient-grating Method," *Phys. Rev. Lett.*, 41, 131 (1978).

7. H. Hervet, W. Urbach, and F. Rondelez, "Mass Diffusion Measurements in Liquid Crystals by a Novel Optical Method," *J. Chem. Phys.*, 68, 2725 (1978).

8. B. A. Smith and H. M. McConnell, "Determination of Molecular Motion in Membranes Using Periodic Pattern Photobleaching," *Proc. Nat. Acad. Sci., U.S.A.*, 75, 2759 (1978).

9. B. A. Smith, "Measurement of Diffusion in Polymer Films by Fluorescence Redistribution After Pattern Photobleaching," *Macromolecules*, 15, 469 (1982).

10. B. A. Smith, E. T. Samulski, L-P Yu, and M. A. Winnik, "Polymer Diffusion in Molten Poly(propylene oxide)," *Macromolecules*, 18, 1901 (1985).

11. F. Lanni and B. R. Ware, "Intensity Dependence of Fluorophore Photobleaching by A Stepped-intensity Slow-bleach Experiment," *Photochem. Photobiol.*, 34, 279 (1981).

12. J. Davoust, P. F. Devaux, and L. Leger, "Fringe Pattern Photobleaching, a New Method for the Measurement of Transport Coefficients of Biological Macromolecules," *EMBO J.*, 1, 1233 (1982).

13. F. Lanni, Center for Fluorescence Research in Biomedical Sciences, Carnegie Mellon University, personal communication.

14. B. A. Smith, W. R. Clark, and H. M. McConnell, "Anisotropic Molecular Motion on Cell Surfaces," *Proc. Natl. Acad. Sci., U.S.A.*, 76, 5641 (1979).

15. R. M. Weis, K. Balakrishnan, B. A. Smith, and H. M. McConnell, "Stimulation of Fluorescence in a Small Contact Region Between Rat Basophil Leukemia Cells and Planar Lipid Membrane Targets by Coherent Evanescent Radiation," *J. Biological Chem.*, 257, 6440 (1982).

16. K. Zero, D. Cyr, and B. R. Ware, "Tracer Diffusion of Counterions Through a Solution of Polyelectrolyte," *J. Chem. Phys.*, 79, 3602 (1983).

17. S. Gorti, L. Plank, and B. R. Ware, "Determination of Electrolyte Friction from Measurements of the Tracer Diffusion Coefficients, Mutual Diffusion Coefficients, and Electrophoretic Mobilities of Charged Spheres," *J. Chem. Phys.*, 81, 909 (1984).

18. J. R. Simon, A. Gough, E. Urbanik, F. Wang, B. R. Ware, and D. L. Taylor, "Analysis of Rhodmine and Fluorescein-labeled F-actin Diffusion in Vitro by Fluorescence Photobleaching Recovery," *Biophysical J.*, 54, 801 (1988).

19. B. A. Smith, E. T. Samulski, and L-P Yu., "Tube Renewal Versus Reptation: Polymer Diffusion in Molten Poly(propylene oxide)," *Phys. Rev. Lett.*, 52, 45 (1984).

20. D. E. Koppel and M. P. Sheetz, "A Localized Pattern Photobleaching Method for the Concurrent Analysis of Rapid and Slow Diffusion processes," *Biophysical J.*, 43, 175 (1983).

21. D. E. Koppel, P. Primakoff, and D. G. Myles, "Fluorescence Photobleaching Analysis of Cell Surface Regionalization," *Applications of Fluorescence in the Biomedical Sciences*, Alan R. Liss, Inc., 477-497, 1986.

22. F. Lanni and B. R. Ware, "Modulation Detection of Fluorescence Photobleaching Recovery," *Rev. Sci. Instr.*, 53, 905 (1982).

23. L. M. Smith, B. A. Smith, and H. M. McConnell, "Lateral Diffusion of M-13 Coat Protein in Model Membranes," *Biochemistry*, 18, 2256 (1979).

24. J. L. R. Rubenstein, B. A. Smith, and H. M. McConnell, "Lateral Diffusion in Binary Mixtures of Cholesterol and Phosphatidylcholines," *Proc. Natl. Acad. Sci. U.S.A.*, 76, 15 (1979).

25. L. M. Smith, J. W. Parce, B. A. Smith, and H. M. McConnell, "Antibodies Bound to Lipid Haptens in Model Membranes Diffuse as Rapidly as the Lipids Themselves," *Proc. Natl. Acad. Sci. U.S.A.*, 76, 4177 (1979).

26. S. Gorti and B. R. Ware, "Probe Diffusion in An Aqueous Polyelectrolyte Solution," *J. Chem. Phys.*, 83, 6449 (1985).

27. B. R. Ware, "Laser Techniques for the Study of Counterion Condensation," in **Advances in Laser Spectroscopy**, Vol. 3, B. A. Garetz and J. R. Lombardi, Eds., Wiley, New York, 171 (1986).

28. F. Lanni, D. L. Taylor, and B. R. Ware, "Fluorescence Photobleaching Recovery in Solutions of Labeled Actin," *Biophysical J.*, 35, 351 (1981).

29. B. R. Ware and J. W. Klein, "Assembly of Actin Filaments Studies by Laser Light Scattering and Fluorescence Photobleaching Recovery," *Biophysical J.*, 49, 147 (1986).

30. A. Monzo-Villarias and B. R. Ware, "Actin Oligomers below the Critical Concentration Detected by Fluorescence Photobleaching Recovery," *Biochemistry*, 24, 1544 (1985).

31. E. A. Walling, G. A. Krafft, and B. W. Ware, "Actin Assembly Activity of Cytochalasins and Cytochalasin Analogs Assayed Using Fluorescence Photobleaching Recovery," *Archives Biochem. Biophys.*, 264, 321 (1988).

32. M. B. Mustafa, D. Tipton, and P. S. Russo, "Temperature Ramped Fluorescence Photobleaching Recovery for the Direct Evaluation of Thermoreversible Gels," *Macromolecules*, 22, 1500 (1989).

8
SURFACE SENSORS FOR MOISTURE AND STRESS STUDIES

L. T. Nguyen

Philips Research Laboratories
Signetics Corporation
Sunnyvale, California

Electronic packages must serve the dual purpose of protecting the delicate circuitry from mechanical damage during processing, and preventing any moisture-induced failures. Hermetic ceramic packages can achieve such goals reliably, albeit at high cost. Plastic packages, on the other hand, can provide considerable savings for somewhat more moderate reliability. This reduced reliability results from the permeability of all polymeric sealants to water, and from the curing shrinkage inherent with polymer systems. The former leads to device failure in the presence of ionic contaminants, while the latter can generate stresses equally harmful to the operation of the device (1). Thus, the ideal future packaging technique for LSI (large-scale integrated circuits), VLSI (very large-scale integrated circuits), and ULSI (ultra large-scale integrated circuits) devices must offer some optimal balance between cost and reliability (2,3).

With the current trend toward high packing density for improved operating speeds, comes the need to provide means for evaluating in situ the performance of a package. Characterization of package performance generally falls within three distinct areas, namely, electrical operation, heat dissipation, and environmental reliability. Test structures can generally be designed directly on a device to evaluate the parameters related to each of the previous areas (4).

This chapter, however, deals only with the environmental aspect of device qualification. It discusses the common sensing elements incorporated into test devices to monitor the effects of moisture and contamination on the operation of a packaged IC, and the influence of encapsulation stresses on the package integrity. Elements for quantifying the extent of moisture permeation into a package fall into two broad categories, namely, conductivity sensors and capacitive sensors. Stress monitoring elements, on the other hand, are usually variants of the familiar strain gages introduced in the early 1950s. Currently, the most successful design is based on piezoelectric semiconductor gages. The moisture and stress sensors described in this chapter exhibit sufficient versatility to be applicable to both thin films and relatively thick coatings.

8.1 MOISTURE MONITORING

Plastic packaging had dominated the electronics consumer market due to improved processing methods and higher purity of the packaging material. Unfortunately, one of the drawbacks placating the generic plastic package is its nonhermeticity. Moisture can readily penetrate the plastic encapsulant, which in most cases, is either a filled epoxy, a silicone gel, or some modified variant of the two materials. Subsequently, water can combine with ionic impurities or processing residues to induce metal corrosion (5-7). Common tests to evaluate the moisture resistance of a plastic package include the temperature/humidity bias (THB), the autoclave, and the pressure/temperature/humidity bias (PTHB).

During the typical THB, devices are subjected to bias at 85°C and 85% relative humidity (RH). The current generation of devices can readily withstand this test for thousands of hours with only minor failure rates. Thus, once a powerful and rapid feedback indicator of moisture resistance, the THB is no longer widely used. Another test that is also being slowly phased out is the autoclave. This is an unbiased test which involves heating parts to 121°C at 100% RH. The low incidence of device failures subjected to the autoclave has downgraded its importance among qualification stress tests. Since no bias is applied, this test is mainly used nowadays to compare the effectiveness of different passivation schemes on full scale wafers. Poor adhesion between interlevel dielectrics, or between glass and metal, shows up either through cracks or delamination. Furthermore, the saturated steam at 15 psig always induces some extent of corrosion around the bond pads. Finally, the test most favored among reliability engineers is the PTHB, sometimes known as HAST (highly accelerated stress testing), which subjects components to supersaturated steam (85% RH and typically 120°C). Operating conditions for PTHB have not been standardized yet, so that experimental correlation is difficult, and acceleration life factors vary widely depending on the failure modes (8,9). Nevertheless, rather high acceleration factors are observed, and thus, small sampling sizes and relatively short test duration can be used.

Regardless of the integrity of the package, moisture eventually diffuses through the plastic to reach the die surface. At this point, the nature of the passivation, the corrosion resistance of the metallic interconnect, and the adhesion of the dielectric interlayers with metal all determine the moisture resistance of the package. Since the amount of water permeating a package is usually minute, special characterization techniques have to be developed. Simple gravimetric techniques, such as thermogravimetry or quartz spring balance, can still be used (10-13). However, when applied to package moisture analysis, thermogravimetry only shows the water vapor content of the device atmosphere. This is accomplished by constant monitoring of the weight of a package which is subjected to some known heating profile. Any moisture vapor released from the package is removed by a stream of dry carrier gas. In this mode, the test is destructive. The quartz spring balance can yield useful weight gain/loss information on polymer films, although thinner films often need backing substrates for support, and require extreme care in handling. Other methods commonly used include mass spectrometry, electrolytic hygrometry, surface conductivity sensors, and capacitive transducers (14).

Mass spectrometry involves piercing the hermetic package with a sharp tool, and analyzing the internal gases escaping into the vacuum chamber and the spectrometer

(15). It reveals the total water content in a package and the presence of other gaseous species. However, it cannot distinguish between absorbed water and water vapor. Furthermore, it does not lend itself readily to process control to track manufactured parts since the test is destructive. Electrolytic hygrometry determines total water content, but cannot analyze other gases. It also suffers from the same drawbacks as mass spectrometry. Surface conductivity sensors are by far the oldest methods of detecting water on surfaces, and will be discussed in more detail later, together with the capacitive transducers. More recent studies have also introduced dielectric film sensors which will also be presented. In situ sensors can be classified into two broad categories, those operating by a *volume effect* (capacitive, dielectric film) compared to the *surface effect* sensors (surface conductivity).

Thus, it can be seen that the appropriate sensing structures incorporated into a test chip can provide in situ information on the amount of water present within an electrical package. In situ moisture sensors can be used in a variety of applications ranging from the evaluation of sealants and die attach materials, long-term hermeticity studies, corrosion failure analysis, measurement of permeation rates in filled encapsulants, correlation of helium leak rates in "hermetic" ceramic packages with moisture levels, to the qualification of new packages or packaging technology.

8.1.1   "Volume Effect" Sensor

A *volume effect* device is based on the principle of physical adsorption of water on the sensor when exposed to a humid environment. This phenomenon is widely known, and has been put to use in a number of applications in biological, pharmaceutical, and industrial process control. Within this category are elements such as *aluminum oxide sensors*, *dielectric film transducers*, and *metal oxides*.

Aluminum Oxide Sensors: The structure is essentially a capacitor with a bottom metal electrode, usually aluminum, covered with an anodized porous oxide (e.g., Al_2O_3 for aluminum electrodes). Generally, a thin and porous gold overcoat is deposited for corrosion protection. The gold and aluminum act as electrodes. Water vapor diffuses through the porous gold film to the oxide surface, and upon adsorption, alters the impedance of the device. Generally, the sensor is attached to a header with a die-attach epoxy, and subsequently wire bonded. The capacitance of the device can be measured with an impedance bridge at various frequencies. Previous industrial sensors are made mainly from aluminum foil, while current fabrication technology uses thin film techniques.

Figure 8.1 illustrates a schematic of such a sensor with two bonding pads (16). The corresponding model and electrical equivalent circuit is depicted in Fig. 8.2. Pores are idealized as cylindrical cavities distributed uniformly and equally spaced from each other. The thin Al_2O_3 layer present on the aluminum base is represented by the parallel R_2C_2 circuit. Similarly, pores can be described mathematically by equivalent RC circuits. Upon exposure to a humid environment, water vapor permeates the gold layer, and equilibrates on the pore walls. The total complex impedance of the sensor is determined by the amount of water adsorbed. This can be seen in Fig. 8.3 which describes the characteristic calibration curve of an industrial type Al_2O_3 device. The meter reading on the scale is directly proportional to the admittance (1/impedance) of the sensor. High moisture levels translate into high admittances. Note that at a given pressure in Fig. 8.3, there is a direct correspond-

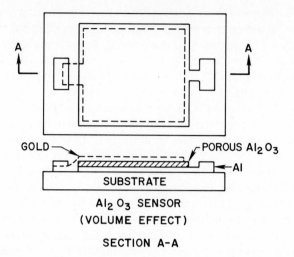

GOLD ⎯⎯   ⎯ POROUS Al_2O_3

←Al

SUBSTRATE

Al_2O_3 SENSOR
(VOLUME EFFECT)

SECTION A-A

Figure 8.1:   Schematic of an aluminum oxide humidity sensor (16). This "volume effect" device has an upper electrode of gold and a lower electrode of aluminum. Moisture uptake changes the impedance of the sensor.

ence between the vapor content (in ppm) and the dew/frost point observed. Water vapor levels down to 1 ppm can be detected with this type of sensor.

The aluminum oxide based sensor is thermally fairly stable. In experiments with hybrid microelectronic packages where bakeouts at 150°C from 4 to 168 h are required, admittance readings remained constant (16). Failure modes tend to be either bad bonding or broken wire bonds which translate into extremely "dry" readings, or shorting which results in the exact opposite. In one example, for instance, open failures near one of the bonding pads are traced back to a flaw in the original device layout. Aluminum in the lower electrode overlaps the gold-over-chrome pad, and readily oxidizes in the humid test environment (17). A redesign of the contact fault involves sintering the aluminum side of the sensor directly to the epitaxial substrate. The other bonding pad is connected to the porous gold side of the dielectric sandwich.

Nevertheless, the sensor suffers from two major drawbacks. First, hysteresis is always observed when the sensor is subjected to a "wet" to "dry" cycle (18). Indeed, for a given fixed relative humidity, the admittance reading is larger during the increasing water concentration phase compared to the decreasing portion of the cycle. Presumably, such results stem from the different rates of adsorption and desorption of water molecules from the pore surfaces. And, second, there is a contin-uous shift of readings toward lower admittance values. Typical variations can be around 25% after storage for one week, and up to 80% of the original calibration value after seven weeks (17). Bakeout at 100°C under vacuum followed by storage under nitrogen does not seem to restore the sensor. It is precisely this gradual shift in the calibration curve which renders aluminum-oxide-based sensors unreliable for any extended studies.

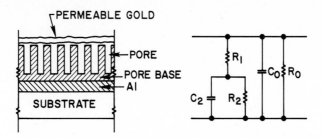

Figure   8.2:   Schematic idealized cross section of the aluminum oxide humidity sensor. The corresponding electrical equivalent circuit is also shown. C_0, C_2, R_0, R_1, and R_2 are the capacitance of the oxide layer, the capacitance of the porous base, the resistance of the Al_2O_3 layer, the resistance of the sides of the pore, and the resistance of the porous base, respectively (16).

Although aluminum-oxide-based sensors possess excellent sensitivity in detecting moisture content, they have not yet been applied to polymer thin film characterization.   The construction of the sensor raises several intriguing possibilities.   For instance, the introduction of a dielectric layer between the gold and the aluminum bottom electrode, or on top of the gold film, would give more information on the permeation nature of the film than simple moisture adsorption characteristics.   A variant of this idea was reported recently in the study of moisture diffusion in polyimide films (19).   In this case, polyimide is spin coated between two aluminum electrodes connected to an impedance bridge.   Three electrode configurations are tested, as shown in Fig. 8.4.   Figure 8.4a depicts a guarded parallel plate capacitor where the upper electrode consists of alternating stripes of aluminum and exposed polyimide of equal width (~50 μm).   This distance defines the path length for water to diffuse in and out of the film.   The cross section through one stripe is shown in Fig. 8.4b.   The second structure tested is another "stripe electrode" with much smaller spacing (5 μm).   This configuration limits the amount of water desorbing from the film during weight measurements, and allows correlation between capacitance and weight changes during moisture uptake.   In the third configuration shown in Fig. 8.4c, the upper electrode consists of an aluminum layer with uniformly distributed square holes.   Diffusion within this structure is inherently two-dimensional which permits precise evaluation of variations of moisture uptake with relative humidity and temperature.   Figure 8.4d illustrates a typical dew point cycle and corresponding capacitance data for a 51-μm stripe electrode sample at room temperature.   It was found that the low-frequency dielectric permittivity of the polyimide is directly proportional to the amount of water absorbed.   Moisture uptake depends on the ambient relative humidity, while the diffusion process can be described as Fickian with a diffusion coefficient in the range of 5 x 10^{-9} cm²/s.   Asymmetry in the sorption and desorption kinetics also suggests some concentration dependency of the diffusion coefficient at a fixed temperature.

The standard parallel plate design has recently been used to determine the dielectric properties of polymer thin films during supercritical fluid extraction (20).   This terminology refers to processes that use fluids above their critical points as ordinary solvents.   These can extract a solute from a mixture, and be removed cleanly by

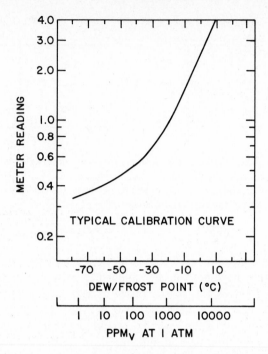

Figure 8.3: Calibration curve obtained for a typical industrial type Al_2O_3 moisture sensor. The meter reading is directly proportional to the admittance (e.g., 1/impedance) of the device. At a given pressure, there is a one-to-one correspondence between the vapor content and the dew/frost point (16).

merely throttling to a pressure below the critical temperature or pressure. Liquid CO_2 is used in this case to extract nonvolatile siloxane components: bis-(4-acryloxybutyl) tetramethyldisiloxane and tris-(chloromethyldimethylsiloxy) methacryloxypropylsilane, from the model polymer: poly(2,3-dichloro-1-propylacrylate). This polymer/siloxane system is used in X-ray microlithography, where incomplete removal of the siloxanes leads to poor pattern definition and high defect formation. By monitoring the capacitance of the system at a set frequency during extraction, dielectric changes would reveal compositional or morphological alterations. Neglecting end effects, the parallel plate capacitor is described by the familiar relation, $C = A\varepsilon/d$, where C is the capacitance of the film of thickness d, and dielectric constant ε, sandwiched between two electrodes of area A. Figure 8.5 illustrates the configuration used. The polymer is spun coated on a glass disk which carries an evaporated comb pattern of aluminum electrodes. The upper electrode is another layer of aluminum evaporated through the comb patterned mask rotated 90° with respect to the lower electrode. After pressurization, the capacitance typically exhibits a steep rise, followed by a gradual decrease. Decompression, on the other hand, resulted in a drop of the capacitance due to out-gassing of CO_2. Such behavior suggests that swelling of the film (up to 4%) separates the electrodes further apart, and accounts for the capacitance decrease during

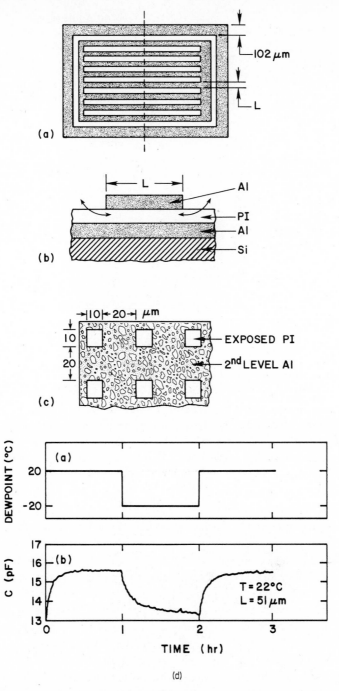

Figure 8.4: Schematic of planar Al/polyimide/Al electrodes. (a)"Stripe-electrode" top view; (b)Cross section showing the shortest paths of moisture diffusion; (c)Top view of upper electrode (19); (d)Dew point and capacitance data for 51-μm strip electrode sample at room temperature.

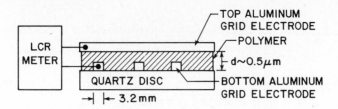

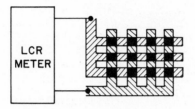

Figure 8.5: Schematic of parallel plate capacitor used in the supercritical fluid extraction of thin polymer films (20).

pressurization rather than modification of the film dielectric constant due to CO_2 absorption. Indeed, once the CO_2 concentration is reduced by decompression, the capacitance loss is accelerated. The technique is self-limiting, however, since repeated decompression pits the film and electrode, and drastically alters the capacitance.

Structures are not simply restricted to the metals used in thin film technology. Sensors have also been made with materials such as traditional *LiCl electrolytes* (21), newer *swelling organic polymers* (22), and promising *metal oxides* (23). Electrolytes are usable in the ambient temperature and humidity range, but display unsatisfactory life time and accuracy. Furthermore, they would also be inappropriate in the study of thin polymer films for microelectronics, where care is required for all handling activities to avoid ionic contamination.

Dielectric Film Sensors: Sensors using a hygroscopic polymer as the dielectric have been available commercially for over 10 years. Such transducers are made up of an acetal polymer such as cellulose acetate butyrate (CAB) coated on both sides with either gold electrodes or carbon-particle plates (24,25). One version consists of a CAB dielectric with integrally bonded conductive layers of carbon particles in a CAB matrix as depicted in Fig. 8.6 (26). The capacitance of the CAB end-mounted element is governed by the area of the second conducting layer and the thickness of the dielectric core layer. Once exposed to a humid environment, water is absorbed by the CAB, which raises its dielectric constant and the corresponding capacitance of the device. The transient response of the CAB transducer for step changes in the relative humidity at room temperature and an air velocity of 10 ft/min is illustrated in Fig. 8.7. Approximately 400 s are required for the element to change by 90%

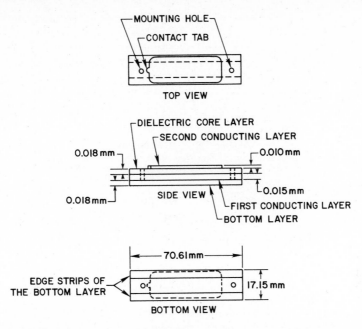

Figure   8.6:   Schematic of dielectric thin film humidity sensor. Cellulose acetate butyrate (CAB) is the hygroscopic dielectric medium. Conductive layers of CAB filled with carbon black form the electrodes (26).

from 90 to 50% RH. Unfortunately, hysteresis is observed, reducing the reliability of the sensor. Furthermore, the CAB loaded with carbon black is brittle, and tends to age with time and exposure to high temperatures. The sensing mechanism is also susceptible not only to moisture, but also environmental contaminants and ionic gases such as CO_2 and NO_x.

Two variant applications of this technology have been reported recently. The first one still relies on a cellulose acetate as the hygroscopic sensor (27). However, in this case, the porous electrode consists of a layer of chromium deposited under conditions that impart tensile stresses in the film, inducing numerous cracks in the polymer. These cracks, spaced a few micrometers apart from each other, turn the polymer into small islets covered by the chromium (100 Å to 1 μm). Thus, permeability is increased tremendously without impairing the mechanical strength or the electrical conductivity of the film. Presumably, this technique can also be extended to polymers such as polyimide, which exhibit low moisture sensitivity without cracks.

To circumvent some of the drawbacks of monolayered film devices, one new twist to the sensing mechanism can be added. A sensor recently introduced has two functional layers sandwiching either a glass or polyimide substrate: an upper layer of Allylamine polymer ($(CH_2 = CHCH_2NH_2)_n$) deposited by plasma polymerization (1000 to 6000 Å thick) and a lower layer of Ni/Cr thin film coated by standard vacuum deposition (28). Strain gage patterns are then formed into the lower layer by selective photoetching. The polymer can either stretch or shrink depending on

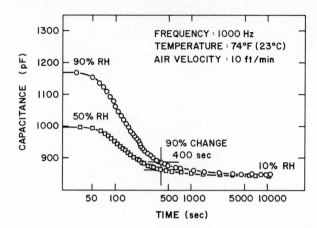

Figure 8.7: Response characteristics of thin film moisture sensor for step changes in the relative humidity (26).

whether moisture is absorbed or desorbed. The corresponding strain is transferred to the metal resistance gages. As long as the organic film is not grossly influenced by trace ionic gases or contaminants, the strain within the metal should remain unaffected. Figure 8.8 compares the sensitivity to moisture of sensors using a quartz (a) or polyimide (b) substrate. The flexible polyimide transmits strain more effectively than the rigid quartz. Indeed, for a given moisture level (and corresponding strain), the output voltage of the polyimide backed sensor is about six times larger than its glass analog. This type of sensor characteristically has extremely fast response to ambient changes in humidity due to the thin polymer. More work certainly needs to be carried out to match the temperature coefficients of both films. Thin film technology can also improve the performance of the sensor by integrating the circuitry associated with the gages. Nevertheless, the decoupling of the humidity sensing function from the strain sensing operation minimizes the influence of contaminants and ionic gases.

Metal Oxide Sensors: Humidity sensors based on the enhanced surface electrical conductivity of metal oxides upon absorption of water vapor have been reported. Two groups are generally recognized. The proton type utilizes the increase in ionic conductivity from either capillary condensation within micropores, or water physisorbed to the surface (29-31). On the other hand, the semiconductor type relies on the changes in electronic conductivity due to water chemisorption. In this case, water chemisorbed to the oxide surface acts as electron-donating gas. For n-type oxide semiconductors, for instance, increased water vapor pressure results in larger electrical conductivity (32).

Considerable research has been devoted to developing durable and reliable *protonic ceramic humidity sensors*. Porous TiO_2/V_2O_5 ceramic devices resulted from efforts to quantify water content in cement (29). A complex metal oxide blend of zinc, chromium, vanadium, and lithium constitutes the matrix for another sensor (30). Porous $MgCr_2O_4/TiO_2$ forms the base for another sensor developed by

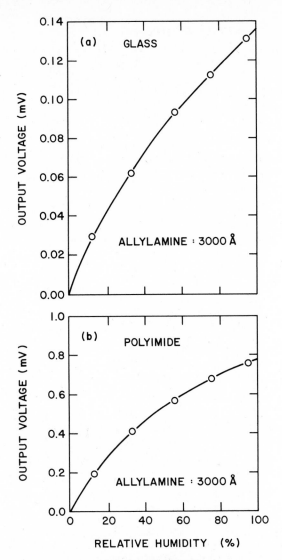

Figure 8.8:   Output sensitivity of a dual-layered film sensor. The device is made of a moisture sensitive plasma polymerized film and a strain sensitive Ni/Cr film. The supporting substrate can be either (a) quartz, or (b) polyimide. Better sensitivity is obtained with polyimide (28).

Matsushita Electric Industrial Co. for microwave process control (31). A schematic of the latter sensor is depicted in Fig. 8.9a. The $MgCr_2/TiO_2$ contains up to 0.35 M of TiO_2, and exhibits a single solid phase solution with a pure $MgCr_2O_4$ type spinel structure. Pore sizes range from 500 to 3000 Å, with an average porosity of 35%. The open pore structure allows easy adsorption or absorption of water vapor, and also relieves stresses from thermal shock.   Fired ruthenium oxide (RuO_2) adhering

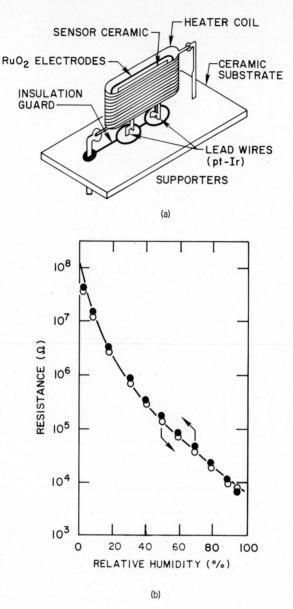

Figure 8.9: (a) Schematic of protonic metal-oxide-based humidity sensor. (b) Device response to moisture exposure (31).

strongly to the ceramic serves as electrodes. Heat cleaning above 500°C is carried out with the coil wrapped around the sensor. Figure 8.9b shows the typical humidity resistance of the $MgCr_2O_4/TiO_2$ sensor. Resistance decreases by almost four orders of magnitude as the relative humidity is raised from 0 to 100%. This particular sensor does not reveal any outward sign of hysteresis.

Semiconductive sensors fall within the perovskite oxide group (ABO_3) because it is possible to modify their water absorption properties and electronic characteristics with the proper selection of cationic components (32). High humidity sensitivity can be found for systems containing calcium and strontium, such as $CaTiO_3$, $CaSnO_3$, $SrTiO_3$, and $SrSnO_3$. Oxides with transition metals at B sites (e.g., $CaMnO_3$, $SrFeO_3$, or $LaCrO_3$), on the other hand, are insensitive to humidity changes. The conductivities of the latter systems are too large to be affected by water absorption.

Ideal sensor materials should have a low surface resistivity which is reversibly responsive to relative humidity, and long-term stability after repeated heat cleaning cycles at high temperatures. Presumably, a polymeric film can be cast or deposited onto one of these substrates to determine its moisture permeation characteristics. However, this raises a host of potentially troubling issues. For instance, the film may adhere poorly to the porous ceramic. Furthermore, the periodic heat cleaning steps required to maintain the sensor in calibrated condition may gradually degrade the polymer. The bakeout of devices at temperatures up to 500°C to remove chemisorbed water would decompose even polyimide. And, finally, metal oxide devices tend to deteriorate irreversibly over extended use due to chemisorbed water, or adhering dust particles and oil residues.

8.1.2 "Surface Effect" Sensor

Interdigitated Comb Structure: Also known as the "condensation type" or "dew point" device, this transducer is composed of a set of interdigitated metal electrodes, as illustrated in Fig. 8.10. Shown are some characteristic dimensions which can vary with the particular application. A cross section of the device with a layer of encapsulation is also illustrated. Traditional usage calls for sealing the sensor in the microelectronic package, which is positioned inside an environmental chamber. The conductivity between the electrodes is monitored continuously as the device is cooled. At the dew/frost temperature, water vapor condenses on the transducer, causing an increase in conductivity. Clearly, one of the obvious drawbacks of this test procedure is the problem arising when the frost point resides below 0°C. In such a case, water vapor would condense directly in the solid rather than liquid phase.

Several design variations can be found depending on the particular application. For instance, a nonglassivated interdigitated pattern of aluminum metal stripes deposited on oxidized silicon is reported to be more dependable than its glass counterpart (33). This is due to the negligible amount of foreign ions present which may induce stray surface leakage currents. Leakage from one metal line to another metal line can be due to various reasons such as bulk conductivity of the insulator substrate, insulating films, or polymer sealant, or electrochemical conductivity from ionic reactions near the biased lines. Such property can be turned into advantage, since the sensor also serves to detect the presence of ionic contaminants. In the typical test to monitor the moisture level in hermetic packages, the sensor undergoes the same processes (e.g., mounting, bonding, sealing) as actual devices. The finished transducer is then placed into a fluorocarbon bath heated to 100°C to ensure water desorption from internal cavity walls. While the bath is cooled down at a known rate, a fixed voltage is applied across the pads. Normal leakage current is around 10^{-11} A. This rises gradually as water starts to condense on the sensor surface. A peak is reached once all the water present has condensed. Moisture content (in ppm) can be

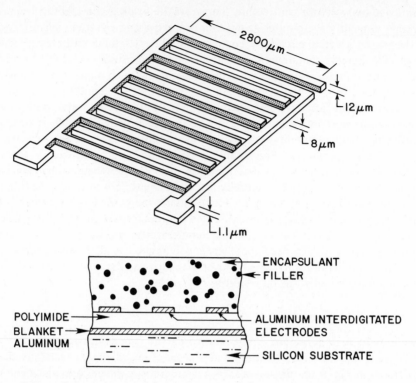

Figure 8.10: Schematic of "surface effect" humidity sensor consisting of interdigitated comb electrodes deposited on SiO_2. A cross section of the electrode configuration and the encapsulating film is also shown (52,53).

determined by plotting leakage current against decreasing temperature on a combined nomograph of dew point, moisture level, and package cavity pressure. The latter can be estimated for a constant cavity volume at a given temperature from the Gay-Lussac law. For a hermetic package, for instance, humidity content should not exceed 5000 ppm corresponding to a dew point less than 0°C (MIL-STD 883).

It has been shown that chemisorption of water molecules to the sensor surface is negligible during the condensation phase (34). Fixation of water molecules to the thermal oxide occurring by physisorption with successive surface coverage following the thermodynamics is described by

$$m = \frac{m_0 c RH}{(1 - RH)\,[1 + (c - 1)RH]} \tag{1}$$

where m, m_0, and RH are the total mass of adsorbed water, the mass of water in one adsorbed monolayer, and the relative humidity, respectively. $c = \exp(\Delta H / RT)$ with ΔH as the difference between the latent heat of evaporation of water and the adsorption energy of the water vapor ($\Delta H \sim 0.7$ eV/molecule), R, the universal gas constant, and T, the absolute temperature. The adsorbed water transforms the interelectrode spaces into an electrolytic bath, depending on whether the test is carried out above or below the dissociation potential of water (~ 1.2 V). The ensuing surface conductivity, σ, of the SiO_2 surface can occur even with only a few layers of water, and can be described by the equation (35)

$$\sigma = \sigma_0 \ m \ \exp\left(-\frac{E_m}{RT}\right) \qquad (2)$$

where σ_0 is the surface conductivity of the substrate in the dry condition and E_m is the activation energy of the conduction. This was later verified experimentally for packages submitted to conditions between 10 and 70% RH (36),

$$\sigma = 2 \times 10^{-3} \ . \ (RH) \ . \ \exp[-11.5(RH)] \qquad (3)$$

It should be noted that the electrolysis of the water adsorbed between two metallization lines modifies the structure of the top layers (36). Such alteration can partly account for difficulties in measurement reproducibility with this type of sensor. In addition, reproducibility is also influenced by the hydration of the oxide under varying conditions (time and temperature) of device storage. Nevertheless, a thermodynamic model of the sensitivity of the sensor can be formulated, which relates the pore size of the oxides to the shift of the triple-point temperature of water molecules physisorbed to these micropores (37).

Studies focus on the critical number of monolayers of water adsorbed for the water to behave like a liquid, and ionic conduction to occur. Among these models, the liquid slab theory and the patch adsorption hypothesis are strongly favored. The liquid slab model assumes van der Waal interaction between neighboring molecules, and attraction between adsorbent and the uniform slab of water (38). This behavior occurs when the adsorption isotherm follows the relation

$$-\ln(RH) = \frac{C_l}{n^s} \qquad 2 \leq s \leq 3 \quad \text{for bulk water} \qquad (4)$$

where C_l is a characteristic interaction constant and n is the number of monolayers adsorbed. Generally, the critical number falls between 3 and 6 for different types of alumina. On the other hand, the patch adsorption model postulates that water molecules initially form isolated islands on the surface. No significant leakage current occurs at this point. However, as humidity increases, these islands start to connect, forming a continuous current path, and leading eventually to high leakage (39). With α-alumina substrates, one complete monolayer is established at about 40% RH.

A theoretical estimation of the capacitance of an interdigitated comb structure can be postulated (40). This arises from the need for a better understanding of the geometrical effects of hybrid circuit elements on microwave integrated circuits. Capacitance values can be calculated from the Schwarz-Christoffel transforms which give the functional relationships between two dimensional geometries. However, the conformal mapping method introduces complex elliptical integrals into the formula. On the other hand, the following simplified equations still yield relatively accurate results at low frequencies. For an interdigitated comb structure with periodic stripes of width W and length L spaced S distance apart, the capacitance is described by

$$C = (\varepsilon_r + 1)\varepsilon_0 L[2A_1(n - 1) + A_2] \tag{5}$$

where ε_r, ε_0, and n are the dielectric constant of the metal, the dielectric constant of the substrate, and the number of fingers per electrode, respectively. A_1 and A_2 are geometrical parameters with the following expressions:

$$A_1 = K_1 \left(\frac{t_{subs}}{S} \right)^m \left(\frac{W}{t_{subs}} \right)^n \tag{6}$$

$$A_2 = \frac{K_2 W}{(2n - 1)(W + S)} + K_3 \tag{7}$$

where t_{subs} is the thickness of the substrate. When curve fitted with experimental data on combs of different geometries (in CGS units), the constants K_1, K_2, K_3, m, and n assume values of 0.614, 0.775, 0.408, 0.25, and 0.4388, respectively. Test data with capacitors of NiCr/Au and Ta_2N metal on alumina substrates indicate that the capacitance can be regarded as constant up to about 5 MHz. The corresponding frequency response is directly dependent on the line width, and inversely dependent on spacing.

More elaborate theoretical expressions for the capacitance and conductor loss of an array of interdigitated microstrips developed for gallium arsenide monolithic circuits were presented recently (41). Size limitations restrict these lumped capacitors to less than 1 pF, compared to the range of 1 to 30 pF for elements discussed previously (40). The lumped capacitor is formed by the fringing field of the interdigital gap between fingers. By including the thickness of the conductor in the analysis, good correlation between theory and experiments is obtained. Results indicate that the elements remain essentially lumped up to 18 GHz. Losses are dependent on finger width, line spacing, and metallization thickness. This suggests that to design an optimum comb structure, width and spacing should be selected first for maximum Q-values and minimum parasitics.

More recently, the interdigitated comb structure was used for real-time monitoring of high-quality dielectric films (42). The need was prompted by the large interest in producing reliable insulating monolayers for GaAs superlattices, MOS devices, and nonlinear optical processing. In this case, the electrode was fabricated on a glass substrate using standard lithography and wet etching techniques. Monolayers of two polymers, polymethyl methacrylate and *n*-docosyl merocyanine, were deposited by the Langmuir/Blodgett (L/B) technique. Since an L/B layer has a known composition and a precise thickness, multiple layers can be coated onto a substrate. A standard impedance bridge was used to record dynamically the residual amount of water during drying between dipping cycles. Using the method of images, a theoretical expression for the capacitance was developed. It predicted a sensitivity of 2 fF/monolayer, which agreed to within 10% of the experimental values. The discrepancy arose from the effect of water and the assumptions underlying the model. In the former case, the sensor was sensitive to water and water vapor in the vicinity, which caused an increase in the sensitivity parameters. On the other hand, the method of images assumed that the electrodes have essentially zero thickness. Better characterization would be obtained if precautions were taken either to ensure complete film drying, or better control of the test environment. A sensor with smaller line width dimensions would also improve the sensitivity of the technique.

More refined and accurate results can also be obtained with finite element modeling (43-46). Previous derivations are based on simple two-dimensional geometries, while three-dimensional problems all consider more realistic planar multiconductor configurations. In one instance, the charge density on the conductors is derived from the method of moments (43). However, the program places a constraint of fixed cell size which imposes unnecessary restrictions in designing, and forces long and costly computer iterations for complex geometries. Variable cell sizes are used in another program which calculates capacitances for both planar and out-of-plane configurations (44). A constant charge on each cell, however, is assumed. A subsequent program circumvents this limitation (45). In this case, singularity functions are included in a function set which allows a better description of the charge density distribution, and solves larger circuit layouts. Improvements in computation speed is later obtained with the use of a mixed-order finite element technique (46). The latter combines the Rayleigh-Ritz method with constant charge approximation and the point-matching method with a singular charge assumption.

Triple Track Structure: Another variant of the interdigitated comb discussed previously is the "triple track" structure for material testing studies (4,47-51). This sensor consists of three parallel aluminum lines forming a meander pattern on the phosphorous-doped SiO_2 deposited on silicon. Dimensions and spacing vary with the applications, but lines are typically 1 μm thick by 5 μm wide (20). As illustrated in Fig. 8.11, a typical surface leakage test involves grounding the outside conductors so that lines are alternately biased similar to the interdigitated electrode configuration. The pads are then connected either to a picoammeter for current detection, or to an impedance meter for dielectric measurements.

In one study, for instance, the track is biased either positively or negatively to observe anodic or cathodic corrosion (47). A variety of plastic resins (e.g., epoxy novolac, epoxy anhydride, silicone, and silicone/epoxy) is used to encapsulate the test devices. Leakage current under a constant 30-V potential is monitored for different conditions, ranging from 30 to 110°C at 10 to 95% RH. At constant RH, ionic conductivity is slowed down at low temperatures. Interestingly, the initially low-surface conductivity increases steadily as moisture reaches the die surface, either

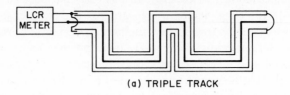

(a) TRIPLE TRACK

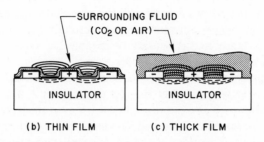

(b) THIN FILM   (c) THICK FILM

Figure 8.11:   Schematic of triple track patterns and impedance meter (dielectric measurements) or ammeter (surface leakage measurements).   Fringe fields in devices covered with thin and thick films are shown (20).

along the plastic/lead frame interface or through the bulk of the plastic. Epoxy anhydride fares better than its Novolac counterpart in retarding metal corrosion. On the other hand, silicone and silicone/epoxy copolymer combine good adhesion to the die and low ionic content to provide the best moisture protection. Similar conclusions are obtained with triple track devices encapsulated into 24-lead premolded chip carrier configurations (48). Another study looks into the effect of silicone gels and overcoats in preventing unwanted leakage in integrated circuits and hybrid integrated circuits (49). Triple tracks with different metallurgies (e.g., Ta_2N and Ti/Pd/Au) on alumina substrates are tested. Although silicone gel is soft and exerts mild stresses on delicate bonds during thermal cycling, its low modulus provides little mechanical protection.   Most often, the gel itself has to be covered with a metal/plastic lid or some harder coating. A judicious selection of polymeric overcoat does improve the moisture resistance of the package, although simply having a harder layer over the gel does not guarantee success.

Another aging study of triple track test vehicles in a more aggressive environment is also reported   (50).   The sampling test includes devices which are bare, encapsulated with silicone gel, covered with silicon nitride caps, and encapsulated with caps. Both positive and negative 10-V bias is considered to determine whether the failure mechanism is anodic or cathodic. Vehicles are subjected to 85°C, 85% RH with 1 ppm of sulfur dioxide. The latter is the main oxidized sulfur component in the atmosphere, and is the principal cause of corrosion in industrial environments. However, results indicate that humid SO_2 is not very aggressive since failures occur only after extended aging. Silicone gel is again found to retard galvanic corrosion. It effectively inhibits, but does not prevent high leakage currents between the metallized tracks.

In another investigation, leakage current is monitored with respect to epoxy molding formulations such as coupling agents, filler particle size, and internal mold release agents (51). Standard pressure cooker conditions, 121°C at 100% RH and 2 atm, are applied to molded metallized tracks under 10-V steady bias. Silane coupling agents with hydrophobic radicals postpone the onset of current leakage by almost 80%. These agents are used routinely to enhance the adhesion between inorganic fillers and the polymeric matrix. Furthermore, increasing the average filler size to about 25 μm also improves the moisture resistance of the coating. Presumably, this is due to the higher content of silane coupling agents coated on the large filler particles. On the other hand, although mold release agents facilitate part removal from the mold cavities, they can also wreak havoc the reliability of the package. Indeed, they tend to migrate toward the film/substrate interface, degrade the adhesion, and leave pathways for moisture to collect. Amide-based release agents hydrolyze easily while their ester counterparts do not. The latter can also effectively repel water.

More recently, a single test chip with some of the structures discussed previously to integrate the reliability monitoring functions has been revealed (4). Four distinct areas are considered:

1. Moisture content is detected by a conventional dew point sensor made up of unpassivated interdigitated aluminum stripes.

2. Corrosion is monitored by a combination of aluminum meander surrounded by an interdigitated comb structure. The resistance of the meander increases as corrosion progresses. By using two of these structures, both cathodic and anodic corrosion can be studied with respect to the grounded comb.

3. Wire bond reliability is studied by measuring wire bond contact resistance.

4. Drift in device parameters such as transistor threshold voltage, diode leakage current, channel leakage current, and frequency shift of a ring oscillator are monitored.

Furthermore, diffused resistors capable of 15-W power output and thermal diodes for temperature sensing also ring the test chip. These elements can induce different temperature gradients across the chip, and help evaluate packaging tasks such as thermal performance, package and board design, and various heat removal schemes. At this writing, further work still needs to be done in calibrating all the structures on the chip. Nevertheless, the test chip does provide a good vehicle for investigating, among other packaging issues, the effects of moisture and contamination on the operation of the encapsulated device.

For dielectric studies with the triple track pattern, the capacitance recorded is governed by the thickness of the polymer coating. When a film thinner than the electrode height and spacing is spin coated directly on the device, the fringe field extends beyond the boundary of the film. Measurements in this case are influenced both by changes in the film and the surrounding medium. A thick coating, on the other hand, would completely shield the field from the exterior. The resulting capacitance should be independent of film thickness and external environment, and sensitive only to the film structural composition.

This is shown in the supercritical fluid extraction study discussed earlier (20). Upon pressurizing a bare triple track device with CO_2, the capacitance increases and stabilizes. Dense CO_2 displaces the atmospheric air and alters the dielectric constant recorded. A similar behavior is observed with a thin (0.7-μm) polymeric layer over the test device, revealing the extent of the fringing field on the surrounding medium. On the other hand, a thicker film (10-μm) sufficiently contains the field that solvent pressurization shows little effect on the capacitance measured.

In another study to investigate the transport characteristics of moisture in encapsulants, interdigitated aluminum comb structures are also used for dielectric measurements (52,53). A cross section of such structure is shown in Fig. 8.10. Metal stripes (1.1 μm x 12 μm x 2800 μm) spaced 8 μm apart from each other are deposited on top of polyimide and a blanket layer of aluminum. Unfilled and filled blends of bisphenol A/cycloaliphatic epoxies, and a solvent-based silicone-modified epoxy are tested. These are encapsulants widely used in either molded packages or in face coatings such as tape automatic bonding. Fused silica particles with an average diameter of 1.5 μm are added to the unfilled formulation at various volume loadings to study the effects of fillers on moisture permeation. An automated gain phase analyzer records the capacitance during temperature/humidity cycling. Since coatings a few mils thick are evaluated, it can be reasonably assumed that the fringing field is contained within the film. Once exposed to the humid test environment, any changes in the capacitance would reflect the presence of water, which has diffused through the encapsulant to the epoxy/chip interface. The study shows that the diffusion coefficients of the epoxies display some weak dependency on water concentration, but are strongly governed by the temperature. Furthermore, increasing filler content lowers the saturated moisture content within the coated sensors. The corresponding diffusion follows a dual-mode absorption with a fraction of the water permeating freely through the polymeric matrix, and the balance immobilized by a combination of microvoids and interaction between moisture, fillers, and resin. Interestingly, water diffusion is dictated by the hygrothermal history of the coating. It is expected, for instance, that a coated sensor at 85% RH first exposed to 25°C, then to 85°C, and back down to 25°C would display the saturated moisture content of a new specimen equilibrated at 25°C. However, Fig. 8.12 indicates that this is not the case. The new equilibrium value is approximately the same as the saturation moisture level of the device at 85°C/85% RH. Thus, once an epoxy film is exposed to a test condition, the matrix sustains structural damage in the form of irreversible interaction such as microcracks and microvoids. The extent of damage remains constant as long as the specimen is not subjected to a more severe condition (52). As a moisture indicator, the comb structure also reveals that the primary route of moisture ingress into a molded package occurs via the bulk surfaces. This holds especially true when adhesion promoters are coated on the lead frame to enhance the quality of the resin/lead frame interface (53).

8.1.3 Other Techniques

Previous techniques for detecting moisture in films or packages, for example, mass spectrometer, Al_2O_3 hygrometers, or special comb or triple track structures, are not ideally suited for production line control. Al_2O_3 hygrometers are also susceptible to aging problems, while the high voltage required in DC conductance measurements

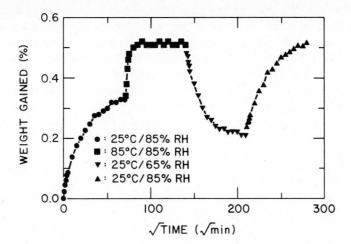

Figure 8.12:   Effect of the hygrothermal history on the permeation characteristics of an encapsulant film (52). The polymer is an unfilled blend of bisphenol-A and cycloaliphatic epoxies.

introduces artifacts from electrolysis of the adsorbed moisture. Operationally simpler tests have been suggested for manufacturing process control. These include the track capacitance technique (54), the capacitance ratio test (55), and the AC conductance method (56).

The first technique, sometimes referred to as the Philips capacitance test, derives from the dew point test described earlier. It is found that the capacitance between a pair of metallized tracks on any integrated circuit has a component dictated by the moisture content adsorbed to the surface of the die. Measurements are typically conducted at low frequencies (e.g., 100 Hz). Slow localized cooling of the die can raise the relative humidity at its surface, until sufficient water is adsorbed to result in a detectable change in capacitance. The temperature corresponding to this increase gives a measure of moisture content within the package (54).

An improved version of the test is subsequently introduced to assess the humidity content of both hermetic and plastic packages (55). The underlying concept is based on the fact that the capacitance between a selected pair of metallized tracks is frequency dependent. While maintaining the die at a constant dew point temperature corresponding to a given moisture level, measurements are conducted at two frequencies (e.g., 100 and 1000 Hz) to provide a ratio of the moisture-induced capacitance to the intrinsic device capacitance. Comparing this ratio with that from a package of calibrated moisture content yields a useful pass/fail criterion. Although the selection of the dew point temperature is somewhat arbitrary, the accuracy of the method is comparable with alternative techniques.

The AC conductance method, like the Philips capacitance test and the capacitance ratio method, uses the chip itself as the sensor. The conductance between any two pins of the chip which are not connected to any electrical component is measured with a lock-in-amplifier. A voltage of 100 mV is applied at 100 Hz. Such low voltage eliminates any possible artifacts from the dissociation of adsorbed water. The conductance is monitored continuously as a function of temperature as the chip undergoes cooling, followed by a heating cycle. Experimental results indicate that moisture-induced capacitance variation is only between 0 and 20%, while the corresponding conductance changes range from 10 to 400%. Thus, the AC conductance method shows better sensitivity than the other capacitance techniques. The integrity of a passivation layer, either inorganic or polymeric, can be checked. Since the die itself acts as a sensor, selective condensation on the device is no major concern. A proper choice of pins allows the use of most of the die surface.

8.1.4 Implications

It should become apparent that each technique has its own advantages and drawbacks. However, when judiciously used in combination, these methods can detect water in plastic packages in a fairly cost-effective manner. For instance, selection of packaging materials with desirable properties such as high-temperature stability or low out-gassing can be accomplished with mass spectroscopy and thermogravimetry. Surface conductivity sensors can be used to compare these materials with and without process contaminants for current leakage in a humid environment. Aluminum oxide sensors, dielectric sensors, or capacitive sensors can also be used concurrently to measure the internal moisture content of sealed or encapsulated parts. Once such tasks or related tests are performed, improved standards on acceptable levels of moisture for long-term reliable device operation can be established.

8.2 STRESS MONITORING

Stresses within a polymer film arise either from the mismatch in the coefficients of thermal expansion between film and substrate, or from curing shrinkage. Generally, the first class of stresses, thermal stresses, is generated throughout the process life of the package. For instance, molding or encapsulation requires up to 175°C for an extended period of time. Typical qualification requirements also submit the package to wave solder reflow temperatures (~215°C), shippability readiness (e.g., -40 to +60°C for at least ten cycles), high-temperature bias device test (e.g., 165°C for 200 h), or just simple thermal cycling (e.g., 0 to 125°C for 3 cycles/h and at least 3000 cycles). On the other hand, shrinkage of a plastic film or coating is an intrinsic property of the polymerization process as a three-dimensional structure is developed. Current resins for encapsulation incorporate inert fillers to minimize volume shrinkage effects, raise the thermal conductivity, and lower the typically high reaction exotherm.

For a simple wafer coated with a thin polymer film such as polyimide, thermal and curing stresses can reach sufficiently high magnitude to bend the wafer. The curvature of such a wafer can be easily measured with an interferometer, a Dektak profilometer, or X-ray diffraction, and the resulting strain can be converted to surface stresses (57,58). These stresses in a multilayer structure such as a plastic package, for instance, can induce a host of reliability problems. Defects stemming from stresses can range from wire bond disconnection, cracking of the passivation, or delamination at the resin/chip interface.

Thus, a good understanding of the origins of stresses is a prerequisite to the design of a more reliable package. This section briefly reviews some of the techniques which have been used to characterize intrinsic stresses in polymers, and describes in detail piezoelectric transducers used in mapping out the stress profile in encapsulated dies. Although these methods are more suitable for bulk polymer and thick coatings, nevertheless, they can give some insight into the possible strains encountered with thin films.

8.2.1 Miscellaneous Techniques

The concern for cracking of potted parts from internal stresses has been a major issue as early as 1950. In order to obtain some information on the magnitude of the stresses, or at least a numerical comparison between different resins, a *strain sensing element* was introduced (59). This consists of a cylindrical stainless steel tube with wire strain gages bonded to its inner surface. Both external and internal surfaces are ground to a wall thickness of 0.035 in. (0.89 mm) for sensitivity. Once encapsulated, the tube is subjected to internal shrinkage and thermal strains which are recorded by the wire gages. Phenolic SR-4 wire gages, stable up to 260°C, are mounted to measure both radial and longitudinal strains. Temperature and hydrostatic calibration is necessary before stress as a function of temperature can be obtained through thermal cycling of the resin. Although somewhat crude, nevertheless, the method provided for the first time quantitative information on catalyst/resin formulations, annealing processes, and even the extent of cure of the resin.

Different strain gage techniques based on the previous method for measuring internal stresses in casting resins are reported, and described in a brief review (60). In one case, for instance, an aluminum transducer with strain gages bonded to the inner wall is potted at the center of the resin mass. Bending strains (perpendicular to the axis direction) are small compared to the axial strains resulting from both longitudinal stresses and shear stresses at the resin/wall interface. Data indicate that compressive stresses developed upon potting the transducer increase rapidly during the gelation of the polymer (over 900 psi or 6.2 MPa after 3 h). Further polymerization during room temperature storage leads to a gradual increase in these stresses (up to 1500 psi or 10.3 MPa after six months). Once the maximum degree of crosslinking is obtained, subsequent changes become smaller. Thermal cycling also reveals that internal stresses are relieved upon heating, and the rate of cooling and the initial temperature of cooling dictate the magnitude of the compressive stresses (60).

A similar configuration is used to evaluate epoxy molding compounds for large-scale integrated circuits and advanced memory devices (61). Encapsulants made by incorporating fused and standard silica on a one-to-one ratio exhibit compressive stresses less than 700 psi (4.8 MPa), compared to the 1500 psi (10.3 MPa) from conventional epoxy compounds. Furthermore, when additives are added to the compounds, both thermal shock and glass transition temperatures are enhanced, resulting in longer life for metal-oxide semiconductor (MOS) devices. The fluids act as internal lubricants to relieve the thermal stresses.

A wide array of wire strain gages is available for this method of quantifying polymeric stresses. Such a gage typically has two basic components, namely, a carrier and a sensing foil. The carrier provides dimensional stability, support, and even some measure of mechanical protection for the sensing element. Carriers can be made from polyimide, glass reinforced epoxy, metal, or even paper. The sensing foil, on the other hand, is a photoetched sheet of metallic alloy with the desired grid pattern to yield strain in uniaxial or biaxial directions. The studies reported are based on constantan (Cu/Ni) or Dynaloy (Ni/Cr) on polyimide- or glass-reinforced epoxy carriers. For instance, such gages can be mounted directly on the paddle of dual-in-line lead frames, and subsequently molded to test different formulations of low-stress encapsulants (62).

A subsequent technique uses a concept similar to that of a *bimetallic strip*. In this case, when a polymer film coated on a strip of metal foil expands or shrinks, the metal strip becomes concave or convex. As long as elastic limits are not exceeded, the curvature can be translated into surface stresses in the resin (63). The coating is applied to a large and thin aluminum foil sheet, which is then cut into smaller strips after precuring. The width of the strips is kept to a minimum to lower any curvature across the strip. Samples are placed in an oven with controllable temperature and humidity. Curvature measurements are carried out using a low power microscope equipped with an ocular micrometer. The stress, σ, can be determined from the strip deflection by the following equation

$$\sigma = \frac{E_2 h_2^3}{12 \, h_1 \rho H} \, F(m, n) \tag{8}$$

where E_1, E_2, h_1, h_2, H, ρ, δ, ℓ are the Young's modulus of the resin, the Young's modulus of the metal substrate, the thickness of the film, the thickness of the metal, the total thickness ($h_1 + h_2$), the radius of curvature ($= \ell^2/8\delta$), the deflection at the strip center, and the length of strip between supports, respectively. Furthermore, $F(m, n)$ is a characteristic parameter described by

$$F(m, n) = \frac{(1 - mn^2)^3}{(1 + mn)^3} + \frac{[mn(n + 2) + 1]^3 + m(mn^2 + 2n + 1)^3}{(1 + mn)^3} \tag{9}$$

with $m = E_1/E_2$ and $n = h_1/h_2$. As indicated in Eqs. (8) and (9), this method requires knowledge of the moduli of both substrate and polymer film. Experiments with two epoxies (Epon® 828 and 1001) and three curing agents (diethylenetriamine, *m*-phenylenediamine, and phthalic anhydride) are conducted as a function of cure temperature and humidity. Results indicate that the stresses within the cured epoxy resins are due mainly to thermal contraction. Surprisingly, no effect from polymerization shrinkage is observed between the different curing agents. Volumetric shrinkage, however, does exist since the resins do not contain any inorganic fillers. This is an indication of the sensitivity limit of the method. Nevertheless, the test shows that thermal stresses are reversible, with a hysteresis loop occurring only at fast temperature changes. Moisture tends to plasticize the matrix and relax the observed stresses. This condition can be reversed upon reheating. Finally, the glass transition temperature of the polymer, T_g, governs these stresses.

Virtually no stresses are observed above T_g. Below T_g, on the other hand, stresses are approximately proportional to the difference between the measured temperature and T_g.

Another simple method to characterize stresses within a polymer is the *thermometer pressure test* (64). The underlying principle resides in the fact that when a thermometer is strained, for example, pressure is applied to the glass bulb, mercury rises in the column to a certain height. This is due partly to the actual temperature of the mercury, and partly to the straining effect of the bulb. Calibration by independently heating the bulb and by applying external hydrostatic pressure would reveal the influence of each effect. The technique is useful in determining the influence of a silicone rubber conformal coat (Dow Corning RTV-731) on the stress profile within a device encapsulated with methylene dianiline filled with 75% (by weight) of silica. The stress relief is found to be directly proportional to the coating thickness. Furthermore, the coating must be quite thick in order to eliminate stresses entirely. The study reveals some alternatives for reducing stresses, such as filling the compound, flexibilizing the material, reducing the exotherm, and protecting with rubbery conformal coatings.

Another technique that deserves some mention is *photoelasticity*. The basic technique for three-dimensional photoelasticity as applied to problems of restrained shrinkage is well established (65,66). It differs from the conventional "frozen stress" technique because, in this case, the loads are self-induced due to the differential shrinkage occurring from uniform temperature changes. As a result, the isochromatic fringes observed depend both on elastic effects and "frozen stresses." When slices are carefully taken from the sample during the cooling period, the magnitude of the stresses changes since stresses normal to the surface are relieved. However, the stress distribution should remain fairly constant. Thus, this mixed character of the fringe pattern must be decoupled before any meaningful conclusions can be inferred. An attempt was made recently to study stress levels in encapsulants with birefringent model compounds used in optoelectronic applications (67). Lead frames with dies already mounted to the die paddles are molded with two types of epoxies and postcured. Each sample is then sliced with a diamond saw, leaving the die intact. Comparison of the ensuing isochromatic fringe pattern with that from blank samples subjected to calibrated loads yields the corresponding stresses. Measurements and results obtained from laminated plate theory and a two-dimensional elastic finite element analysis agree only for the upper layer of material. However, a large discrepancy exists for the lower layer, and cannot be accounted for by strictly geometrical effects. Differences between the temperatures of the mold halves, for instance, may explain the variation. Finally, the technique also indicates that postcuring relaxes the stress concentrations in the vicinity of objects with sharp corners such as the edges of the die or the burrs from the stamped die paddle.

Recent advances in etching single crystal silicon into complex mechanical structures have created opportunities for new characterization techniques. For instance, orientation-dependent etchants can dissolve crystal planes differentially to yield a wide array of angular surfaces (68). As a result, various microdevices have been produced commercially, ranging from accelerometers, flow sensors, to chemical sensors. One reported application of silicon machining in determining thin film mechanical properties used cantilever microbeams (69). Young's modulus and

breaking stresses of silicon dioxide and silicon nitride were measured from their resonant frequencies. The concept was subsequently extended to both metallic and
polymeric films on [100] silicon substrates (70,71). In this case, after standard pat-
terning, portions of the silicon were removed selectively, yielding in the process
microbeams coated with the desired films. Distinct curvatures were obtained, either
from the thermal mismatch between the substrate and the dielectric medium, or in
the case of metal films, from the intrinsic growth stresses. The state of stress can be
readily deduced from such curvatures. Another version of the cantilever beam structure is the so-called "release structure" (72). This is an asymmetric structure patterned into the film prior to release from the substrate. Once the silicon backing was
removed, the film was free to deform under the action of its internal stresses. This
geometry is sensitive to residual tensile stresses.

8.2.2 Piezoelectric Gages

Among the techniques listed previously to quantify polymer film stresses, the
piezoresistivity in silicon must count as one of the most successful methods (73-84).
This effect has been used to make strain, force, torque, and pressure sensors. Only
recently has it been extended to the characterization of stresses in plastic packages.
The need arises when it is found that stress plays a major role in the reliability of IC
devices. For instance, passivation defect density is directly dependent on the stress
produced by different resins (73). Encapsulation strain and thermal cycling also
account for the deformation of aluminum metal lines (74). Commercial
piezoresistive sensors are therefore used to delineate stress contours across the
surface of a typical die (75). Such sensors also reveal cracks in the dielectrics
between metal layers. Good adhesion between the encapsulant and the die surface is
necessary for stress transfer (75). Furthermore, resistor orientation dictates the sensitivity of the sensor. Rather low signals are obtained with p-type grids parallel to
the [100] crystal direction. When similar sensors are incorporated into reliability test
devices, package stresses with certain molding compounds can reach as high as 60%
of the silicon fracture strength (77,78).

The resistivity of silicon becomes anisotropic under external loading. The change
in resistivity, however, is dependent on the direction of current flow with respect to
the crystal lattice. The three-dimensional aspect of the lattice translates into six
fractional resistivity changes. When related to stresses, an ensuing 6 x 6 matrix of
coupling coefficients is obtained, although due to the symmetry of the lattice, only 3
of the 36 coefficients are independent. These are typically denoted by Π_{11}, Π_{12}, and
Π_{44}, and are measured experimentally to be a function of doping density and carrier
(n or p) type. The choice of the coordinate system with respect to the silicon crystal
lattice dictates the coupling coefficient matrix. These aspects were reviewed recently
in a study which uses finite element techniques to evaluate piezoresistive effects on
sensors located at different orientations on a [100] die surface (79). The modeling
suggests a reorientation of the substrate with respect to the photomask so as to lower
the sensitivity of the resistors to stress.

Determination of the stress requires recording the resistance of two nonparallel
sensors. When the axes of a silicon die are parallel to the [010] and [001] directions
in the [100] plane of the silicon surface, piezoresistivity effects are minimized. In

this case, the fractional change of each resistor is derived from two simultaneous equations such as

$$\frac{\Delta R_x}{R_x} = \Pi_{11}\sigma_x + \Pi_{12}\sigma_y + \Pi_{12}\sigma_z \tag{10}$$

$$\frac{\Delta R_y}{R_y} = \Pi_{12}\sigma_x + \Pi_{11}\sigma_y + \Pi_{12}\sigma_z \tag{11}$$

where ΔR_i and σ_i are the resistivity change in the i direction due to current flow in that direction and the normal stress in the i direction, respectively. On the other hand, were the die edges and resistor elements parallel to [011], as is most IC fabrication on [100] wafers, maximum piezoresistivity effects would be obtained. The corresponding fractional resistance changes are described by

$$\frac{\Delta R_x}{R_x} = \frac{1}{2}(\Pi_{11} + \Pi_{12} + \Pi_{44})\,\sigma_x + \frac{1}{2}(\Pi_{11} + \Pi_{12} - \Pi_{44})\,\sigma_y + \Pi_{12}\sigma_z \tag{12}$$

$$\frac{\Delta R_y}{R_y} = \frac{1}{2}(\Pi_{11} + \Pi_{12} + \Pi_{44})\,\sigma_x + \frac{1}{2}(\Pi_{11} + \Pi_{12} + \Pi_{44})\,\sigma_y + \Pi_{12}\sigma_z \tag{13}$$

Note that with the proper selection of orientation, no shear stresses are involved.

Although the sensors are designed to quantify stresses from both thin films of passivation and thick films of encapsulants, reported literature deals mainly with the latter. This probably results from the more severe reliability issues related to encapsulation stresses. Piezoelectrically induced parameter shifts from encapsulation stresses change with time and ambient conditions. These stresses are not uniform from one package to another, even though the packages can be molded simultaneously. Factors that account for these variations can include a thermal gradient within the mold, a mismatch in the coefficients of thermal expansion of the materials, a different postcure history, or a brittle die attach adhesive. By subjecting the sensors through each step of the manufacturing line, a measure of the stresses imparted to the devices by each process can be determined. Compressive stresses from the molding steps readily distort the die and affect the surface structures.

Piezoresistive sensors start to play a major role in reliability programs throughout the industry. Such devices may facilitate either a preliminary screening of compound formulations (80) or spur on more systematic studies. Residual stresses after die attachment can be monitored (81). In this case, for instance, different die attach method/lead frame metal combinations are tested, namely, Au-Si/Alloy 42, epoxy/copper 194, and polyimide/copper 194. Eutectic Au-Si with an Alloy 42 lead frame yields the lowest stress profile. Thick layers of die attach material also tend to relax bonding stresses imparted to the die. A recent study investigates the effect of an overcoat on die surface stress and package cracking (82). Silicone RTV, silicone gel, siloxane modified polyimide, and polyimide are evaluated. Stressing according to MIL-STD 883B conditions B (-55 to +125°C) and C (-65 to +150°C) does not produce any cracking in packages with overcoats. Figure 8.13a shows results of finite element modeling of a PLCC (plastic leaded chip carrier) configuration for different conditions. All die coats reduce the compressive stresses, with silicone gel exhibiting the best improvement and polyimide the smallest. The simulation does not quite agree with the data obtained from a rosette of piezoresistive sensors, as shown in Fig. 8.13b.

It seemed that both the silicone gel and modified polyimide affected the in-plane stresses, while the standard polyimide and silicone rubber raised the compressive stresses on the die. The discrepancy could be due to two main reasons. First, the finite element model was rather simplistic, with a coarse two-dimensional mesh size throughout the cross section simulated and finer mesh only at the die corner. Elastic material properties were assumed with no provision for viscoelastic stress relaxation, which may have an effect especially since the temperature cycled (150°C) was close to the glass transition temperature of the molding compound. And, second, good adhesion between the various interfaces was assumed throughout the package. This does not necessarily apply since delamination between the molding compound and either the lead frame or the die coat, or both, may have already occurred during molding. As a result of this poor adhesion, stresses may only be partially transferred from the molding compound to the die/lead frame structure, and would drastically affect the influence of the die coat in preventing passivation cracking.

A more systematic use of piezoresistive gages is reported recently. The study focuses on lead frame material, molding compounds, die overcoats, chip size, and package geometry (83). The piezoresistive elements are made on p-doped [111] silicon substrates, with accompanying resistors for power dissipation, and thermal diodes for temperature measurements. Mixed results are obtained. For instance, Alloy 42 causes less mechanical stress, but increases the risk of cracking the package. Similarly, silicone rubber as die attach or protective overcoat lowers the stress level, albeit at higher risk of wire bond fatigue from thermal cycling. The highest stresses are obtained during the low temperature portion of the thermal cycle. Stresses are reduced at higher temperatures. As expected, Fig. 8.14 reveals that high-stress levels are obtained with larger die sizes. Thus, the current push toward large dies for higher integration must seriously consider such implications. An improvement over previous sensors is obtained through the use of digitally addressable multiple strain rosettes (84,85). The sensing element, in this study, is an arsenic implanted pattern consisting of four resistors located at 0°, 45°, 90°, and 135° from each other. The decoding logic used to address individual resistors is incorporated directly into the chip. By a judicious distribution of these rosettes on the die surface, surface stress contours can be mapped out, as depicted in Fig. 8.15. This shows a two-dimensional plot of average stress on a die encapsulated with a rubber modified epoxy compound. Such information is useful in studying the thermomechanical behavior of molding compounds, and locating the high stress concentration areas within a package.

A caveat should be pointed out with piezoresistive sensors. The independent coupling coefficients are shown to be related to each other by the following equations (86):

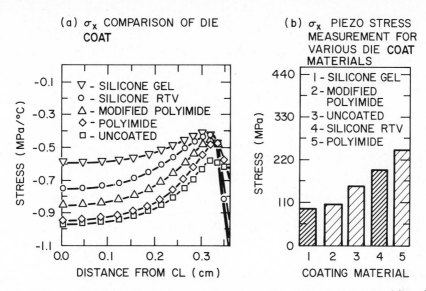

Figure 8.13:   In-plane surface stress of various die coating materials determined from finite element analysis. Stress levels are plotted as a function of distance from the package center line. Similar surface stress levels as recorded by piezo-stress chips with different coatings (82).

$$\Pi_{11} = 21\Pi_{44} \qquad \Pi_{12} = -125\Pi_{44} \qquad (14)$$

However, there is still a large uncertainty on the absolute values of the constants, especially Π_{44}. Consequently, absolute values of stress derived from the sensors should be used with care in predicting cracking of the passivation, die, or package. Rather, such techniques should be used for comparing stress levels obtained for different coating applications.

Piezoresistivity data are very sensitive to temperature fluctuations. Usually, resistance measurements need to be corrected for temperature shifts, either by incorporating temperature compensation schemes directly into the bridge setup, or by determining the chip temperature and adding a correction factor based on the device thermal sensitivity.

Principal stresses can easily be measured with these devices. Shear stresses developing between the plastic and the die, on the other hand, are more difficult to detect, since they are generally about an order of magnitude less than the normal stresses. Nevertheless, shear stresses can cause various reliability problems such as cracked passivation, lifted ball bonds, or shifted metal lines. Such damage does not appear until after several thermal test cycles, although in some severe cases, damage can occur immediately after postcuring the plastic packages. Damage is greatest in the vicinity of the edges of the die where the highest shear stresses are encountered. Simple structures have been designed to measure the extent of shear damage imparted by the plastic (81,82). The typical test would detect shorting between a wide metal line and a nearby narrower sense line, both positioned along edges of the

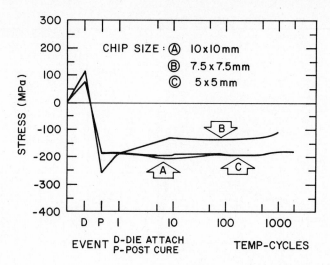

Figure 8.14: Stress level for three different chip sizes as a function of temperature cycling (83).

die. The wide line is designed to simulate the worst case of a large power bus layout. Both microscopic and electrical data can thus be obtained. Only qualitative information is obtained since the results describe the effects of shear stresses, rather than give some absolute value of these stresses. The experimental data obtained usually must be verified with finite element modeling of the package. Shear stresses should run parallel to the die surface, exhibiting maximum at the corners and approaching zero at the center.

8.3 CONCLUSIONS

Sensors have been subjected to the same microfabrication techniques that produced the microprocessor, which in turn, created a revolution in digital process control systems. As a result, they have gained wider acceptance and usage in various applications related to electronic packaging.

To detect the presence of moisture in packages, for instance, two types of sensors can be selected, either a "volume effect" sensor (e.g., Al_2O_3, dielectric, or metal-oxide-based transducers) or a "surface effect" sensor (e.g., interdigitated comb structures or triple track patterns). Both can be used to study the moisture diffusion characteristics in thin and thick polymer films. However, each technique has its inherent advantages and drawbacks, and must be used in conjunction with other independent methods to obtain more information on moisture permeation into packages.

Moisture-induced failure is not the only major issue affecting the reliability of a package. The stresses encountered during encapsulation can also drastically influence its lifetime. Major efforts have thus been devoted to characterizing the nature of these stresses. Within this context, piezoresistive gages have been the preferred

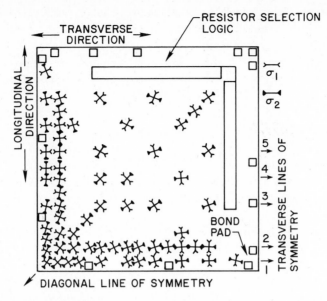

Figure   8.15:   Two-dimensional stress plot of the average surface stress of a die coated with a rubber-modified epoxy-based molding compound. The average is recorded from a sampling group of 16 packages prior to thermal shock testing (85).

means for screening molding compounds, evaluating die overcoats, analyzing lead frame configurations, or studying die attach materials for large area chips.

8.4 REFERENCES

1.   P. V. Robock and L. T. Nguyen, "Plastic Packaging," Chapter 8, in **Microelectronics Packaging Handbook**, R. R. Tummala and E. J. Rymaszewski, Eds., Van Nostrand Reinhold, New York (1989).

2.   E. R. Winkler, "Integrated Circuit Packaging Trends," *Solid State Technol.*, 94, June (1982).

3.   E. C. Blackburn, "VLSI Packaging Reliability," *Solid State Technol.*, 113, January (1984).

4.   S. C. O'Mathuna and P. L. Moran, "A Single IC to Evaluate the Electrical, Thermal, and Hermetic/Corrosion Resistant Properties of IC Packages," *I.M.C. Proc.*, 378, Kobe, Japan, May 28-30 (1986).

5.   R. W. Thomas, "Moisture, Myths, and Microcircuits," *I.E.E.E. Trans. PHP-12*, 167 (1976).

6.   L. Gallace and M. Rosenfield, "Reliability of Plastic Encapsulated Integrated Circuits in Moisture Environments," *R.C.A. Rev.*, 45, 249 (1984).

7.   R. B. Comizzoli, R. P. Frankenthal, P. C. Miller, and J. D. Sinclair, "Corrosion of Electronic Materials and Devices," *Science*, 234, 341 (1986).

8.   J. Gunn, S. Malik, and P. M. Mazumdar, "Highly Accelerated Temperature and Humidity Stress Test Technique (HAST)," *I.E.E.E. Proc. Int. Reliab. Phys. Symp.*, 48 (1981).

9.   J. W. Osenbach and J. L. Zell, "Humidity-Temperature-Voltage Acceleration Model for Corrosion of Thin Film Aluminum," *Electrochem. Soc. Meeting*, Paper 312, Chicago, Oct. 11-12 (1988).

10. R. M. Felder, C. J. Patton, and W. J. Koros, "Dual-Mode Sorption and Transport of Sulfur Dioxide in Kapton Polyimide," *J. Polym. Sci.*, 19, 1895 (1981).

11. S. A. Jenekhe and J. W. Lin, "Diffusion and Permeation of Vapors in Composite Polymer Thin Films," *Thin Solid Films*, 105, 331 (1983).

12. E. Khor, M. S. Gay, L. T. Taylor, and J. W. Wightman, "Moisture Effects in Lithium-Doped Polyimide Films," *J. Polym. Sci.: Polym. Chem. Ed.*, 23, 175 (1985).

13. D. K. Yang, W. J. Koros, H. B. Hopfenberg, and V. T. Stannett, "Sorption and Transport Studies of Water in Kapton Polyimide," *J. Appl. Polym. Sci.*, 30, 1035 (1985).

14. Semiconductor Measurement Technology, *ARPA/NBS Workshop on Moisture Measurement Technology for Hermetic Semiconductor Devices*, Washington, D.C., May 1981.

15. D. R. Carley, R. W. Nearhoof, and R. Denning, "Moisture Control in Hermetic Leadless Chip Carriers with Silver-Epoxy Die-Attach Adhesive," *R.C.A. Rev.*, 45, 279 (1984).

16. M. G. Kovac, D. Chleck, and P. Goodman, "A New Moisture Sensor for *In Situ* Monitoring of Sealed Packages," *I.E.E.E. Proc. Int. Reliab. Phys. Symp.*, 85 (1977).

17. J. B. Finn and V. Fong, "Recent Advances in Al_2O_3 *In Situ* Moisture Monitoring Chips for CERDIP Package Applications," *I.E.E.E. Proc. Int. Reliab. Phys. Symp.*, 10 (1980).

18. S. Hasegawa, "Performance Characteristics of a Thin-Film Aluminum Oxide Humidity Sensor," *I.E.E.E. Proc. Electron. Comp. Conf.*, 30, 386 (1980).

19. D. D. Denton, D. R. Day, D. F. Priore, S. D. Senturia, E. S. Anolick, and D. Scheider, "Moisture Diffusion in Polyimide Films in Integrated Circuits," *J. Electron. Mat.*, 14, 119 (1985).

20. D. H. Ziger, "*In Situ* Capacitance Studies of Thin Polymer Films During Compressed Fluid Extraction," *J. Mater. Res.*, 2, 884 (1987).

21. F. W. Dunmore, "An Improved Electric Hygrometer," *J. Res. Nat. Bur. Stand.*, 23, 701 (1939).

22. K. Otsuka, S. Kinoki, and T. Usui, "Organic Polymer Humidity Sensor," *Denshi-Zairyo*, 19, 68 (1980).

23. T. Seiyama, N. Yamazoe, and H. Arai, "Ceramic Humidity Sensors," *Sensors Actuators*, 4, 85 (1983).

24. R. E. Ruskin, Ed., *Humidity and Moisture: Principles and Methods of Measuring Humidity in Gases*, Vol. 1, Reinhold, 597 (1965).

25. K. W. Misevich, "Capacitance Humidity Transducer," *I.E.E.E. Trans. Ind. Electron. Contr. Instrum.*, Vol. IECI-16, No. 1, July (1969).

26. P. E. Thoma, J. O. Colla, and R. Stewart, "A Capacitance Humidity-Sensing Transducer," *I.E.E.E. Trans. Comp., Hybrids, Manuf. Tech.*, Vol. CHMT-2, 321 (1979).

27. G. Delapierre, H. Grange, B. Chambaz, and L. Destannes, "Polymer-Based Capacitive Humidity Sensor: Characteristics and Experimental Results," *Sensors Actuators*, 4, 97 (1983).

28. T. Tani, K. Horiguchi and T. Miwa, "Thin Film Humidity Sensor," *I.E.E.E. Proc. Electron. Comp. Conf.*, 163 (1983).

29. Y. Sudo, "Humidity Sensitive Ceramic Element," *Denshi-Zairyo*, 19, 74 (1980).

30. T. Yuki and Y. Yokomizo, "Highly Efficient Humidity Sensor," *Sensa-Gijutsu*, 1, 23 (1981).

31. T. Nitta, "Ceramic Humidity Sensor," *Ind. Eng. Chem. Prod. Res. Dev.*, 20, 669 (1981).

32. H. Taguchi, Y. Takahashi, and C. Matsumoto, "The Effect of Water Adsorption on $(La_{1-x}Sr_x)MnO_3$ $(0.1 \leq x \leq 0.5)$," *Yogyo-kyokai-shi*, 566 (1980).

33. R. K. Lowry, L. A. Miller, A. W. Jonas, and J. M. Bird, "Characteristics of a Surface Conductivity Moisture Monitor for Hermetic Integrated Circuit Packages," *I.E.E.E. Proc. Int. Reliab. Phys. Symp.*, 97 (1979).

34. D. Kane, R. Gauthier, M. Brizoux, and J. Perdrigeat, "Physical Characterization of Surface Conductivity Sensors in the Aim of an Absolute Moisture Measurement in Electronic Components," *I.E.E.E. Proc. Int. Reliab. Phys. Symp.*, 441 (1984).

35. H. Koelmans, "Metallization Corrosion in Silicon Devices by Moisture-Induced Electrolysis," *I.E.E.E. Proc. Int. Reliab. Phys. Symp.*, 168 (1974).

36. R. P. Merret and S. P. Sim, "Assessment of the Use of Measurement of Surface Conductivity as a Means of Determining Moisture Content of Hermetic Semiconductor Encapsulation," *ARPA/NBS Workshop on Moisture Measurement Technology for Hermetic Semiconductor Devices*, Washington, D.C., 94 (1978).

37. D. Kane and M. Brizoux, "Recent Developments on Moisture Measurement by Surface Conductivity Sensors," *I.E.E.E. Proc. Int. Reliab. Phys. Symp.*, 69 (1986).

38. G. D. Halsey, "Physical Adsorption on Non-Uniform Surfaces," *J. Chem. Phys.*, 16, 931 (1948).

39. B. D. Yan, S. L. Meilink, G. W. Warren, and P. Wynblatt, "Water Adsorption and Surface Conductivity Measurements on α-Alumina Substrates," *I.E.E.E. Proc. Int. Reliab. Phys. Symp.*, 95 (1986).

40. L. Binotto and G. F. Piacentini, "Analysis of Interdigitated Thin Film Capacitors," *Thin Solid Films*, 12, 325 (1972).

41. R. Esfandiari, D. W. Maki, and M. Siracusa, "Design of Interdigitated Capacitors and Their Application to Gallium Arsenide Monolithic Filters," *I.E.E.E. Trans. Microwave Theory Tech.*, 31, 57 (1983).

42. G. G. Zou, S. T. Kowel, and M. P. Srinivasan, "A Capacitance Sensor for On-Line Monitoring of Ultrathin Polymeric Film Growth," *I.E.E.E. Trans. Comp., Hybrids, Manuf. Tech.*, 11, 184 (1988)

43. P. D. Patel, "Calculation of Capacitance Coefficients for a System of Irregular Finite Conductors on a Dielectric Sheet," *I.E.E.E. Trans. Microwave Theory Tech.*, Vol. MTT-19, 862 (1971).

44. A. E. Ruehli and P. A. Brennan, "Efficient Capacitance Calculations for Three-Dimensional Multiconductor Systems," *I.E.E.E. Trans. Microwave Theory Tech.*, Vol. MTT-21, 76 (1973).

45. P. Balaban, "Calculation of the Capacitance Coefficients of Planar Conductors on a Dielectric Surface," *I.E.E.E. Trans. Circ. Theory*, Vol. CT-20, 725 (1973).

46. P. Benedek, "Capacitances of a Planar Multiconductor Configuration on a Dielectric Substrate by a Mixed Order Finite Element Method," *I.E.E.E. Trans. Circ. Syst.*, Vol. CAS-23, 279 (1976).

47. S. P. Sim and R. W. Lawson, "The Influence of Plastic Encapsulants and Passivation Layers on the Corrosion of Thin Aluminum Films Subjected to Humidity Stress," *I.E.E.E. Proc. Int. Reliab. Phys. Symp.*, 103 (1979).

48. J. H. Martin and L. D. Hanley, "Humidity Test of Premolded Chip Carriers," *I.E.E.E. Trans. Comp., Hybrids, Manuf. Tech.*, Vol. CHMT-4, 210 (1981).

49. R. G. Mancke, "A Moisture Protection Screening Test for Hybrid Circuit Encapsulants," *I.E.E.E. Trans. Comp., Hybrids, Manuf. Tech.*, Vol. CHMT-4, 492 (1981).

50. M. Iannuzzi and R. P. Kozakiewicz, "Bias Humidity Performance and Failure Mechanisms of Nonhermetic Aluminum SIC's in an Environment Contaminated with SO_2," *I.E.E.E. Trans. Comp., Hybrids, Manuf. Tech.*, Vol. CHMT-5, 345 (1982).

51. O. Nakagawa, I. Sasaki, H. Hamamurra, and T. Banjo, "The Relationship Between Moisture Resistance and Epoxy Molding Compounds in Integrated Circuits," *J. Electron. Mat.*, 13, 231 (1984).

52. L. T. Nguyen and C. A. Kovac, "Moisture Diffusion in Electronic Packages: I. Transport within Face Coatings," *S.A.M.P.E. Electron. Mat. Proc. Conf.*, 574 (1987).

53. L. T. Nguyen, "Moisture Diffusion in Electronic Packages: II. Molded Configurations vs. Face Coatings," *S.P.E. 46th Ann. Tech. Conf.*, 459 (1988).

54. N. Bakker, "In-line Measurement of Moisture in Sealed IC Packages," *Philips Telecom. Rev.*, 37, 11 (1979).

55. R. P. Merrett, S. P. Sim, and J. P. Bryant, "A Simple Method of Using the Die of an Integrated Circuit to Measure the Relative Humidity Inside its Encapsulation," *I.E.E.E. Proc. Int. Reliab. Phys. Symp.*, 17 (1980).

56. N. K. Annamalai and S. M. R. Islam, "Moisture Determination in IC Packages by Conductance Technique," *I.E.E.E. Proc. Int. Reliab. Phys. Symp.*, 61 (1986).

57. L. T. Nguyen and I. C. Noyan, "X-Ray Determination of Encapsulation Stresses on Silicon Wafers," *Polym. Eng. Sci.*, 28, 1013 (1988).

58. I. C. Noyan and L. T. Nguyen, "Residual Stresses in Polymeric Passivation and Encapsulation Materials," *Polym. Eng. Sci.*, 28, 1026 (1988).

59. A. J. Bush, "Measurement of Stresses in Cast Resins," *Modern Plastics*, 143, February (1958).

60. D. V. Steele, "Internal Stresses Developed in an Epoxy Resin Potting Compound During Long-Term Storage," *Polym. Eng. Sci.*, 280, October (1965).

61. H. W. Rauhut, "Low Stress and Live Device Performance of Experimental Epoxy Encapsulants," *S.P.E. Reg. Tech. Conf.*, Toronto, Canada, March (1983).

62. B. Jordan, "Lower Stress Encapsulants," *I.E.E.E. Proc. Electron. Conf.*, 130 (1981).

63. H. Dannenberg, "Determination of Stresses in Cured Epoxy Resins," *S.P.E. J.*, 669, July (1965).

64. F. D. Swanson, "The Use of Conformal Coatings to Relieve Encapsulation Stresses," *Insulation*, 42, February (1966).

65. R. C. Sampson, "A Three Dimensional Photoelastic Method for Analysis of Differential Contraction Stresses," *Exp. Mech.*, 3, 225 (1963).

66. "Standard Test Method for Shrinkage Stresses in Plastic Embedment Materials Using a Photoelastic Technique for Electronic and Similar Applications," *A.S.T.M. Standard F 100-71*, Reapproved (1981).

67. K. M. Liechti, "Residual Stresses in Plastically Encapsulated Electronic Devices," *Exp. Mech.*, 226 (1985).

68. K. E. Petersen, "Silicon as a Mechanical Material," *M.R.S. Symp. Proc.*, 76, 99 (1987).

69. K. E. Petersen and C. R. Guarnieri, "Young's Modulus Measurements of Thin Films Using Micromechanics," *J. Appl. Phys.*, 50, 6761 (1979).

70. T. P. Weihs, S. Hong, J. C. Bravman, and W. D. Nix, "Measuring the Strength and Stiffness of Thin Film Materials by Mechanically Deflecting Cantilever Microbeams," *M.R.S. Fall Meeting*, Boston, MA, November 28 - December 3 (1988).

71. D. E. Pulver, P. J. Ficalora, and E. W. Maby, "ULSI Metallization: The Effect of Stress as a Controlled Variable," *M.R.S. Fall Meeting*, Boston, MA, November 28 - December 3 (1988).

72. M. G. Allen, M. Mehregany, R. T. Howe, and S. D. Senturia, "Microfabricated Structures for In Situ Measurement of Residual Stress, Young's Modulus, and Ultimate Strain of Thin Films," *Appl. Phys. Lett.*, 51, 241 (1987).

73. H. Inayoshi, K. Nishi, S. Okikawa, and Y. Wakashima, "Moisture-Induced Corrosion and Stress on the Chip in Plastic Encapsulated LSIs," *I.E.E.E. Proc. Int. Reliab. Phys. Symp.*, 17, 113 (1979).

74. M. Isagawa, Y. Iwasaki, and T. Sutoh, "Deformation of Al Metallization in Plastic Encapsulated Semiconductor Devices Caused by Thermal Shock," *I.E.E.E. Proc. Int. Reliab. Phys. Symp.*, 18, 171 (1980).

75. S. Komatsu, K. Suzuki, N. Iida, T. Aoki, T. Ito, and H. Sawasaki, "Stress Sensitive Diffused Resistor Network for a High Accuracy Monolithic D/A Converter," *Int. Electron. Dev. Conf.*, 144 (1980).

76. R. J. Usell, Jr. and S. A. Smiley, "Experimental and Mathematical Determination of Mechanical Strains Within Plastic IC Packages and Their Effect on Devices During Environmental Tests," *I.E.E.E. Proc. Int. Reliab. Phys. Symp.*, 19, 65 (1981).

77. J. L. Spencer, W. H. Schroen, G. A. Bednarz, J. A. Bryan, T. D. Metzgar, R. D. Cleveland, and D. R. Edwards, "New Quantitative Measurements of IC Stresses Introduced by Plastic Packages," *I.E.E.E. Proc. Int. Reliab. Phys. Symp.*, 19, 74 (1981).

78. W. H. Schroen, J. L. Spencer, J. A. Brian, R. D. Cleveland, T. D. Metzgar, and D. R. Edwards, "Reliability Tests and Stress on Plastic Integrated Circuits," *I.E.E.E. Proc. Int. Reliab. Phys. Symp.*, 19, 81 (1981).

79. K. M. Schlesier, S. A. Keneman, and R. T. Mooney, "Piezoresistivity Effects in Plastic Encapsulated Integrated Circuits," *R.C.A. Rev.*, 43, 590 (1982).

80. D. R. Edwards, G. Heinen, G. A. Bednarz, and W. H. Schroen, "Test Structure Methodology of IC Package Material Characterization," *I.E.E.E. Trans. Comp., Hybrids, Manuf. Tech.*, Vol. CHMT-6, 560 (1983).

81. C. G. M. van Kessel, S. A. Gee, and J. J. Murphy, "The Quality of Die Attachment and its Relationship to Stresses and Vertical Die Cracking," *I.E.E.E. Trans. Comp., Hybrids, Manuf. Tech.*, Vol. CHMT-6, 414 (1983).

82. F. Shoraka, C. A. Gealer, and E. Bettez, "Research Reveals Differences in Coating Effects on Die Stress," *Semicon. Int.*, 110, October (1988).

83. P. Lundstrom and K. Gustafsson, "Mechanical Stress and Life for Plastic Encapsulated Large Area Chip," *I.E.E.E. Proc. Electron. Comp. Conf.*, 396 (1988).

84. S. A. Gee, V. R. Akylas, and W. F. van den Bogert, "The Design and Calibration of a Semiconductor Strain Gage Array," *I.E.E.E. Int. Conf. Microelectron. Test Structures* (1988).

85. S. A. Gee, W. F. van den Bogert, V. R. Akylas, and R. T. Shelton, "Strain Gage Mapping of Die Surface Stresses," *I.E.E.E. Proc. Electron. Comp. Conf.*, 343, Houston, May 22-24 (1989).

86. C. S. Smith, "Piezoresistance Effect in Germanium and Silicon," *Phys. Rev.*, 94, 42 (1954).

9
PHOTOTHERMAL ANALYSIS OF
THIN POLYMER FILMS

H. Coufal

Almaden Research Center
IBM Research Division
San Jose, California

Exposure of a sample to any form of radiation causes local heating via absorption and subsequent thermalization of energy. If the incident radiation is modulated or pulsed, a transient heat source is generated in the sample. Thermal waves and, due to thermal expansion, acoustic waves are launched into the sample under study and interact with the sample before being detected by one of the numerous detection schemes available nowadays. By changing the incident wavelength, spectra of the sample can be recorded readily. Scanning of the excitation across the sample allows mapping of optical, thermal, and acoustical properties of the sample under investigation. Any type of radiation can be utilized for the excitation of these phenomena. When using electromagnetic radiation, lasers are the preferred light source since they provide convenient access to an energy source of high spectral resolution, power, and, in the case of time-resolved experiments, well-characterized pulse shape. In addition, lasers with their parallel beam facilitate focusing of the energy into a small area. Other sources of energy that can be advantageous are particle beams, such as electron or ion beams.

9.1 PHOTOTHERMAL AND PHOTOACOUSTIC EFFECT IN THIN FILMS

In spectroscopy, a sample is excited with a tunable light source. In conventional spectroscopy, the intensities of the incident, transmitted, and reflected light are recorded as a function of wavelength. From the numerical difference, the light absorption in the sample is calculated. If the sample is a weakly absorbing thin film, incident and transmitted light intensities have to be measured with high precision to determine the small absorbed fraction of the incident light. For light-scattering samples, a conventional measurement of the diffuse reflectivity and the transmitted light for deriving the absorbed fraction of the incident energy is subject to a number of systematic errors. In photothermal detection schemes, however, only the energy absorbed in the sample contributes toward the signal, making photothermal techniques the method of choice for spectroscopic studies of thin films or samples that scatter light. Due to thermal diffusion, only the heat deposited within a thin layer of the surface of the sample will be detectable at the surface. This feature allows spectroscopy of weakly absorbing thin films on an absorbing substrate as well as optically opaque films to be studied using photothermal techniques. Spectroscopy of thin films is, therefore, an extremely important area of application for photothermal analysis.

Since thermal diffusion, sound propagation, and possibly, other energy transport processes, such as charge transport or excitonic processes, are involved in the generation of the signal, the associated material parameters can be studied.

Compared to more conventional thermal analysis methods, photothermal analysis has the advantage of noncontact generation of a well-defined heat source at the surface or in the volume of the sample of interest. With the high time resolution of some photothermal detection schemes, the transient temperatures at a sample surface or within the volume of the sample can be determined in real time with very high time resolution. This allows studies of thermal properties and the temperature dependence of other physical properties such as phase transitions or charge distributions. Due to the high sensitivities of many photothermal detection schemes, only very low intensities are needed for excitation. Because sample heating is negligible, temperature sensitive sample properties can be studied.

Laser-induced ultrasound allows the convenient noncontact generation of high-frequency ultrasonic waves in a sample. Compared to conventional ultrasonic transducers, photothermal generation is always noncontact and free of transducer ringing. Coupling of the sound pulse from the transducer to the sample (subjected to acoustic impedance matching, etc.) in the case of a transducer is no longer a problem with laser-generated ultrasound since the sound is generated in the sample! This feature might be convenient in studying bulk samples. For the ultrasonic analysis of thin films, the wavelength of the probing pulse should be substantially less than the sample thickness. In this case, techniques based on conventional ultrasonic transducers are cumbersome, to say the least, due to the above mentioned acoustical problems. With a sufficiently short and powerful laser pulse, absorbed in a thin sample, ultrahigh ultrasonic frequencies are generated easily, making photoacoustic analysis of thin films relatively straightforward.

In imaging, a large number of techniques compete with photothermal imaging and microscopy. Optical microscopy is an extremely well-developed, powerful, and mature analytical technique. The signal-to-noise ratio in microscopy is excellent and parallel recording and processing of a frame is state of the art. Quite clearly, photothermal imaging with its relatively low signal to noise ratio and the complicated signal generation (and hence contrast mechanism) will never be able to compete with light microscopy. Thermal wave and ultrasonic imaging of thin films are, however, two niches where photothermal imaging is becoming a powerful tool.

One of the key features of photothermal excitation is the fact that it does not require extensive sample preparation or physical contact with the sample. The same is true for a number of other detection schemes. In many applications, excitation and detection of the signal can be accomplished on the same side of the sample enabling single-sided, remote monitoring of the sample under study. This feature makes photothermal analysis a good candidate for industrial and even military applications.

A number of comprehensive books (1-4) have reviewed the field of photoacoustics and photothermal phenomena. In addition, excellent review articles (5-9) have surveyed the field. Moreover, updated information on the status of the field can be found in recent special issues of journals that were dedicated to photoacoustics (10,11) and in the proceedings of international conferences on photoacoustics (12,13). The focus of this chapter will be to complement this substantial body of literature and to critically assess the potential of photothermal and photoacoustic methods for the analysis of thin polymer films. References will be limited to most recent developments in this rapidly growing field.

The physical principles of the signal generation process and selected detection schemes are briefly reviewed. Various detection schemes are compared in detail. Typical applications in thermal analysis, ultrasonic testing, and imaging of thin films are discussed. Examples highlighting the advantages and drawbacks of photothermal and photoacoustic techniques are then analyzed, showing the reader how the potential of photothermal techniques can be assessed for the analysis of his or her thin film materials.

9.1.1 Signal Generation Process

In a typical photothermal experiment (Fig. 9.1), a homogeneous thin film of thickness T is excited with a modulated or pulsed light source. The light absorption in the sample can be characterized by a wavelength-dependent optical absorption length $A(\lambda)$

$$A(\lambda) = \frac{1}{\beta(\lambda)} \tag{1}$$

with $\beta(\lambda)$ being the commonly used optical absorption coefficient of the sample. Due to radiationless processes, part of the absorbed energy is released as heat. With the incident energy being either modulated or pulsed, the heat generation will show a corresponding time dependence. Via heat diffusion, then, a temperature profile develops in the sample. For a heat source with a modulation frequency f, the heat diffusion can be described by the thermal diffusion length D

$$D(f) = \sqrt{\frac{k}{\pi \rho c f}} \tag{2}$$

where k is the thermal conductivity, ρ the density, and c the specific heat of the sample. Defining the thermal diffusivity as

$$\alpha = \frac{k}{\rho c} \tag{3}$$

the thermal diffusion length can be expressed in terms of this material parameter and the modulation frequency:

$$D(f) = \frac{\sqrt{2\alpha}}{2\pi f} \tag{4}$$

The amplitude of a thermal wave is attenuated by a factor of $e^{-2\pi}$, that is, 2×10^{-3}, when diffusing one diffusion length (14). Therefore, only the heat generated within one thermal diffusion length of a sample surface will be able to reach this surface. If the sample is in contact with another medium, thermal waves are reflected and transmitted at the interface. Heat diffusion extends into the second medium according to Eq. (2) or (4) with the material parameters of that medium instead of the sample

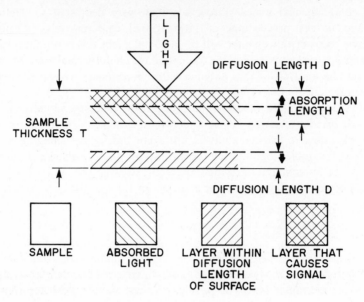

Figure 9.1: Schematic of the photothermal signal generation process. For definitions of the absorption and diffusion length, please see the text.

properties. For pulsed excitation, the thermal transit time τ is the relevant parameter. The temperature at a distance ℓ from a pulsed heat source reaches its maximum at a time τ after the excitation. For one-dimensional heat flow, the thermal transit time is given by

$$\tau = \frac{\ell^2}{2\alpha} \tag{5}$$

The thermal diffusion length, $D(t)$, which is defined as the square root of the mean-square distance of thermal energy from the location of a transient point source of heat, is given by

$$D(t) = \sqrt{2\alpha t} \tag{6}$$

at a time t after the heat pulse is generated. For modulated continuous wave (CW) radiation, the thermal diffusion length $D(f,t)$ in its frequency dependent definition, Eq. (2) or (4), is the most convenient way to analyze the observed signals. For pulsed excitation, Eq. (6) is the equivalent time domain formula.

The transient temperature profile in the sample (and, when applicable, in the adjacent medium as well) is accompanied by a stress profile due to thermal expansion. Sound waves are hence generated in the sample, at the surfaces or interfaces as well as in the adjacent materials. These sound waves, being essentially unattenuated in the frequency range considered here, can then propagate over long distances. Additionally, if the heat generation is sufficiently localized, the surfaces or interfaces buckle slightly.

All detection schemes that detect the temperature at the sample surface directly will have to deal with the peculiar character of thermal waves. Due to the fact that thermal waves are critically damped,

- the detector has to probe the temperature within one thermal diffusion length of the excited area and

- only the energy deposited within this distance is effective in signal generation.

The first restriction can be overcome by indirect detection of the surface temperature change. Because the sound waves generated photothermally propagate essentially unattenuated over large distances, an acoustic detector can therefore be placed far away from the illuminated area of the sample. For moderate light intensities, that is, below the threshold intensity required to ablate the sample material, the amplitude of the observed acoustic signal will depend on the temperature distributions in the heated sample volume and the adjacent medium. Such distributions are determined by the thermal diffusion length $D(f, t)$ and the optical absorption length $A(\lambda)$. The sound wave serves as a carrier of the thermal information. Due to this diffusive character of thermal waves, the information obtained from the radiation-induced thermal and acoustic waves is completely different from conventional optical spectra. Some of the unique features and problems of photothermal and photoacoustic spectroscopy are due to this difference.

In an optically opaque sample of thickness $T > A(\lambda)$, such as shown in Fig. 9.1, the radiation-induced heating can always be observed at the illuminated side of the sample. However, a temperature increase on the back side is only noticeable when the light is absorbed within one thermal diffusion length of the back side, that is, when the condition $T - A(\lambda) < D(f, t)$ is met. This condition is a function of the material constants of the sample such as $\beta(\lambda)$, k, ρ, and c and of an experimental variable, the modulation frequency f. As mentioned above, an acoustic transducer will be able to detect a signal on either side of the sample. On the back side, this can be done even if there is no sizable temperature fluctuation reaching that side! For photoacoustic spectroscopy, sample thickness is, therefore, only a minor concern. It is, however, important to remember that the sound wave serves only as a carrier of the thermal information. The amplitude of the observed acoustic signal will depend on the temperature distributions in the heated sample volume and the adjacent medium, both determined again by the thermal diffusion length $D(f, t)$ and the optical absorption length $A(\lambda)$. For spectroscopy, therefore, the relation between these two parameters has to be analyzed.

When the thermal diffusion length $D(f, t)$ is larger than the optical absorption length $A(\lambda)$ $[D(f, t) > A(\lambda)]$, and when the sample thickness T is larger than the absorption length $[T > A(\lambda)]$, all the incident energy is absorbed and contributes toward the observed signal. In this case, a small variation of the absorption coefficient does not change the observed signal! The signal is *saturated* and proportional to the incident energy and $(1-R)$ with R being the reflectivity of the sample. This condition can be avoided by using a thin sample or by increasing the modulation frequency until $D(f, t) < A(\lambda)$ is achieved.

A sample with an optical absorption length smaller than the sample thickness $A(\lambda) < T$ absorbs all the incident light, hence no light is transmitted. Such an opaque sample is not accessible to conventional transmission spectroscopy. If the thermal diffusion length $D(f, t)$ is, however, smaller than the optical absorption length $A(\lambda)$, $D(f, t) < A(\lambda)$, the energy deposited within this thermal diffusion length is proportional to the absorption coefficient β of the sample. Due to the fact that the thermal diffusion length decreases with increasing modulation frequency according to Eq. (2), this condition can, at least in principle, be fulfilled by modulating at a sufficiently high frequency. Photothermal spectroscopy thus allows recording of *spectra for opaque samples.*

At very high modulation frequencies, only the light absorbed in the surface layer contributes to the signal. When using a lower modulation frequency, the same surface layer and the adjacent layer generate the signal. By comparing spectra at various modulation frequencies, surface absorption can be readily distinguished from bulk absorption. In principle, the *depth profile* of a layered structure can be obtained in a nondestructive manner. The penetration depth or range of this profiling technique is determined by the thermal diffusion length at the lowest modulation frequency. The resolution, however, is limited by the highest modulation frequency. The bandwidth of the detection scheme together with the thermal diffusivity of the sample under study limits the depth profiling capability of this technique.

In conventional optical transmission spectroscopy, the path length along which absorption occurs has to be known to allow determination of the absorption coefficient, a task that can be cumbersome for rough samples, fibers, or powders. In photothermal and photoacoustic spectroscopy, however, it is sufficient to assure that the thermal diffusion length is smaller than the relevant sample dimension such as grain size or fiber diameter, to obtain a qualitative absorption spectrum. For many practical applications, a qualitative analysis is sufficient and *no sample preparation* is required. However, extensive calibration procedures are required for quantitative analysis.

In spectroscopy, samples with an absorption coefficient that is a function of wavelength are, of course, of particular interest. Different parts of the spectrum might have an absorption length $A(\lambda)$, possibly varying from larger than the selected thermal diffusion length to much shorter (Fig. 9.2). In this case, parts of the spectrum where $D(f, t) > A(\lambda)$, that is, with a large absorption coefficient β, will be saturated. On the other hand, the thermal diffusion length $D(f, t)$ is a function of the modulation frequency and the thermal properties of the sample. According to the above consideration, spectra of the same sample will be dependent on the modulation frequency (Fig. 9.3). Furthermore, spectra of samples with identical optical but different thermal properties might be quite different. The same holds true for samples with identical light absorption but different fluorescence yields (and therefore different heat generation). Thus, photothermal and photoacoustic spectroscopy thus requires particular care in the selection of the modulation frequency, the consideration of thermal sample parameters and thickness, and an understanding of the signal generation process.

9.1.2 Detection Methods

Radiation-induced transient effects, thermal as well as mechanical effects, have been studied for quite some time. Determining the temperature or the energy of a sample has traditionally been the domain of thermometry and calorimetry. Radiation-induced mechanical effects were studied using interferometry or ultrasonic transducers. In the past few years, interest in radiation-induced thermal and acoustic

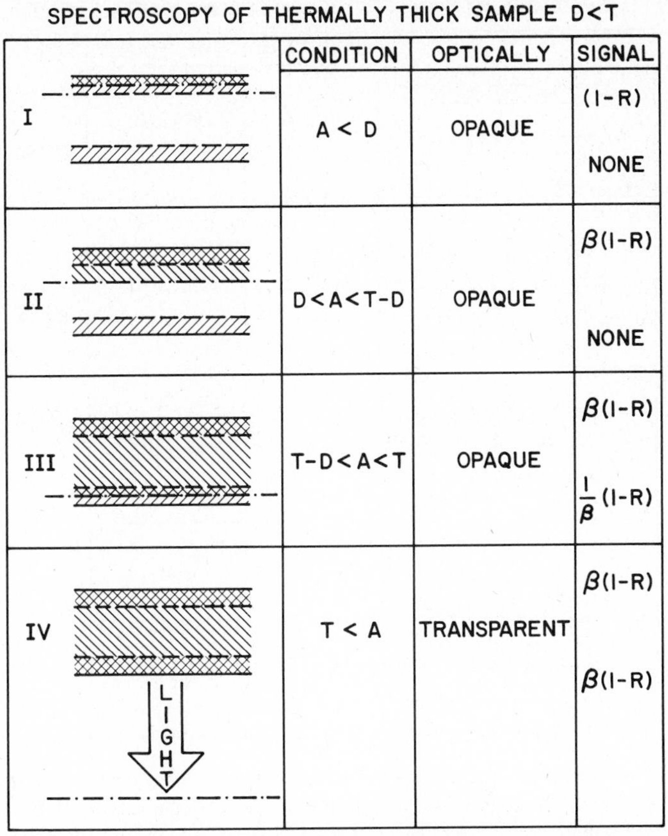

SPECTROSCOPY OF THERMALLY THICK SAMPLE D<T

		CONDITION	OPTICALLY	SIGNAL
I		$A < D$	OPAQUE	$(1-R)$ NONE
II		$D < A < T-D$	OPAQUE	$\beta(1-R)$ NONE
III		$T-D < A < T$	OPAQUE	$\beta(1-R)$ $\frac{1}{\beta}(1-R)$
IV		$T < A$	TRANSPARENT	$\beta(1-R)$ $\beta(1-R)$

LIGHT

Figure 9.2: Schematic of the photothermal signal generation process in a thermally thick sample as a function of absorption length A. Shown is the relationship between thermal diffusion length D and sample thickness T for an increasing A (I-IV). For these cases, characterized by the relationship between the parameters, the variation of the photothermal signal with optical reflection coefficient R and absorption coefficient β of the sample is given for detection on front and back side, respectively. See Fig. 9.1 for the definition of the symbols and the text for more details.

processes increased substantially, largely due to scientific and industrial applications of lasers. A number of classical detection schemes were adapted for this particular type of application and new detection methods were also developed. Detectors tie into various stages of the signal generation chain described above. In photothermal detection schemes, the radiation-induced temperature increase in the sample or at the sample surface is monitored by measuring either the temperature directly or a temperature-dependent property of the sample or of the adjacent medium. Photo- and optoacoustic methods detect the acoustic waves caused by the radiation-induced heating of the sample itself or a gas or a liquid that is in thermal and/or acoustical contact with the sample. For experiments with the emphasis on surface temperature, classical temperature sensors such as thermocouples are clearly the method of choice

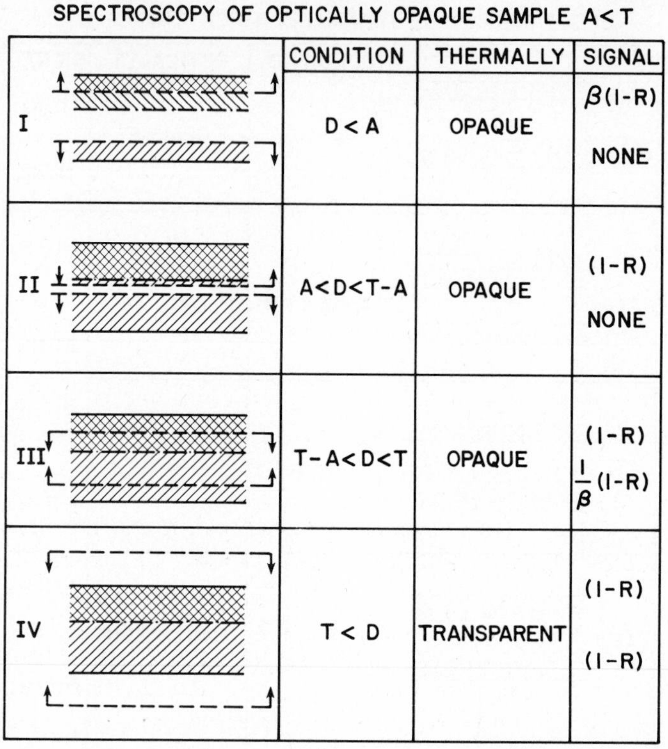

SPECTROSCOPY OF OPTICALLY OPAQUE SAMPLE A < T

		CONDITION	THERMALLY	SIGNAL
I		D < A	OPAQUE	$\beta(1-R)$ NONE
II		A < D < T - A	OPAQUE	$(1-R)$ NONE
III		T - A < D < T	OPAQUE	$(1-R)$ $\dfrac{1}{\beta}(1-R)$
IV		T < D	TRANSPARENT	$(1-R)$ $(1-R)$

Figure 9.3: Schematic of the photothermal signal generation process in an optically opaque sample as a function of thermal diffusion length D. Shown is the relationship between optical absorption length A and sample thickness T for an increasing D (I-IV). For these cases, characterized by the relationship between the parameters, the variation of the photothermal signal with optical reflection coefficient R and absorption coefficient β of the sample is given for detection on front and back side, respectively. See Fig. 9.1 for the definition of the symbols and the text for more details.

because they are easy to calibrate and convenient to use. For spectroscopic measurements, however, sensitivity and convenient coupling of the detector to the sample under study are the main concern.

The transient heating of a pyroelectric material can, for example, be detected via the induced electrical charge or current. In fact, many commercially available laser power meters function on this basis. A variation of this technique uses a *pyroelectric calorimeter* as a substrate for the sample of interest. The thermal wave, generated at the front side of the sample, is detected on the back side with the pyroelectric detector. The sample has to be thermally thin [i.e., $T - A(\lambda) < D(f, t)$] to take advantage of this detection scheme. For a very thin sample, extremely short thermal transit times can be obtained according to Eq. (5). Utilizing a pyroelectric thin film calorimeters (15), a sensitivity of nanojoules combined with a time resolution of nanoseconds was achieved recently. This technique is, however, restricted to pyroelectric samples or thin films that can be directly deposited onto a pyroelectric substrate. Work currently under way will eventually allow the deposition of a thin pyroelectric detector film on top of the substrate under study. *Thin film resistance thermometers* which are directly deposited onto the sample of interest (16) compete with pyroelectric temperature sensors as far as time resolution is concerned. *Thin film thermocouples* are another type of temperature sensors for photothermal applications (17,18). These three sensors have the advantage that they are well understood and can be readily calibrated in absolute units. Their main drawback is that they require thermal contact with the sample and have to be deposited directly onto the sample of interest.

In contrast, *photothermal radiometry* can be utilized with bulk samples and is in addition a noncontact technique. Blackbody radiation from the sample is imaged onto a suitable infrared detector (19). A change in surface temperature then effects a change of the observed signal. Due to the Stephan-Boltzmann law, the total power of the blackbody radiation is proportional to the fourth power of the temperature of the sample. The radiometry signal, therefore, increases dramatically with increasing base temperature of the sample. This unique feature together with the fact that remote, single ended, noncontact probing of a moving sample can be readily implemented make photothermal radiometry the ideal choice for process monitoring of thin films in vacuums systems or in hostile atmospheres or at high temperatures. Infrared detectors require, however, cryogenic cooling to reach acceptable sensitivities. High sensitivity of the detector is typically accompanied by low time resolution. A sizable number of detection schemes are based on particles emitted from the sample surface. Electrons (20) and atoms (21) emitted from a surface that is heated by a short laser pulse have been analyzed to derive the time-dependent surface temperature of the sample. Laser-induced desorption of adsorbates (22,23) from a surface or the analysis of the products of a photothermal reaction during laser-assisted chemical deposition of a metal (24) have been used successfully to measure surface temperatures. Most of these methods are of limited general utility: most of them require ultrahigh vacuum conditions, are restricted to few types of samples, and are surface specific.

A large number of techniques utilize a *probe laser* to detect thermal effects caused by another light source. The power of the probe laser is typically orders of magnitude smaller than that of the excitation source and different wavelengths are commonly used. A change in temperature of the sample or the adjacent medium is associated with a change in the refractive index of that material. This change in refractive index causes a change in the reflection or transmission of the probe laser (25). Recently, variations of this technique (26,27) achieved a time resolution of the order of 10 ps! The refractive index gradient caused by the temperature gradient in the sample or the adjacent medium forms a transient thermal lens capable of deflecting a probe laser (28). The same is true for the surface buckling due to localized heating (29). In a similar fashion, acoustic wave fronts give rise to refractive index gradients. These as well as surface displacement due to acoustic waves can be probed by lasers or other optical techniques (30). All probe laser techniques have the advantage of optical excitation and probing, and are therefore noncontact techniques. They require, however, careful alignment of two lasers, and optically flat samples, which are fairly easy to meet with thin films vacuum deposited on flat substrates. A large number of publications in the semiconductor area have been using photothermal deflection schemes to determine properties of thin films on wafer substrates.

Piezoelectric transducers (31) attached to the sample under study convert the sound waves that are generated in the sample into an easily recorded electrical signal. Their main drawbacks are the requirement for good mechanical contact with the sample, and problems inherent in acoustic detection such as matching of the acoustic impedance of the sample and transducer, susceptibility to acoustic noises and trade-offs between sensitivity and time resolution. Transducers using a change in capacitance or inductance (32) between the sample surface and a reference electrode or inductor to monitor the surface displacement overcome many of the disadvantages of conventional piezoelectric ultrasonic transducers. The main area of application of ultrasonic detection of photothermal transient signals is in ultrasonic material testing for weakly absorbing samples such as adsorbates or transparent materials. The same holds true for special transducers employed for the detection of surface acoustic waves (33). If the sample can be in contact with a gas atmosphere, *microphones* are frequently used for detection (see, for example, Ref. 2). Besides requiring a gas-filled cell to contain the sample, the frequency range of microphones is rather limited, and suppression of acoustic noises poses restraints on such systems.

An evaluation of the pros and cons of the major photothermal and photoacoustic detection schemes shows that each of the detectors has its merits, making it the prime choice for certain applications or a particular type of sample. As should be evident from the above discussion, the signal generation and detection process can be rather complicated and may involve a large number of individual processes. A substantial loss in sensitivity and time resolution is associated with each diffusion or conversion process. These losses become particularly significant when detection occurs at the end of the signal generation chain. Microphone detection, in particular, has low signal generation efficiency and time resolution. It requires the most elaborate theoretical models for the quantitative interpretation of the observed signals. However, even with these limitations, high signal-to-noise spectra can be obtained using a microphone. The spectra can be readily interpreted due to the long history and large body of literature associated with this approach. If the highest time resolution is required for studies of extremely thin films, pyro- and piezoelectric detection schemes excel. Photothermal deflection schemes seem to be most popular in all applications dealing with semiconductor wafers and thin films on these wafers.

9.1.3 Instrumentation

A typical photothermal setup comprises a suitable light source, the detector, and signal recovery electronics (Fig. 9.4). It is important to keep in mind that the signal generation process involves optical, thermal, and possibly acoustic properties of the sample.

In spectroscopy, it is important to eliminate thermal and acoustic artifacts in the recorded spectra. To account for the wavelength dependence of the source, the thermal parameters of the sample, and the characteristics of the detector, a reference sample is normally employed. Reference data are obtained in a single beam spectrometer before or after the sample spectrum is taken or, in a dual beam arrangement, simultaneously with the sample spectrum. Due to the strong influence of the modulation frequency, it is imperative to record both data sets at the same

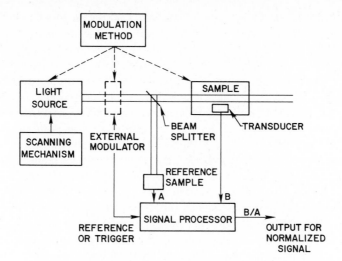

Figure   9.4:   Schematic of a photothermal spectrometer.

modulation frequency with samples of identical or well-known thermal characteristics.

For thermal or acoustic analysis of the sample, it is necessary to eliminate the influence of optical sample parameters and to distinguish between thermal and acoustical signals. The thermal diffusion process is quite different from sound propagation: thermal waves are critically damped within one wavelength to less than 1% of their initial amplitude while sound waves propagate over long distances at the frequencies considered here. The thermal signal can therefore be readily suppressed by a simple delay line. When this is not feasible, the well-defined transit time of a sound pulse can then be utilized to discriminate electronically against the thermal signal by time-gating techniques. Thermal diffusion across an interface depends on the thermal diffusivities of both materials. Sound transmission, on the other hand, is a function of their respective acoustic impedances. Couplers and filters can be designed that transmit only thermal energy. Another important difference between thermal and sound waves of the same frequency is the much longer wavelength of sound waves. For most thin film applications, the wavelength of the interrogating wave should be of the order of the film thickness. The ultrasonic analysis of a film requires, therefore, considerably higher modulation frequencies than the thermal analysis of the same film. The cost of high-frequency electronics increases dramatically with increasing frequency. In comparison, photothermal analysis is a less costly alternative.

The light source in a typical experiment can be intensity modulated by suitable means such as current modulation of a lamp or a mechanical chopper. Wavelength or polarization modulation is advantageous when small absorptions superimposed on a large background absorption are of interest. When applicable, one can use modulation of the absorption properties of the sample (Stark modulation) or other physical parameters (pressure) that allow modulation of the sample characteristics.

Mostly, sinusoidal or square wave modulation of the incident light intensity is employed (1-4). In this case, a lock-in amplifier is the adequate tool for signal recovery. It should be emphasized that if a source of constant intensity is modulated this way the energy per excitation cycle decreases with increasing modulation frequency. The signal at high modulation frequencies, which are desirable because of the above considerations on saturation, will then be extremely weak. Excitation with a short light pulse has the advantage of generating a higher surface temperature than with the same energy in a longer pulse or a periodic excitation, due to the fact that heat loss during the short pulse can be neglected. This results in a superior signal-to-noise ratio but might damage the sample irreversibly or affect temperature sensitive sample properties. With pulsed excitation, box-car integrators are frequently used to monitor the signal in a small time window (34). Transient digitizers allow acquisition of the complete time domain signal and, if desired, transformation into the frequency domain (35). Recently, another modulation scheme is emerging. It is based on noise modulation (36) of the light source and transient digitizers for data recording. Time domain type results can be obtained from the data by cross correlation (37) and frequency domain data by Fourier analysis (35). The advantage of this new technique is that many modulation frequencies are probed simultaneously and therefore depth profiles can be obtained in a small fraction of the time required for recording the spectra subsequently at several modulation frequencies. Assuming a light source of constant intensity that is modulated by an external modulator, noise modulation makes more efficient use of the light intensity than any other modulation scheme.

Photoacoustic detection is frequently employed in FT-IR studies of thin films or powders. Here, the intensity modulation due to the interference fringes can be utilized when a commercial rapid scanning FT-IR spectrometer is used as the light source. One should, however, be aware of the fact that this results in a much lower modulation frequency for the long wavelength side of the spectrum as compared to the short wavelength part of the same spectrum. Therefore, a spectrum might be partially distorted for high wave numbers due to saturation!

With conventional light sources, only a limited spectral, time and, depth resolution can be obtained. Nevertheless, high sensitivity, instrumental simplicity, and simple sample preparation procedures make photothermal detection an interesting alternative to more established techniques, such as diffuse or internal reflectance spectroscopy (38). When combined with the high spectral and time resolution possible with laser excitation, photothermal detection, however, offers a unique combination of advantages which are not available with other techniques This is evident for spectroscopy, but is also true for thermal and ultrasonic analysis of thin films when using photothermal effects to generate the probe that interrogates the sample under study. In the following paragraph, examples will be presented along with some of the above features.

9.2 SPECTROSCOPY OF THIN FILMS

Due to the peculiar signal generation and detection process, photothermal spectroscopy has several unique features. It is quite different from conventional transmission spectroscopy. Photothermal spectra represent only the heat released within the thermal diffusion length of the sample surface. As outlined above, this makes photothermal spectroscopy the method of choice for thin film studies. Additional advantages of photothermal detection methods are that they are not affected by scattered light and the depth-profiling capability. If these techniques are so powerful, why are they not dominating in spectroscopy? Quite clearly, the fact that thermal and acoustic properties of the sample affect the observed spectra complicate the interpretation of the data considerably. Reference samples with well-defined optical, thermal, and acoustical properties are required (39) and photothermal spectra of two samples with identical optical absorption are identical only if the samples have identical thermal and acoustic properties, the same depth profile, as well as the same quantum efficiency for radiationless deexcitation, all recorded at the same modulation frequency. These complications let photothermal spectroscopy complement conventional techniques but not replace them (40).

9.2.1   Homogeneous Films

Spectroscopy of thin polymer films is an area where photothermal detection methods can be advantageous. An example is the first direct measurement of the optical absorption of polyacetylene (41). As a result of its complicated fibrillar morphology, polyacetylene is a very complex optical medium. Optical absorption spectra, which can be compared with various theoretical models, had to be inferred from transmission spectra of thin films. It was found that these films scatter several percent of the incident radiation diffusely. Transmission measurements cannot distinguish between scattering and absorption. Therefore, the absorption constants derived from transmission measurements, even when using specular or diffuse reflectance data for correction, can exhibit systematic errors that make a comparison between various theoretical models almost arbitrary. Photothermal deflection spectra provided the first reliable measurement of the absorption edge of polyacethylene and showed a Urbach type behavior.

Electrode-supported films are of considerable interest in electrochemical studies. One problem with this type of sample is the fact that the sample is deposited onto a metal. Since the amplitude of the light intensity vanishes close to the metal, conventional spectroscopies suffer from lack of sensitivity. This effect was clearly seen in FT-IR studies of Prussian Blue and cupric hexacyanoferrate on various electrodes (42). In another experiment (43), the distance between a silver surface and the chromophore was varied using a varying number of transparent Langmuir-Blodgett spacer layers. Thus, the amplitude of the electric field vector of the standing light wave in front of the silver was actually mapped!

Many of the problems in electrochemistry are usually addressed ex situ with surface scientific methods requiring ultrahigh vacuum. Ex situ transmission and reflection measurements complement these experiments. To ensure that sample preparation is not introducing artifacts, in situ experiments are desirable. Recently, a group was successful in recording photoacoustic spectra in situ for both growing and fully grown electrochemical films (44).

9.2.2   Spectroscopy of Layered Films

Photographic color reversal films with their intricate but well-defined layered structures have served as model systems for the demonstration of the depth-profiling capability of photoacoustic (45) and photothermal (46) detection schemes. As shown in

Fig. 9.3 with a decreasing modulation frequency, a thicker layer of the sample contributes toward the observed signal. With a sample such as a color reversal film, the observed spectra depend strongly on the modulation frequency and would, at least in principle, allow the reconstruction of the depth profile, that is, optical density as a function of wavelength and depth, of the sample.

9.2.3 Nonradiative Quantum Yield

Absorption spectroscopy is the prevalent application of photothermal and photoacoustic spectroscopy. Since radiationless deexcitation of the electronic excited state is a key step in the signal generation chain, the relevant quantum yield for nonradiative processes can be determined. To demonstrate the determination of nonradiative yields and to highlight the frequency dependence of the spectra, let us consider a trilayered sample such as the one shown in Fig. 9.5, indicating a weakly absorbing layer 1 separated from another absorbing layer 3 by a transparent spacer layer 2. In the example discussed here, layer 1 contained Nd_2O_3 molecules, whereas layer 3 was a silver film. The spectra shown in Figs. 9.6 and 9.7 are recorded at two different modulation frequencies with a thermal detector at the back side of the sample (47). At a low modulation frequency, an absorption-like spectrum is observed (Fig. 9.6). Light absorbed by the Neodymium ions is converted into heat. This heat diffuses across the sample and is detected on the back side. At high modulation frequencies (Fig. 9.7), the much shorter thermal diffusion length does not allow heat generated by the Neodymium ions to reach the detector. Only light that is transmitted by the Nd_2O_3 layer and then absorbed by the silver film will contribute to the observed signal. The observed spectrum is, therefore, a transmission spectrum of the Neodymium film detected by a silver absorber with subsequent thermal detection. These two extreme cases correspond to Fig. 9.3 which involves a homogeneous sample. If the structure of the film or its thermal properties are known, additional information can be derived from these spectra. The thermal diffusivity or thickness of the spacer layer can be derived from a comparison of the signals. A more important parameter is the quantum yield for radiationless deexcitation. In the transmission spectrum, Fig. 9.7, light that has been **absorbed** by the Neodymium ions is missing. In the absorption spectrum, only that fraction of the light that has been **absorbed and converted into heat** contributes to the signal. The ratio of both signals represents the probability for nonradiative decay of the optically excited state. In the above example, this quantum yield turned out to be of the order of 90%. Since both signals are measured with the same detector, no calibration of the detector is required for the determination of absolute quantum yields. This technique allows the measurement of quantum yields of thin films and even monolayers (43).

9.3 THERMAL ANALYSIS OF THIN FILMS

Photoacoustic and photothermal spectroscopy can be cumbersome due to the thermal diffusion step in the signal generation process. In the thermal analysis of thin polymer films, however, the same process is advantageous. Photothermal excitation of a light source facilitates the implementation of a well-defined heat source for the probing of the thermal properties of thin films. The heat source does not require mechanical or thermal contact with the sample and can be tailored to the application by simply changing the temporal shape or wavelength of the light source and by

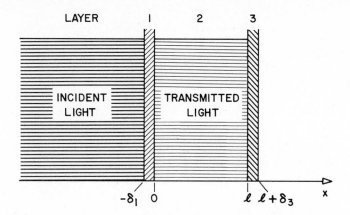

Figure   9.5:   Schematic of composite trilayered sample.

changing the size and location of the illuminated area. Combined with the high sensitivity and time resolution of many photoacoustic and photothermal detection
schemes, thermal analysis of temperature sensitive properties of extremely thin films
becomes viable.

9.3.1 Thermal Diffusivity

The thermal diffusivity of a thin film can be determined by generating a transient
heat source at the front side of the sample and by monitoring the subsequent temperature increase on the back side. When using amplitude modulation of a CW light
source to generate the heat source, the amplitude or phase of the transmitted temperature wave can be analyzed (48). For a film with thickness ℓ, an amplitude decrement $\tilde{D}$

$$\tilde{D} = \ln \frac{\Theta(0)}{\Theta(\ell)} \tag{7}$$

is observed where $\Theta(0)$ and $\Theta(\ell)$ are the amplitudes of the temperature oscillation on
the front and back side of the sample. For a thin film of diffusivity α, the frequency
dependence of the amplitude decrement is given by

$$\tilde{D} = \frac{\ell}{\sqrt{\alpha}} \times \sqrt{\pi f} \tag{8}$$

In addition, a phase lag

$$\Delta = \Phi(\ell) - \Phi(0) \tag{9}$$

is observed between the front and the back sides, with Φ being the phase angle
between the modulated light source and the observed temperatures. In here, $\Delta = \tilde{D}$.
Figure 9.8 shows the typical result for the modulation frequency response of the
amplitude decrement of a thin plastic film. From the slope of the curve, the product
of thermal diffusivity times ℓ^2 can be derived. If the film thickness is known, the

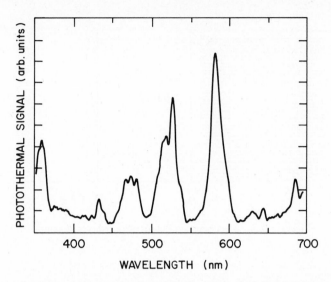

Figure 9.6: Normalized photothermal absorption spectrum of 0.8×10^{15} Nd_2O_3 molecules in a 1-μm-thick PMMA film coated on top of a 0.1-mm undoped PMMA cast on a silver substrate. Recorded by out-of-phase detection at 2.2-Hz modulation frequency (69).

thermal diffusivity is determined. The data in Fig. 9.8 show also the typical increase of the statistical error at higher modulation frequencies. Phase lag measurements (Fig. 9.9) have considerably better signal-to-noise ratio. For very thin films, however, their resolution is insufficient. The answer to this problem is to measure at higher frequencies. Data for the same films as in Fig. 9.9 but recorded in the time domain with pulsed excitation are shown in Fig. 9.10. From these data, the thermal diffusivities of thin films can be derived using numerical models. Experimental requirements for photoacoustic measurements of the thermal diffusivities of thin films, first introduced by Adams and Kirkbright (49), are modest. Due to the limited bandwidth of microphone detection, photothermal methods are currently dominating the field. With the sample directly coated onto a pyroelectric sensor, the highest time resolution can be obtained for extremely thin films (50-52). This technique involves, however, special sample preparation techniques. Other methods typically require optical access to the front and back sides of the sample (53). This problem can be overcome by measuring the lateral heat flow in the sample (52). Still another way is to determine the surface temperature as a function of time while heat diffuses from the surface of the sample into the interior. Photothermal radiometry (54,55) and photothermal deflection schemes (56,57) have demonstrated their utility in this type of non-contact, single-ended measurement. Photoacoustic schemes can even be adapted for samples with inaccessible back sides (58). Corrections for the geometry of the cell improve the accuracy considerably (59).

Laser excitation combined with one of the photothermal detection schemes facilitate measurements of thermal transport coefficients in thin films. Whereas frequency domain experiments are very much straightforward in their interpretation,

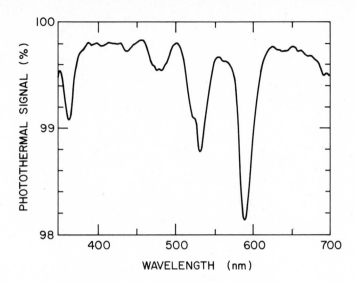

Figure 9.7: Normalized photothermal transmission spectrum of the sample described in Fig. 9.6. Recorded at 88-Hz modulation frequency (69).

time domain experiments with pulsed light sources for excitation require extensive analytical or numerical efforts to derive transport coefficients from the observed data.

9.3.2 Film Thickness

The above discussion on thermal diffusivity assumed that the thickness of the sample can be derived with other methods. But quite clearly, Eq. (8) can be utilized to determine the thickness of a sample with known thermal diffusivity. A large number of more conventional techniques, many of them mechanical or optical, are available to measure film thickness conveniently with the desired precision. Most of the reported thermal thickness measurements demonstrate only the validity of models underlying the derivation of Eq. (8) and show that the results determined with this technique are consistent with other measurements. Why bother with a technique whose accessible thickness range is severely limited by the thermal diffusion length, that is, by Eq. (2) or (6)? One application might be the optical measurement of thin films on opaque substrates of low reflectivity such as SiO_2 on Si (60). All of the optical measurements require an independent measurement of the refractive index of the material to derive film thickness. By a combination of thermal and optical interferometric techniques, the thickness of the film can be determined in situ without an off-line measurement of the refractive index (61).

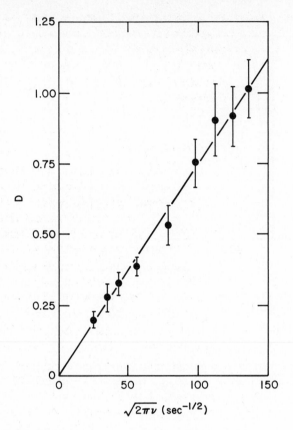

Figure 9.8: Modulation frequency dependence of the amplitude decrement D of a 2.1-μm-thick Novolac® film (72).

9.3.3 Phase Transitions

Heat deposited by a laser beam in a sample can increase the temperature of the sample, as discussed in the paragraph on thermal diffusivity above, or if the sample undergoes a first-order phase transition, provide the latent heat of that transition without changing the temperature at all. This well-known phenomenon was first observed in the context of photoacoustics by Florian et al. (62). Anomalies in the amplitude and phase of the observed signal (63-65) were interpreted (66) as an oscillation of the liquid/solid phase boundary in the sample during the periodic illumination. Photoacoustic experiments have been complemented by photopyroelectric measurements of phase transitions (67). These experiments on model systems qualified photothermal methods and the associated theory for the study of phase transitions. Most recently, these methods were applied to study the glass/crystal phase transition of semiconductors (68) and laser annealing (69). Figures 9.11 and 9.12 show representative data from this study. In response to the laser pulse, the sample, a thin Tellurium film, is heated by the laser and cools down subsequently. With increasing laser fluence, the peak temperature of the film does not follow that

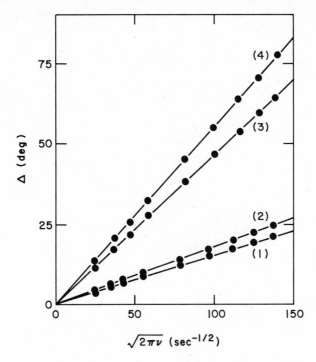

Figure   9.9:   Modulation frequency dependence of the phase lag Δ for four Novolac® films of thickness: (1) 0.65 μm, (2) 0.80 μm, (3) 2.1 μm, and 2.45 μm (72).

increase due to the latent heat of the melting and boiling transition. When the time resolution is high (Fig. 9.11d), the sample melts during the first part of the laser pulse and is heated all the way up to the boiling point by the remainder of the pulse. During the cool down period (Fig. 9.12), the latent heat is released as a result of crystallization, keeping the sample temperature constant for an extended period of time. The amount of latent heat is proportional to the amount of material left after boiling off part of the film. The crystallization process is therefore finished earlier for samples of smaller masses, that is, samples that stayed longer at the boiling temperature due to larger laser fluence. Compared with frequency domain experiments, time-resolved data do not require sophisticated models for the interpretation of the data. Frequency domain experiments fostered the understanding of the photoacoustic signal generation process but have only limited appeal to users of conventional thermal analysis equipment. Time domain studies might, however, be very attractive tools for studies of laser-induced thermal processes, such as annealing, and ablation. Their high time resolution combined with excellent sensitivity enable real-time studies of single transient events crucial for that type of research.

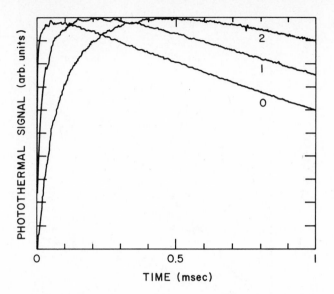

Figure 9.10: Photothermal signal for a pyroelectric thin film calorimeter excited by a 10-μJ laser pulse with 15-ns pulse width: (0) bare calorimeter, (1) coated with a 0.65-μm, and a (2) 0.80-μm thick Novolac® film (72).

9.4 ULTRASONIC ANALYSIS OF THIN FILMS

Pulsed lasers have been used by many authors as a convenient and flexible means for the generation of ultrasound in thin films or solids. Recent reviews on this subject are available elsewhere (70-72). For the ultrasonic analysis of a thin film, the wavelength of the sound wave used to monitor the thickness of the sample and its acoustic properties should be less than the sample thickness. With a sound velocity of the order of 5×10^3 m/s, a film thickness of the order of 1 μm requires a pulse of less than 200 ps to meet this criterion. Conventional techniques are very cumbersome at the corresponding frequencies. In comparison, photothermal techniques are relatively straightforward.

Similar to the above discussed photothermal generation of a heat source, the key advantage of photoacoustic generation of a sound source is that the photoacoustic method requires no mechanical contact with the sample. Timing and location of the sound source can be selected by simple optical techniques and arrays of coherent sources can be readily implemented. A number of photothermal detection schemes is based on optical probing of the sample surface (30). Detection of the ultrasonic pulse therefore does not require mechanical contact with the sample. For semiconductor applications, this means among other things that no contamination will be introduced by a coupling medium. In addition, the excitation source as well as the detector can scan readily across the sample. As long as the intensity of the excitation stays well below the thresholds for evaporation or ablation of the sample, no irreversible processes are anticipated and the method can be considered truly nondestructive.

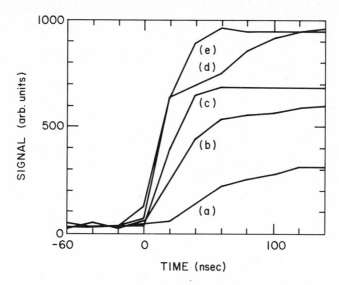

Figure 9.11: Initial temperature during annealing of 30-nm-thick Te films with one pulse from a XeCl excimer laser for five different fluences. The pulse energy is approximately tripled between pulses: (a) 0.7 mJ/cm^2, (b) 2 mJ/cm^2, (c) 7 mJ/cm^2, (d) 24 mJ/cm^2, and (e) 70 mJ.cm^2 (91).

Conventional piezoelectric ceramic and single crystal transducers are excellent mechanical resonators. When stimulating such transducer with a short pulse, its mechanical resonance is excited causing the transducer to ring for quite some time after the pulse. The time resolution of the transducer and hence the thickness of the sample that can be studied is limited by this effect. Numerical corrections or electronic filters can compensate partially for this transducer transfer function. Piezoelectric polymers are not haunted by this problem due to the large internal damping of these materials (73) associated with a much lower conversion of the mechanical energy into an electrical signal.

Coupling of the transducer to the sample of interest is another issue that can be eliminated in photothermal studies. Coupling of the acoustic energy is a function of the acoustic impedances of sample and transducer. Impedance matching can be accomplished by selecting appropriate materials and geometries. Since the choice of piezoelectric materials is rather limited, it might be not possible to match the acoustic impedance of the sample of interest by this technique. Antireflection layers, the other alternative, have limited bandwidth and thus cause severe distortions in a pulsed mode. At the high frequencies desirable for thin film studies, attenuation in the coupling medium can become a serious problem. Light, however, can be readily coupled into virtually any material.

Lasers delivering nanosecond pulses have been available for quite some time at reasonable costs. Generation of ultrasonic pulses with these laser has been dominating the literature, Refs. 74-76 serve as a few examples for these applications. High-power picosecond lasers are just becoming available commercially. The costs

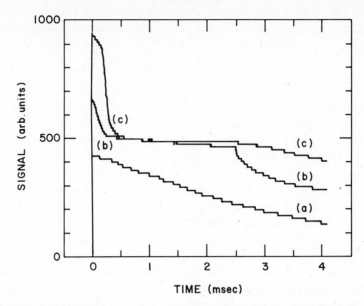

Figure 9.12: Pyroelectric signal during annealing of 30-nm-thick Te films with one pulse from a XeCl excimer for three different fluences. The pulse energy is increased by one order of magnitude between pulses: (a) 0.7 mJ/cm^2, (b) 7 mJ/cm^2 and 70 mJ/cm^2 (91).

of picosecond lasers and the fast electronics required to take advantage of the ultrashort pulses for the analysis of thin films are, however, still rather prohibitive. Maintenance of these lasers is labor intensive. At this time, only few authors (27,77) have been able to take advantage of this emerging diagnostic tool.

9.4.1 Nondestructive Evaluation of Thin Films

In spectroscopy, thermal and ultrasonic analysis of a thin film are, rigorously speaking, nondestructive evaluation of the material if excitation and detection of the probing thermal and ultrasonic waves are accomplished in a nondestructive way. Evaluation in a wider sense refers here to well-developed applications that are potentially of industrial interest. The unique feature of photothermal and photoacoustic techniques is that they allow nondestructive mapping of geometrical, optical, thermal, and, when applicable, acoustic sample properties (78).

9.4.2 Depth Profiling

The depth-profiling capability of photothermal and photoacoustic methods has been mentioned several times in this chapter. As far as thin films are concerned with semiconducting thin films in particular, the key issue is to distinguish surface absorption from bulk and interface absorption. This problem can be addressed by recording photothermal or photoacoustic spectra of the sample of interest at two or more modulation frequencies [Eq. (4)] or at different times after pulsed excitation [Eq. (6)]. Instead of recording these spectra subsequently, data can be obtained by modulating the light source with several frequencies at the same time and using Fourier transform techniques to retrieve amplitudes and phases of the individual frequency components (35). With pulsed excitation, individual time domain signals can be recorded and analyzed to distinguish contributions due to the surface and bulk of the sample. From these data then the location of the absorber can be inferred in analogy to the procedure used in Figs. 9.6 and 9.7. These methods have been utilized to determine, for example, the surface passivation of amorphous silicon films (79), and to separate volume absorption from interfacial absorption in TiO_2/SiO_2 multilayered films (80) or from the influence of subsurface defects on plasma-sprayed coatings (81).

Thermal and acoustic waves can be utilized to measure or profile temperature- or pressure-dependent properties of a thin film. Pyroelectric (82) and piezoelectric properties of thin films can be determined readily with these techniques. Using time-resolved measurements the depth profile of charge or field distributions can be determined. Techniques employing laser-induced thermal (83) or pressure pulses (74) are the standard methods for measurements of charge or field distributions in dielectric films (84). Here the fact that only the excited layer of the sample induces an electrical signal and that the excitation sweeps across the sample are utilized to derive the depth profile.

Polymer electrets such as β-polyvinylidene fluoride (PVDF) are frequently used in pyro- and piezoelectric transducers for the detection of radiation induced transients. The polarization profile in these materials determines the ultimate time resolution of the transducer and serves as an example to highlight some of the differences between thermal and ultrasonic analysis of a sample. PVDF films from two manufacturers were used for the data shown in Figs. 9.13-9.16. Due to the completely different manufacturing processes, the polarization profiles are quite different; bulk properties, however, are almost identical. Using a 500-ps laser pulse absorbed in a tungsten rod, a shock wave is generated. The piezoelectric response of the two foils to this shock wave is quite different. The pulse generates a response of approximately twice the halfwidth, but only half the amplitude in the second foil (Fig. 9.14). Photothermal analysis of the same foils, excited with the same laser directly on the surface of the transducer, shows that the first foil is well poled. The rise time of the signal is fast and excitation from the front and back of the film results in almost identical signals (Fig. 9.15). This behavior would be expected for a homogeneously poled film. The photothermal response of the second film (Fig. 9.16) shows that one side of the transducer has a fast rise time, but the back side has a fairly thick unpolarized layer and the polarization reaches the bulk value only far away from the surface.

For a qualitative interpretation such as the above, photothermal analysis has the advantage that for the same spatial resolution a much lower time resolution suffices. With the bandwidths of commercial electronics limited to approximately 6 GHz, and that at very substantial expense, photothermal analysis might be the method of choice over photoacoustic analysis. Qualitative interpretation of acoustic signals are, however straightforward. While the pressure pulse traverses the sample, it induces a signal that is a direct measure of the polarization. The depth of the sample is directly mapped into the time dependence of the observed signal. Since thermal pulses diffuse into the sample the main contribution to the signal is always caused by the surface. While the pulse diffuses into the sample, successive layers of the sample add up to this surface signal. The deconvolution of the observed signal into a depth profile is,

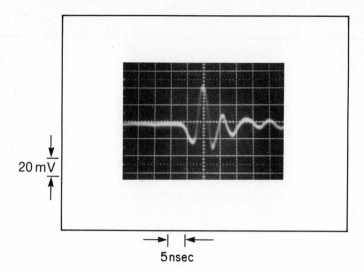

Figure 9.13: Piezoelectric response of an ultrasonic transducer built from 9-μm-thick β-PVDF from Kureha to an ultrasonic pulse generated by absorption of a 500-ps pulse with an energy of 1 mJ.

however, possible by solving the heat diffusion equation for this geometry. A number of algorithms have been developed to accomplish this deconvolution (84).

9.4.3 Imaging

Maps of the geometrical, optical, thermal, and acoustical features can be obtained by scanning the excitation across the sample or by using masks to illuminate different combinations of pixels subsequently (85,86). With optical spectroscopy being a well-developed technique with the parallel processing capability of all pixels (i.e., viewing or recording of one complete picture), photothermal and photoacoustic techniques are not competitive in the area of geometry and optical imaging.

Due to the complicated signal generation process, interpretation of photothermal and photoacoustic images can be ambiguous (87,88). The capability of these microscopes to image subsurface features is, however, of considerable technological interest. Defect maps in thin dielectric films (89), maps of radiation-induced defects (90), or thickness measurements of Si membranes (91) are just a few examples of thin film applications. The bulk of the application is in the area of integrated circuits and thin film devices. Two machines have been marketed successfully that use either an electron beam (92) or a laser for excitation (93). Detection is accomplished with an ultrasonic transducer inserted into a modified electron microscope, or laser beam deflection or changes in reflectivity to probe the temperature change.

The power of this imaging technique is evident in Fig. 9.17. A commercial PVDF film is imaged in an electron microscope. Clearly visible in the thermal wave image are unpolarized domains of PVDF that do not cause a signal when heated by the electron beam. In addition, a 1-μm square poled by exposure to the electron beam is visible as a dark area.

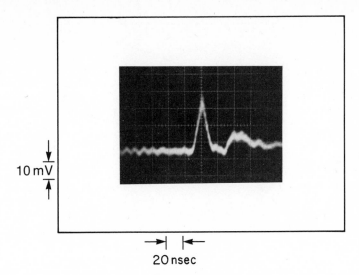

Figure 9.14: Piezoelectric response of an ultrasonic transducer built from 24-μm-thick SOLEF® foil to the same pulse used in Fig. 9.13.

Due to the large potential of this imaging technique, it would justify considerable efforts to improve equipment and the interpretation of images.

9.5 DIELECTRIC PROPERTIES OF THIN FILMS

Besides the already discussed applications, a number of publications have been dealing with the photoacoustic or photothermal detection of surface plasmons. Plasmons are collective oscillations of electrons which have been extensively studied with conventional spectroscopic techniques for quite some time. The plasmon surface polariton is extremely sensitive to surface roughness and the thickness and dielectric constant of the adjacent materials. Using conventional cell-microphone techniques to detect the nonradiative decay, it was possible to demonstrate that the sensitivity of this method is superior to conventional attenuated total reflection measurements and that the influence of dye monolayer on the amplitude and width of the resonance was readily observed (94). With the same technique it was shown that radiative and nonradiative deexcitation channels complement each other (95). With diffraction gratings on the sample surface instead of a prism coupler, the photothermal detection of plasmons was further enhanced (96) and even free-standing Ag films studied (97). The field enhancement at a silver surface due to resonant surface plasmon excitation has been observed (98). The influence of Langmuir-Blodgett monolayer assemblies on the resonance has been studied in detail (99). From these data the dielectric properties of a monolayer can be calculated. The thickness of the layer and hence the orientation of the molecules in the monolayer are derived. The attention in this field has been focused most recently on the properties of silver island films (100), the influence of dye overlayers on these island films (101), and the optical absorption of small, oblate silver spheroids (102). From demonstrating that photoacoustic methods

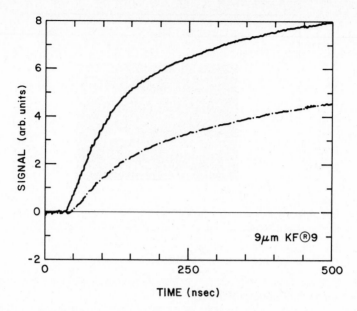

Figure 9.15: Pyroelectric response of the sample of Fig. 9.13 excited directly with a 0.5-mJ pulse from a XeCl excimer laser with a duration of approximately 40 ns. Shown are the signals for excitation from both sides.

duplicate results obtained with other techniques, the field developed into experiments that provide data which can not be observed with any other spectroscopic tool.

9.6 CONCLUSION

Using well-prepared and characterized test samples, photoacoustic and photothermal techniques have been established. The thermal diffusion process makes these techniques unique. It limits the access to a thin layer at the sample surface. This is a severe drawback with bulk samples. For thin films thermal diffusion poses, however, no major problem. It allows the monitoring of subsurface features and the derivation of depth profiles. Due to the complex signal generation and detection process, the interpretation of photothermal and photoacoustic data is rarely unambiguous. Nevertheless, many of the applications are maturing and in the foreseeable future, photothermal and photoacoustic analysis might be able to complement conventional thin film analytical tools.

9.7 REFERENCES

1.   Y. H. Pao, **Optoacoustic Spectroscopy and Detection**, Academic Press, New York (1977).

2.   A. Rosencwaig, **Photoacoustics and Photoacoustic Spectroscopy**, Chemical Analysis, Wiley, New York (1980).

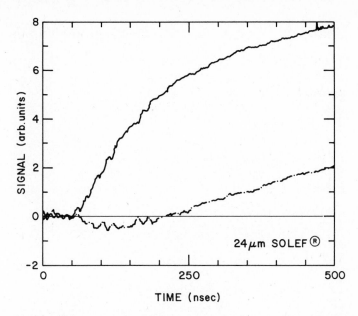

Figure 9.16:   Pyroelectric response of the sample of Fig. 9.14. The conditions of the experiment are otherwise identical to Fig. 9.15.

3.   V. P. Zharov and V. S. Letokhov, **Laser Optoacoustic Spectroscopy**, Springer, Berlin (1986).

4.   A. Mandelis, Ed., **Photoacoustic and Thermal Wave Phenomena in Semiconductors**, Elsevier, Amsterdam (1987).

5.   C. K. N. Patel and A. C. Tam, "Pulsed Optoacoustic Spectroscopy of Condensed Matter," *Rev. Mod. Phys.*, 53, 517 (1981).

6.   J. B. Kinney and R. H. Staley, "Applications of Photoacoustic Spectroscopy," *Ann. Rev. Mater. Sci.*, 12, 295 (1982).

7.   G. A. West, J. J. Barrett, D. R. Siebert, and K. V. Reddy, "Photoacoustic Spectroscopy," *Rev. Sci. Instr.*, 54, 797 (1983).

8.   A. C. Tam, "Applications of Photoacoustic Sensing Techniques," *Rev. Mod. Phys.*, 58, 381 (1986).

9.   H. Vargas and L. C. M. Miranda, "Photoacoustic and Related Photothermal Techniques," *Phys. Rep.*, 161, 43 (1988).

10.   *IEEE Trans. Ultrason., Ferroelectrics, Freq. Contr.*, UFFC-33 (5) (1986).

11.   *Appl. Phys. B*, 43 (1) (1987), whole issue.

12.   *Can. J. Phys.*, 64 (9) (1986), whole issue.

13.   P. Hess and J. Pelzl, Eds., **Photoacoustic and Photothermal Phenomena**, Proc. 5th Int. Conf., Heidelberg, Germany, July 27-30, 1987; Springer Ser. Opt. Sci., Vol. 58, Springer, Berlin, Heidelberg (1987).

Figure 9.17: Secondary electron microscopy (SEM) images of a inductive thin film head for magnetic recording compared with secondary electron induced acoustic images (SEAM) for increasing magnification.

14. H. S. Carslaw and J. C. Jaeger, **Conduction of Heat in Solids**, Clarendon, Oxford (1960).

15. H. Coufal, "Pyroelectric Detection of Radiation Induced Thermal Wave Phenomena," *IEEE Trans. Ultrason., Ferroelectrics, Freq. Contr.*, UFFC-33, 507 (1986).

16. R. Zenobi, J. H. Hahn, and R. N. Zare, "Surface Temperature Measurement of Dielectric Materials Heated by Pulsed Laser Radiation," *Chem. Phys. Lett.*, 150, 361 (1988).

17. P. Baeri, S. U. Campisano, E. Rimini, and J. P. Zhang, "Time Resolved Temperature Measurement of Pulsed Laser Irradiated Germanium by Thin Film Thermocouple," *Appl. Phys. Lett.*, 45, 398 (1984).

18. K. Tanaka, R. Satoh, and A. Odajima, "Photothermal Spectroscopy of Thin Films," *Jpn. J. Appl. Phys.*, 22, L592 (1983).

19. P. E. Nordal and S. O. Kanstadt, "Photothermal Radiometry," *Physica Scripta*, 20, 659 (1979).

20. R. T. Williams, M. N. Kabler, J. P. Long, J. C. Rife, and T. R. Royt, in **Laser and Electron Beam Interactions with Solids**, B. R. Appleton and G. K. Keller, Eds., North-Holland, New York (1982).

21. B. Stritzker, A. Pospieszczyk, and J. Tagle, "Measurement of Lattice Temperature of Silicon During Pulsed Laser Annealing," *Phys. Rev. Lett.*, 47, 356 (1981).

22. I. Hussla, H. Coufal, F. Träger, and T. Chuang, "Pulsed Laser-Induced Thermal Desorption of Xenon," *Ber. Bunsenges. Phys. Chem.*, 90, 240 (1986).

23. M. Buck, B. Schafer, and P. Hess, "Photothermal Desorption Spectroscopy with IR Lasers," *Surf. Sci.*, 161, 245 (1985).

24. P. B. Comita and T. T. Kodas, "Modulated Laser Beam Relaxation Spectrometry of Laser Induced Chemical Vapour Deposition," *Appl. Phys. Lett.*, 51, 2059 (1987).

25. A. Rosencwaig, J. Opsal, W. L. Smith, and D. L. Willenborg, "Detection of Thermal Waves Through Optical Reflectance," *Appl. Phys. Lett.*, 46, 1013 (1985).

26. S. V. Bondarenko, E. V. Ivakin, A. S. Rubanov, V. I. Kabelka, and A. V. Mikhailov, "Acoustic Wave Excitation by Degenerate Four-Wave Mixing of Picosecond Pulses in a Dye Solution," *Optics Comm.*, 61, 155 (1987).

27. G. L. Eesley, B. M. Clemens, and C. A. Paddock, "Generation and Detection of Picosecond Acoustic Pulses in Thin Metal Films," *Appl. Phys. Lett.*, 50, 717 (1987).

28. D. Fournier, A. C. Boccara, N. M. Amer, and R. Gerlach, "Sensitive In situ Trace-Gas Detection by Photothermal Displacement Spectroscopy," *Appl. Phys. Lett.*, 37, 519 (1980).

29. M. A. Olmstead, N. M. Amer, and S. Kohn, "Photothermal Displacement Spectroscopy: An Optical Probe for Solids and Surfaces," *Appl. Phys. A*, 32, 141 (1983).

30. For a review see, for example, J. P. Monchalin, "Optical Detection of Ultrasound," *IEEE Trans. Ultrason., Ferroelectrics, Freq. Contr.*, UFFC-33 pp. 485-499 (1986).

31. A. C. Tam and C. K. N. Patel, "Ultimate Corrosion-Resistant Optoacoustic Cell for Spectroscopy of Liquids," *Opt. Lett.*, 5, 27 (1980).

32. H. M. Frost, "Electromagnetic-Ultrasound Transducers: Principles, Practice and Applications," in **Physical Acoustics**, Vol. 14, W. P. Mason and R. N. Thurston, Eds., 179, Academic Press, New York (1979).

33. R. E. Lee and R. M. White, "Excitation of Surface Elastic Waves by Transient Surface Heating," *Appl. Phys. Lett.*, 12, 12 (1968).

34. A. C. Tam and C. K. N. Patel, "Optical Absorption of Light and Heavy Water by Laser Optoacoustic Spectroscopy," *Appl. Opt.*, 18, 3348 (1979).

35. H. Coufal, "Fourier Analysis of Frequency Domain Multiplexed Photothermal Signals," *J. Photoacoust.*, 1, 413 (1984).

36. A. Mandelis, "Time-Delay-Domain and Pseudorandom-Noise Photoacoustic and Photothermal Wave Processes: A Review of the State of the Art," *IEEE Trans. Ultrason., Ferroelectrics, Freq. Contr.*, UFFC-33, pp. 590-614 (1986).

37. Y. Sugitani, A. Uejima, and K. Kato, "Correlation Photoacoustic Spectroscopy," *J. Photoacoust.*, 1, 217 (1982).

38. P. R. Griffiths and J. A. de Haseth, **Fourier Transform Infrared Spectrometry**, Wiley, New York (1986).

39. H. Coufal, "Self-Supporting Carbon Glass Films: A PAS Standard?," *Appl. Opt.*, 21, 104 (1982).

40. J. M. Bennett, "Optical Scattering and Absorption Losses at Interfaces and in Thin Films," *Thin Solid Films*, 123, 27 (1985).

41. B. R. Weinberger, C. B. Roxlo, S. Etemad, G. L. Baker, and J. Orenstein, "Optical Absorption in Polyacethylene: A Direct Measurement Using Photothermal Displacement Spectroscopy," *Phys. Rev. Lett.*, 53, 86 (1984).

42. M. D. Porter, D. H. Karweik, T. Kuwana, W. B. Theis, G. B. Norris, and T. O. Tiernan, "The Detection Capabilities of Fourier Transform External Reflection and Photoacoustic Spectroscopy for Electrode-Supported Films," *Appl. Spectrosc.*, 38, 11 (1984).

43. W. Knoll and H. Coufal, "Standing Light Wave in Front of a Silver Mirror Investigated by Photothermal Spectroscopy," *Appl. Phys. Lett.*, 51, 892 (1987).

44. J. K. Dohrmann and U. Sander, "In Situ Photoacoustic Spectroscopy of Electrochemically Grown PbO_2 Films: Optical Constants from Photoacoustic Interference Signals," *Ber. Bunsenges. Phys. Chem.*, 90, 605 (1987).

45. P. Helander and I. Lundstrom, "Photoacoustic Study of Layered Samples," *J. Appl. Phys.*, 52, 1146 (1981).

46. J. Pelzl, R. Grygier, and H. Coufal, "Nondestructive Depth Profile of Color Reversal Film by Photothermal Technique," in *Progress in Basic Principles of Imaging Systems,*, F. Grazer and E. Moisar, Eds., Vieweg, Braunschweig (1986).

47. H. Coufal, "Photothermal Spectroscopy of Weakly Absorbing Samples Using a Thermal Wave Phase Shifter," *Appl. Phys. Lett.*, 45, 516 (1984).

48. C. Starr, "An Improved Method for the Determination of Thermal Diffusivities," *Rev. Sci. Instr.*, 8, 61 (1937).

49. M. J. Adams and G. F. Kirkbright, "Thermal Diffusivity and Thickness Measurements for Solid Samples Utilizing the Optoacoustic Effect," *Analyst*, 102, 678 (1977).

50. H. Coufal and P. Hefferle, "Thermal Diffusivity Measurements of Thin Films with a Pyroelectric Calorimeter," *Appl. Phys. A*, 38, 213 (1985).

51. P. K. John, L. C. M. Miranda, and A. C. Rastogi, "Thermal Diffusivity Measurement Using the Photopyroelectric Effect," *Phys. Rev. B*, 34, 4342 (1986).

52. C. C. Ghinzoni and L. C. M. Miranda, "Photopyroelectric Measurement of the Thermal Diffusivity of Solids," *Phys. Rev. B*, 32, 8392 (1985).

53. O. Pessoa, Jr., C. L. Cesar, N. A. Patel, H. Vargas, C. C. Ghinzoni, and L. C. M. Miranda, "Two-Beam Photoacoustic Phase Measurement of the Thermal Diffusivity of Solids," *J. Appl. Phys.*, 59, 1316 (1986).

54. W. P. Leung and A. C. Tam, "Thermal Diffusivity in Thin Films Measured by Noncontact Single-Ended Pulsed-Laser-Induced Thermal Radiometry," *Opt. Lett.*, 9, 93 (1984).

55. R. E. Imhof, F. R. Thornley, J. R. Gilchrist, and D. J. S. Birch, "Opto-Thermal Study of Thermally Insulating Films on Thermally Conducting Substrates," *J. Phys. D*, 19, 1829 (1986).

56. A. Skumanich, H. Dersch, M. Fathalla, and N. M. Amer, "A Contactless Method for Investigating the Thermal Properties of Thin Films," *Appl. Phys. A*, 43, 297 (1987).

57. J. P. Roger, F. Lepoutre, D. Fournier, and A. C. Boccara, "Thermal Diffusivity Measurement of Micron-Thick Semiconductor Films by Mirage Detection," *Thin Solid Films*, 155, 165 (1987).

58. G. Benedetto and R. Spagnolo, "Photoacoustic Measurement of the Thermal Effusivity of Solids," *Appl. Phys. A*, 46, 169 (1988).

59. T. Hashimoto, J. Cao, and A. Takaku, "Thermal Diffusivity Measurements for Thin Films by the Photoacoustic Effect," *Thermochim. Acta*, 120, 191 (1987).

60. A. Mandelis, E. Siu, and S. Ho, "Photoacoustic Spectroscopy of Thin SiO_2 Films Grown on (100) Crystalline Si Substrates," *Appl. Phys. A*, 33, 153 (1984).

61. M. E. Abu-Zeid, A. E. Rakhshani, A. A. Al-Jassar, and Y. A. Youssef, "Determination of the Thickness and Refractive Index of Cu_2O Thin Film Using Thermal and Optical Interferometry," *Phys. Status Solidi A*, 93, 613 (1986).

62. R. Florian, J. Pelzl, M. Rosenberg, H. Vargas, and R. Wernhardt, "Photoacoustic Detection of Phase Transitions," *Phys. Status Solidi A*, 48, K35 (1978).

63. C. Pichon, M. Le Liboux, D, Fournier, and A. C. Boccara, "Variable-Temperature Photoacoustic Effect: Application to Phase Transitions," *Appl. Phys. Lett.*, 35, 435 (1979).

64. M. A. A. Sigueira, G. G. Ghinzoni, J. I. Vargas, E. A. Menez, H. Vargas, and L. C. M. Miranda, "On the use of the Photoacoustic Effect to Investigate Phase-Transitions in Solids," *J. Appl. Phys.*, 51, 1403 (1980).

65. P. S. Bechthold, M. Campagna, and T. Schober, "Phase Transitions in Metal-Hydrogen Interstitial Alloys by Temperature Dependent Photoacoustic Measurements," *Solid State Commun.*, 36, 225 (1980).

66. P. Korpiun, J. Baumann, E. Luscher, E. Papamokos, and R. Tilgner, "Photoacoustic Effect at First Order Phase Transitions at Increasing and Decreasing Temperature," *Phys. Status Solidi A*, 58, K13 (1980).

67. A. Mandelis, F. Care, K. K. Chan, and L. C. M. Miranda, "Photopyroelectric Detection of Phase Transitions in Solids," *Appl. Phys. A*, 38, 117 (1985).

68. A. L. Glazov, S. B. Gurevich, N. N. Il'yashenko, N. P. Kalmykova, K. L. Muratikov, and N. A. Rogachev, "Study of Phase Transition in Thin Amorphous Semiconductor Films by a Photoacoustic Method," *Sov. Tech. Phys. Lett. (USA)*, 12, 59 (1986).

69. H. Coufal and W. Lee, "Time Resolved Calorimetry of Te-Films During Pulsed Laser Annealing," *Appl. Phys. B*, 44, 141 (1987).

70. C. B. Scruby, R. J. Dewhurst, A. A. Hutchins, and S. B. Palmer, "Laser Generation of Ultrasound in Metals," in **Research Techniques in Nondestructive Testing**, R. S. Sharpe, Ed., Vol. 5, 281, Academic Press, New York (1982).

71. G. Birnbaum and G. S. White, "Laser Techniques in NDE," in **Research Techniques in Nondestructive Testing**, R. S. Sharpe, Ed., Vol. 7, 259, Academic Press, New York (1984).

72. D. A. Hutchins, "Ultrasonic Generation by Pulsed Lasers," in **Physical Acoustics**, W. P. Mason and R. N. Thurston, Eds., Vol. 18, 21, Academic press, New York (1986).

73. A. C. Tam and H. Coufal, "Photoacoustic Generation and Detection of 10 ns Acoustic Pulses in Solids," *Appl. Phys . Lett.*, 42, 33 (1983).

74. G. M. Sessler, J. E. West, and G. Gerhard, "High-Resolution Laser-Pulse Method for Measuring Charge Distributions in Dielectrics," *Phys. Rev. Lett.*, 48, 563 (1982).

75. H. Sontag and A. C. Tam, "Optical Monitoring of Photoacoustic Pulse Propagation in Si Wafers," *Appl. Phys. Lett.*, 46, 725 (1985).

76. H. Sontag and A. C. Tam, "Optical Detection of Photoacoustic Pulses in Thin Silicon Wafers," *Can. J. Phys.*, 64, 1330 (1986).

77. G. L. Eesley, B. M. Clemens, and C. A. Paddock, *Appl. Phys. Lett.*, 50, 717 (1987).

78. C. Thomsen, H. J. Maris, and J. Tauc, "Picosecond Acoustics as a Non-Destructive Tool for the Characterization of Very Thin Films," *Thin Solid Films*, 154, 217 (1987).

79. A. Rosencwaig, "Nondestructive Evaluation with Thermal Waves," *J. Photoacoust.*, 1, 371 (1983).

80. R. C. Frye, J. J. Kumler, and C. C. Wong, "Investigation of Surface Passivation of Amorphous Si Using Photothermal Displacement Spectroscopy," *Appl. Phys. Lett.*, 50, 101 (1987).

81. H. G. Walther and E. Welsch, "Calculation and Measurement of the Absorption in Multilayer Films by Means of Photoacoustics," *Thin Solid Films*, 142, 27 (1986).

82. S. Aithal, G. Rousset, and L. Bertrand, "Photoacoustic Characterization of Subsurface Defects in Plasma-Sprayed Coatings," *Thin Solid Films*, 119, 153 (1984).

83. H. Coufal, R. Grygier, D. Horne, and J. Fromm, "Pyroelectric Calorimeter for Photothermal Studies of Thin Films and Adsorbates," *J. Vac. Sci. Technol A*, 5, 2875 (1987).

84. G. Li, Q.-R. Yin, W.-G. Luo, and Z.-W. Yin, "Measurement of Piezoelectric Coefficients of ZnO Thin Film with Photoacoustic Technique," *Jpn. J. Appl. Phys. Suppl.*, 24-2, 425 (1985).

85. G. M. Sessler and R. Gerhard-Multhaupt, "A Review of Methods for Charge- or Field-Distribution Studies on Radiation-Charged Dielectric Films," *Radiat. Phys. Chem.*, 23, 363 (1984).

86. H. Coufal, U. Moller, and S. Schneider, "Photoacoustic Imaging Using a Hadamard Transform Technique," *Appl. Opt.*, 21, 116 (1982).

87. H. Coufal, U. Moller, and S. Schneider, "Photoacoustic Imaging Using a Fourier Transform Technique," *Appl. Opt.*, 21, 2339 (1982).

88. J. C. Murphy, J. W. Maclachlan, and L. C. Aamodt, "Image Contrast Processes in Thermal and Thermoacoustic Imaging," *IEEE Trans. Ultrason., Ferroelectrics, Freq. Contr.*, UFFC-33, 529 (1986).

89. L. D. Favro, P.-K. Kuo, and R. L. Thomas, "Thermal Wave Propagation and Scattering in Semiconductors," in **Photoacoustic and Thermal Wave Phenomena in Semiconductors**, A. Mandelis, Ed., 69, Elsevier, Amsterdam (1987).

90. W. C. Mundy and R. S. Hughes, "Photothermal Deflection Microscopy of Dielectric Thin Films," *Appl. Phys. Lett.*, 43, 985 (1983).

91. M. Guardalben and A. Schmid, "Photothermal Analysis of Synergistic Radiation Effects in ThF_4 Optical Thin Films," *Phys. Rev. B*, 35, 4026 (1987).

92. J. I. Burov and D. V. Ivanov, "Thermodisplacement Imaging of Silicon Membranes," *J. Phys. (France)*, 47, 549 (1986).

93. A. Rosencwaig and J. Opsal, "Thermal Wave Imaging with Thermoacoustic Detection," *IEEE Trans. Ultrason., Ferroelectrics, Freq. Contr.*, UFFC-33, 516 (1986).

94. A. Rosencwaig, "Thermal Wave Characterization and Inspection of Semiconductor Materials and Devices," in **Photoacoustic and Thermal Wave Phenomena in Semiconductors**, A. Mandelis, Ed., 97, Elsevier, Amsterdam (1987).

95. H. Taalat and H. D. Dardy, "Photoacoustic Study of the Interaction of Surface Plasmons in Monolayers of Dye Molecules," in **Ultrasonics Symposium Proceedings**, B. R. McAvoy, Ed., 700, IEEE, New York (1983).

96. B. Rothenhäusler, J. Rabe, P. Korpiun, and W. Knoll, "On the Decay of Plasmon Surface Polaritons at Smooth and Rough Ag-Air Interfaces: A Reflectance and Photoacoustic Study," *Surf. Sci.*, 137, 373 (1984).

97. T. Inagaki, M. Motosuga, E. T. Arakawa, and J. P. Goudonnet, "Coupled Surface Plasmons in Periodically Corrugated Thin Silver Films," *Phys. Rev. B*, 32, 6238 (1985).

98. T. Inagaki, M. Motosuga, E. T. Arakawa, and J. P. Goudonnet, "Coupled Surface Plasmons Excited by Photons in a Free-Standing Thin Silver Film," *Phys. Rev. B*, 31, 2548 (1985).

99. C. S. Jung, G. Park, and Y. D. Kim, "Photoacoustic Determination of Field Enhancement at a Silver Surface Arising from Resonant Surface Plasmon Excitation," *Appl. Phys. Lett.*, 47, 1165 (1985).

100. R. K. Grygier, W. Knoll, and H. Coufal, "Detection of Plasmon Surface Polaritons on Periodic Silver Gratings with a Pyroelectric Calorimeter," *Can. J. Phys.*, 64, 1067 (1986).

101. T. Inagaki, J. P. Goudonnet, P. Royer, and E. T. Arakawa, "Optical Properties of Silver Island Films in the Attenuated-Total-Reflection Geometry," *Appl. Opt.*, 25, 3635 (1986).

102. V. N. Rai, "Optical Properties of Silver-Island Films Having an Overlayer of RhB Dye," *Appl. Opt.*, 26, 2395.

103. P. Royer, J. P. Goudonnet, T. Inagaki, G. Chabrier, and E. T. Arakawa, "Photoacoustic Study of the Optical Absorption of Oblate Silver Spheroids in

Attenuated-Total-Reflection Geometry," *Phys. Status Solidi A*, 105, 617 (1988).

10
THERMALLY STIMULATED DISCHARGE CURRENT MEASUREMENT FOR THIN POLYMER FILMS

B. Chowdhury*

M & T Chemicals, Inc.
Rahway, New Jersey

Among the different methodologies currently available for characterization of thin polymer films, measurement of thermally stimulated discharge current (TSDC) is to be considered an emerging technique. The principle of TSDC measurement is not new. The methodology itself has evolved on a sound theoretical and experimental basis over more than half a century for solid dielectrics. Most work has been performed with polymers of sufficient thickness to create a nondissipative interelectrode capacitance which allows measurement of short-circuit current, following a voltage soak of the thermally activated solid.

Polymers generally undergo a degree of polarization and acquire a volume charge and a surface charge when subjected to a suitable voltage field. Such voltage-induced polarization provides a means of bringing about internal rearrangements in the polymer matrix whereby discrete charge accumulation occurs in the various domains of the polymer. The total charge, however, remains largely dormant unless a discharge pattern is developed through thermal stimulation.

The discharge current is entirely characteristic of the charging process. The experimental conditions for the latter require, among other considerations, application of a DC voltage whose value depends on the thickness of the specimen. For very thin films for potential electronic and semiconductor applications, the need to distinguish between the properties of the near surface and the various interfaces becomes more important than the bulk properties from a functional point of view. Experimentally, this requires an enhanced deposition technique to produce intimate electrode contact with micrometer-thick films, eliminating any air inclusion, and an accurate differentiation of the temperature regions for the discharge current, to facilitate interpretation.

The usefulness and sensitivity of the TSDC measurement technique lie in the fact that polarization due to individual molecular processes is associated with their low equivalent frequencies. Essentially, the technique provides information on a molec-

---

* Present address: Matech Associates, 150 E. Grove Street, Scranton, Pennsylvania 18510.

ular level. For this reason, TSDC of thin polymer films is a promising field. It is believed that the minimal thickness of the films is advantageous for the polarization process.

The purpose of this chapter is to provide the reader with a historical perspective of the development of the experimental principles of the TSDC technique, and its present applications for polymer electrets, and to lead into an assessment of its suitability for measurement of surface and interfacial properties of thin polymer films. References will be provided in the text and for additional general reading at the end of the chapter.

The recognition of the electret effect, that is, the development and retention of charge by dielectrics on exposure to a voltage field, can be traced to the work of Gray (1) and to the experimental studies by Faraday (2). The term "electret" was coined by Heaviside (3) as an electrical analog of the properties of a magnet. However, the development of the technique in its current form owes its origin to Eguchi's pioneering studies in 1919 (4). The electret effect has been experimentally demonstrated by others since that time (5-7). Serious attempts to produce theoretical understanding and practical applications of the TSDC method did not appear in the literature till after 1960 (8-14), although related studies of the nature of the thermally induced current discharge phenomenon had been published by Frei and Groetzinger (15) in 1936.

The current state of knowledge of the TSDC theory, the basis of its experimental parameters, and its use for investigation of structure and properties of polymers have emerged since the late 1960s (16-25).

10.1 BASIS OF TSDC MEASUREMENT FOR POLYMERS

10.1.1 Interactive Properties of Polymers

When a polymer is thermally activated, that is, its temperature is gradually raised, the motions of the polymer side chains can attain a high degree of freedom with respect to the motions of the main polymer backbone. The change of state of the polymer from the glassy to the rubbery is marked by a pronounced relaxation at the glass transition temperature, T_g, indicating the initiation of the conformal rearrangements of the main chains arising from the cooperative movement of the backbone with its side chains. If a thermogram is taken by the use of dynamic mechanical analysis of temperature versus damping, a loss peak is obtained at this transition temperature. This peak is termed the α-transition. At a low (generally subambient) temperature, the cooperative movement of the main backbone and its branches becomes frozen, but local motions within terminal groups can remain active. At a somewhat higher temperature than this, the groups or segments can undergo rotational changes and assume different conformation states. Transitions from the above two changes are termed γ- and β-transitions in ascending order of temperature on a thermogram, corresponding to the designations of mechanical spectroscopy. It is the polymer's ability to interact with an electric field through this spectrum of changes that forms the basis of the polarization phenomenon.

10.1.2 Consequences of Interaction with Electric Field

A variety of effects may result from the interaction between an impressed voltage and internal polymer motions. Thus, dipoles may form in the polymer matrix (23), surfaces can acquire charge (23), real charges trapped in the polymer (26) can drift, and considerable charge accumulation can take place at intra- or extrapolymer interfaces through dissociative and occlusive processes in the surrounding atmosphere (27) and through electrode material, respectively (19,28). Charge crowding at the interface between dissimilar neighboring entities (29,30) in the polymer is of much interest because of the possibilities this offers for producing charge storage and charge transfer devices.

An amount of mobile electrical charge is associated with each of the above events. These charges can be immobilized by cooling the polymer in the presence of the electric field. Charge release is controlled by molecular movements which are overwhelmingly temperature dependent (31). The charge components can be sequentially discharged by thermal displacement and recorded as a TSDC thermogram, while shortcircuiting the charging loop after turning off the voltage.

10.1.3 Polarization and Discharge Current

The above events can be described in terms of experimental parameters as follows. For a single dipole relaxation, the polarization, P, can be described as a function of time, t, by the well-known Debye equation

$$\frac{dP}{dt} \ = \ -\alpha(T)P \ + \ \varepsilon_0(\varepsilon_s - \varepsilon_\infty)\,\alpha(T)\,E \tag{1}$$

where $\alpha(T)$, ε_s, ε_∞, and E   are the temperature-dependent relaxation frequency (reciprocal of time), the dielectric constant of the sample at the low (static) frequency, the dielectric constant of the sample at the high frequency, and the field strength, respectively.

It can be readily seen that the dielectric constant ε_0, of the unpolarized specimen is modified by the relaxation strength $(\varepsilon_s - \varepsilon_\infty)$, as a result of polarization. Since dielectric constant, ε_0, is also influenced by heating and the TSDC pattern is based on a continuum of changes in dielectric dispersion due to polarization, it is desirable to test the correspondence between the temperature at which a given TSDC peak maximum occurs and the measured value of ε_0 at that temperature. This correspondence has been found to be linear by the author (32) for the TSDC α -transition of several formulations of epoxy resins. The α-transition is graphically the sharpest transition, and it is suitable for this test, because the dielectric values remain constant over a few degrees of temperature at the peak maximum, and this constant value can be interpolated from separately measured values over a range of temperature.

The frequencies associated with thermally activated molecular processes are typically of the order of 10^{-3} to 10^{-2} Hz. TSDC operates on this low equivalent frequency which accounts for its high sensitivity, that is, its ability to provide information on individual molecular processes. The temperature dependence of this frequency is not straightforward, since multiple dipole relaxations are feasible in polymers. However, a reasonable account of this dependence can be based on the Arrhenius rate equation

$$\alpha(T) = \alpha_0 \exp(-\frac{\Delta H_a}{KT})$$ (2)

where α_0 is the preexponential factor, K is Boltzmann constant, T is the absolute temperature, and ΔH_a is the activation energy.

While the merit of the Arrhenius equation is in its ability to provide a means of calculating the activation energy of thermally dependent rate processes, the temperature dependence of TSDC frequency can also be expressed in terms of "WLF" shift (33)

$$\alpha(T) = \alpha_g \exp C_1(T - T_g) (C_2 + T - T_g)^{-1}$$ (3)

where α_g is the relaxation frequency at the glass transition temperature, T_g, and both C_1 and C_2 are constants. For amorphous polymers, $\alpha_g \sim 7 \times 10^{-3}s^{-1}$, $C_1 = 40$, and $C_2 = 52°C$. The practical significance of the above approximation is that it takes into account the relaxation shift due to the free volume concept in polymers for movement of bulky main chain segments.

The time-dependent discharge current, i(t), ensuing from depolarization is not only related to the original polarization, it is also a function of the total surface area, A, of the dielectric specimen that is polarized in contact with the electrodes

$$i(t) = -A \frac{dP(t)}{dt}$$ (4)

In practice, discharge current is a function of the temperature-dependent frequency for any given relaxation process contributing to the release of part of the time-dependent polarization in the specimen

$$i(t) = -\alpha(T) P(t)$$ (5)

Thus, depolarization in different polymer domains manifests itself in different temperature regions on the TSDC thermogram.

10.2 INTERFACIAL DEPOLARIZATION AND CALCULATION OF RELEVANT PARAMETERS

Consideration of interfacial depolarization requires taking into account charge transfer efficiency across heteroboundaries. The initial polarization due to applied voltage is brought about by an induced charge shift throughout the material (34). Part of the energy of the field, however, is dissipated as heat. The ratio of heat dissipated to the emerging current upon discharge varies with the nature of the interface because drift mobilities of charge carriers (characteristic of the interface) produce an additional amount of Joule heat in an applied field. Experimentally, activation energy of the discharge process can be calculated from the initial rise of the current (35) discharging from interfacial polarization, whence the magnitude of the relaxation time can be calculated. Additionally, drift mobility and conductivity values can be derived from first principles of charge transport in solid systems.

The following relationships hold for step by step calculation of the interfacial parameters discussed above.

- *Activation energy* can be calculated by using the method of Perlman (36)

$$\log_e i(T) = C - \frac{\Delta H_a}{KT} \tag{6}$$

where i(T) is the discharge current at temperature T and C is a constant. Activation energy, ΔH_a, is found by multiplying the slope of the straight line plot of $\log_e$ i against 1/T by the Boltzmann constant, K (in eV/°C).

- *Relaxation time* (τ) for the discharge current at the maximum temperature can be calculated by using the following equation

$$\tau = \frac{KT_{max}^2}{H_r \Delta H_a \; \exp(\Delta H_a/KT_{max})} \tag{7}$$

where H_r and T_{max} are the heating rate and the maximum temperature at which relaxation time is calculated, respectively.

- *Drift mobility* (μ_D), in cm²/V.s, of charge carriers across interface can be calculated from the equation

$$\mu_D = \frac{J}{NeE} \tag{8}$$

where J, N, e, and E are the current density in A/cm² of interface, the number of charge carriers/cm³, the charge on the electron in coulombs, and the field strength in V/cm of sample thickness, that is, the voltage difference across the sample divided by the sample thickness, respectively.

- Finally, *volume conductivity*, $\lambda_{vol.}$ (in Ω^{-1} .cm⁻¹) can be calculated by the formula

$$\lambda_{vol.} = Ne\mu_D \tag{9}$$

where the symbols have the same meaning as before.

The above interfacial parameters have been calculated by the author (37) from experimental results for the silicone/epoxy interface to demonstrate that a more than seventy fold increase in charge - compared to that in depolarized specimen - can be achieved from interfacial polarization for certain silicone/epoxy compositions. Phenomena of this nature hold promise for technological applications. The glass transition temperature, T_g, mentioned previously, is a thermal property that manifests itself in the viscoelastic region and is central to important mechanical properties of polymers, mainly because of its relationship to the time/temperature equivalence of viscoelasticity.

A correlation between the time scale of the viscoelastic response to that of the thermal decay of interfacial polarization for the silicone/epoxy interface has been found by the author (38). In that study, the following equation was devised by the author to determine depolarization rate

$$\text{depolarization rate} = \frac{4\pi di(T)}{\alpha(T)VA} \tag{10}$$

where d, i(T), α(T), V, and A are the sample thickness, the discharge current as a function of temperature, the temperature-dependent transport frequency (reciprocal of relaxation time), the voltage drop across the sample, and the surface area of the sample, respectively. Voltage drop was calculated by multiplying the measured current by sample resistance, R:

$$R = \frac{\rho d}{A} \tag{11}$$

with $\rho = 1/$conductivity.

Relaxation time and conductivity values, required for the above calculations, were derived from experimental parameters, as explained before. The correspondence found between viscoelastic response and interfacial depolarization in epoxy resins having soft segments in a rigid backbone is an important step toward establishing methodologies for studying interfacial properties of polymer composites.

A number of factors, both internal and external to the polymer, can affect the nature of polarization in polymers and thus alter the position and magnitude of TSDC peaks. Factors internal to the polymer such as the degree of crystallinity, and the extent of crosslinking and steric configuration, have been studied extensively. Other factors such as prior history or exposure of the polymer to the deleterious effects of high-temperature aging or moisture absorption can significantly influence TSDC results. Additives that are necessary ingredients in polymer formulations for enhancement of properties, can alter the course of depolarization current. It is for these reasons that TSDC measurement has remained largely an experimental technique and theoretical prediction of results is uncommon. Some of the factors influencing polarization in polymers cited in the literature are described below.

The degree of crystallinity has different effects on TSDC peaks of different polymers. Vanderschueren (39) has found that the intensity of the β-peak decreases considerably with the increase in crystallinity in polycarbonate, whereas the work of Murayama and Hashizuma (40) has shown that the α-peak is influenced by the crystalline phases in polyvinylidenefluoride. Ionic agents, such as NaCl, have been found to cause an upward shift in temperature for the α-peak in polyethylene (41) due, probably, to an increase in the energy barrier for its α-relaxation. Antioxidants added to polymers affect the β- and γ-peaks in polyethylene (41), because these additives raise the activation energy for these relaxations. Specific-purpose dopants, such as copper or iodine, have been reported to affect the dipolar and space charge relaxations (42).

Presence of water in polymers is known to degrade their dielectric properties and cause a downward shift in temperature for their relaxation peaks. Vanderschueren and Linkens (43) have shown that in a wide variety of polymers, the α- and β-relaxations are significantly affected by the presence of water. They attributed this to the weakening of intermolecular forces due to the formation of hydrogen bonds, which lowers the activation energy for these relaxations.

10.3 CONSIDERATIONS OF SURFACES AND INTERFACES IN THIN POLYMER FILMS

10.3.1 Physical Aspects

The usual structural considerations for surfaces and interfaces apply equally to thin films. Dimensional considerations for interfaces in thin polymer films arise from blended composites in which different microsurfaces make contact through heterogeneous phases. Polymer composites derived from incompatible blends have a near surface, whose properties are different from those of the bulk and possess interfacial regions between the phases. This affords an opportunity to achieve combination of properties, such as strength from the bulk and protective properties from the surface.

When a thin film surface is created on a substrate, such as by coating or deposition, or when interfaces are formed by interaction between immiscible polymers in a blend, the properties of the surface and interface can each play a substantially different role for the material. Judgment of compatibility in a polymer mixture is usually based on its possessing a single glass transition temperature, T_g (44), which can be determined by thermal analysis. Auger electron microscopy (AES) has proven useful in the determination of underlying chemical states in thin coatings of either pure composition or mixed phases (45). A practical way of preparing very thin polymer films for thermally stimulated discharge (TSD) measurement is to either deposit a thin layer of the polymer on to the electrode or to carry out in situ polymerization on the electrode surface by such means as uv radiation. Conversely, electrodes may be intimately deposited on the sample specimen, or the sample may be separated from the electrodes by air gaps. The air gap method is particularly suitable for charge deposition on polymer surfaces. The deposited electrode method is used for volume charging which results in dipole alignment.

A number of methods are available for producing surface and interfacial regions between a thin coating and a substrate. These include:

- Chemical deposition from the vapor phase
- Physical vapor deposition, such as by evaporation, ion plating, or sputtering
- Electrolytic deposition
- Plasma spraying
- In situ coating by polymerization

The above processes requiring vacuum techniques have been classified by Schiller, Heisig, and Goedicke (46). In this connection, the article by Weissmantel (47) is worth consulting. A more materials-oriented classification, based on physical dimensions of deposition, has been provided by Bunshah and Mattox (48).

Since the scope of this chapter does not allow a discussion of these techniques, the reader is referred to the book by Bunshah, et al. (49) for more complete information on deposition technologies.

10.3.2 Charging Characteristics of Thin Polymer Films

Charging of thin polymer films by applied voltage can be carried out by forming a capacitor in a parallel plate configuration of the film sandwiched between two metallic elements (50). The capacitance, C, of such a configuration is given by

$$C = \frac{K\varepsilon_{air}A}{d} \qquad (12)$$

where d, K, ε_{air}, and A are the film thickness, the dielectric constant of the film, the dielectric constant of free space, and the film surface area, respectively.

The resultant charge, Q, is the product of the capacitance, C, and the applied voltage, V_A

$$Q = CV_A \tag{13}$$

For TSDC experiment on thin films, the applied voltage is adjusted to the rated voltage, V_R, which is some fraction, K, of the breakdown voltage, V_B

$$V_R = KV_B \tag{14}$$

Charge storage is dependent on the capacitance density, that is,

$$\text{charge storage} = V_R\left(\frac{C}{A}\right) \tag{15}$$

Under short-circuit condition, the charge decays giving rise to a displacement current.

10.3.3 Origin of TSDC in Thin Polymer Films

During charging, part of the energy of the field is dissipated as heat. The ratio of heat dissipated to the emerging current upon discharge varies with the magnitude of the current density in the film, because drift mobilities of charge carriers produce an amount of frictional heat. This heat is dependent on both the heat capacity of the film and the thermal impedance between the film and the electrode contact. For a thin coated polymer surface on a substrate, such as in deposited or coated electrodes, the heat capacity of the polymer material is related to its thermal relaxation time constant, which is the ratio of the heat capacity of the coating to the thermal conductance of the substrate. Therefore, thermal characteristics of polymers have a direct bearing on the charging process.

The charge in the polymer can remain dormant indefinitely only if the polymer has a low electrical conductivity and a relatively high T_g. Dielectric films have a finite parallel resistance and capacitors formed from them are not ideal charge storage devices. When the discharge temperature exceeds the formation temperature of these charge elements, as the temperature is gradually raised, current release follows the short-circuit pathway.

The release of current from the various domains of the polymer is thermally selective and, thus, charge accumulation in the surface and interfaces can be mapped via the discharge phenomenon. Charge crowding at the interface arises from the fact that the polymer may have partially crystalline regions, the conductivity of which is higher than that of the amorphous regions. In such a case, charge piling at the crystal boundaries preserves continuity of current.

10.3.4 An Example of the Usefulness of TSDC for Very Thin Films

Current technology of very thin film fabrication makes it possible to extend the application of such films to uses in devices whose operation depends on quantum mechanical tunneling of charge carriers. For example, a variety of metal/insulator/semiconductor (MIS) structures have been exploited for a wide range of applications (51-53) using Langmuir-Blodgett (L-B) film (<10 nm) (54,55) as insulator. The uniformity of formation and transfer of L-B films has been demonstrated by Batey (56).

Although most applications have been with nonpolymeric L-B films, current interest in polymeric L-B films, particularly on semiconducting substrates, is likely to increase. It is axiomatic that the L-B films used as insulators in MIS devices must be sufficiently thick and stable to withstand a voltage, yet thin enough not to obstruct the passage of charge carriers. Application of the TSDC technique for evaluation of voltage withstanding and charge transfer capabilities of such film barriers will offer an ideal method for optimization of insulator thickness.

10.4 PRINCIPLE OF OPERATION AND EXPERIMENTAL SETUP

The methodology of TSDC measurement involves the following steps:
1.   Thermal activation of the polymer sample by a linear heating program.
2.   Application of a high DC voltage when the temperature reaches a characteristic value for the specimen while continuing the temperature program to a preselected maximum value.
3.   Maintaining constant temperature at the maximum value for a period of time (determined by experiment) while still under voltage field.
4.   Cooling down to ambient or to a lower temperature while maintaining the applied voltage.
5.   Discontinuing the voltage input and shorting the electrodes to a recorder.
6.   Reheating the sample at a slow program rate to the desired maximum temperature and recording the discharge current.

The above sequential steps perform the following functions. In Step 1, the polymer is made structurally responsive to polarization by applied thermal energy prior to voltage application. Step 2 is initiated when the glass transition temperature, T_g, of the polymer is reached. At this temperature, relaxation times for characteristic molecular motions in the polymer are likely to be short and rapid conformational rearrangements begin to occur. The applied voltage can interact with the polymer chain terminal groups and reorient dipoles to bring about specific dipole/dipole interactions. The magnitude of the applied voltage is best determined by experiment. It can range anywhere between 0.1 and 50 kV or even higher per cm of sample thickness, depending on the nature of the polymer. Step 3 ensures alignment of oriented dipoles in the polarized film which are then "frozen-in" by cooling, in Step 4.

In Steps 5 and 6, a plot of TSDC against time is obtained. The heating rate is usually 2 or 3°/min or lower. The amount of charge, Q, frozen at each transition can be found by integrating current with respect to time:

$$Q = \int_0^\infty i\, dt \tag{17}$$

where, as before, i is the current and t is the time. The relationships of voltage field, temperature, and charge in the electret formation process are shown schematically in Fig. 10.1.

The field is applied at time t_0 when the temperature reaches the glass transition value, marked T_g on the temperature axis in Fig. 10.1. T_s and T_{max} are the starting and maximum temperatures, respectively. At time t_f the polymer is cooled and

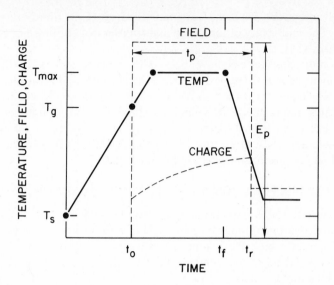

Figure 10.1:   Schematic representation of the events in the polarization process.

voltage is turned off at time t_r. t_p is the duration of voltage application and E_p is the field strength. The charge increases with time but drops to a stable value after losing the instantaneous charge when the voltage is turned off. A practical arrangement for the apparatus for implementation of the above events is shown in Fig. 10.2.

The heating of the sample between two electrodes can be carried out in a programmed temperature bath of silicone oil or in a gaseous medium under programmable thermal control. The method of placement of the sample between the electrodes has been discussed earlier. Cooling can be achieved by circulating a coolant, such as ethylene glycol/water mixture, from a compressor through the sample chamber or by passing liquid nitrogen through the thermostat jacket. The different accessories shown in Fig. 10.2 can be readily obtained from various manufacturers. A diagrammatic representation of a typical TSDC thermogram showing separation of relaxation peaks in the different temperature regions for an unspecified polymer is shown in Fig. 10.3.

The α, β, and γ-transitions, as already described, embody events internal to the polymer. The high-temperature region (ρ-peak area) originates from polarization at heterogneous polymer interfaces due to external charging [Maxwell-Wagner effect (29,30)].

10.5 REVIEW OF SOME APPLICATIONS OF TSDC MEASUREMENT

The first description of an application of the electret effect was in 1928 in the form of a microphone (57). This microphone used a Wax electret which did not have the required electrical stability under working conditions. It was not until after 1962 that this microphone was made workable by Sessler and West (58,59) by the introduction of a thin film polymer electret. This device was made commercially available by the SONY Corporation in 1969. From this functional beginning, the actual meas-

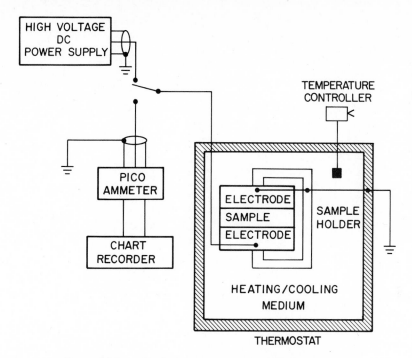

Figure 10.2:   Typical arrangement for TSDC apparatus.

urement of TSDC has quickly found widespread applications in polymer physics. The technique has been found to be useful for elucidation of polymer structures and their influence on polymer reactivity. Additionally, coordination of the experimental results derived from TSDC measurement with corroborative electrical and mechanical techniques has aided in the understanding of the low-frequency behavior of polymer dielectrics and the latent effects of frozen-in mechanical stresses in polymeric materials. Such work, however, has not necessarily been done with very thin (<10 μm) films. Nevertheless, TSDC measurement is characteristically well suited for thin films (6-25 μm) whose increasing applications in modern electronics are likely to produce further refinements in the sample-handling techniques and in their analytical applications.

Selected examples from the literature of the utility of TSDC measurement for polymers are quoted below. Additional references are given at the end of the chapter for general reading on applications.

10.5.1 Structural Parameters

The effects of substitution and steric hindrance on molecular reactivity are best exemplified by the early work on methacrylate polymers and their cyclic substituents.

Figure 10.4 shows the TSDC thermograms of poly(ethylmethacrylate), PEMA, and poly(methylmethacrylate), PMMA (24). The results for methacrylic polymers with   cyclic   substituents   poly(cyclohexylmethacrylate),   PcHMA,   and

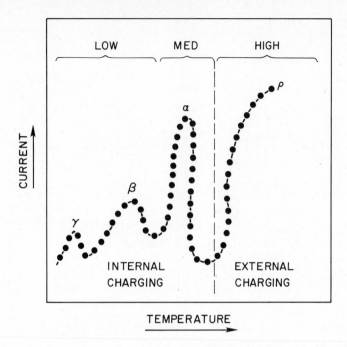

Figure 10.3:   Diagrammatic representation of a TSDC thermogram with separation of peaks.

poly(phenylmethacrylate),   PphMA,   and   for   two   copolymers   with methylmethacrylate are shown in Fig. 10.5 (24).

The α-peak corresponds to the glass transition temperature in each case. VanTurnhout (24) interprets this peak at 66°C for PEMA and at 102°C for PMMA as due to the forced motion of the side groups together with the main chains. The larger ethyl groups push the main chains further apart, causing internal plasticization. Consequently, the ethyl α-maximum is displaced to a lower temperature. The ρ-peak appearing at 85°C for PEMA and at 115°C for PMMA is attributed to drifting of space charges. This peak increases with the conductivity of the polymer. According to Heijboer (60,61), the β-peak occurring at -45°C for PEMA and at -51°C for PMMA can be ascribed to the local motion of the polar side group. Because the ethyl substituent is larger and more sterically hindered, the ethyl β-peak is smaller than the methyl β-peak (24). In Fig. 10.5, a γ-peak at -118°C for PcHMA was assigned by Heijboer (62) to a chair/chair transition of the 6-membered ring. A β-peak is hardly observed because of the steric hindrance to any sweeping motion of the bulky side group in PcHMA (24). The rigid phenyl group, on the other hand, gives no intra-alkyl relaxation, but gives a small β-peak at -13°C, which is partly due to its higher polarity (24). The ρ -peaks for the copolymers (60% MMA-co-40% cHMA and 50% MMA-co-50% PhMA) are found to be larger than expected from the mixing rule which is due to the fact that copolymers are more conductive than homopolymers.

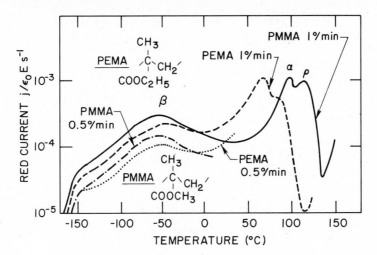

Figure 10.4:   TSDC thermograms of PEMA and PMMA.   The temperature shift with heating rate is also shown (from Ref. 24).

In another study, Vanderschueren and Linkens (63) have found that the β-relaxation decreases in going from PMMA to PEMA to poly(tertiarybutylmethacrylate) to PcHMA, due to increasing steric hindrance for side group motions of these polymers. This phenomenon occurs identically in the dielectric spectra of these polymers at 60 Hz (64). The spectra are shown in Figs. 10.6 and 10.7.

Vanderschueren and Linkens (63) had concluded that the β -peaks for these polymers by the two techniques originate from the same molecular relaxation phenomena. Although correlation of dipolar motions from dielectric studies with TSDC analysis is possible, locating exactly where the role of the ionic species is manifested in a thermogram is not well established (65). This uncertainty is based on the consideration (66) that a relaxation process, such as that at glass transition, T_g, may well coincide experimentally with the dielectric dispersion, giving credibility to the assumption that it is the dipolar nature of the side chains or the terminal groups that acquire mobility. However, high values of activation energy for T_g (1-3 eV) suggest that the dielectric effect might actually be due to the transport of real charges. Nevertheless, Stupp and Carr (66) have demonstrated by direct measurement of dipolar orientation that the decay of preferential orientation of the somewhat polar nitrile side groups in polyacrylonitrile occurs over exactly the same temperature range as the major relaxation process. It had also been shown earlier (67) that the peak due to glass transition in methacrylate polymers does obey the Arrhenius relationship.

10.5.2 Other Physical Parameters

TSDC measurements have been reported for the behavior of carriers and traps contributing to electrical conduction in polymers (13), and significant progress has been made in the understanding of the charging process involving dipoles. The effect of nondipole polarization at the polymer/filler interface on the behavior of carriers has

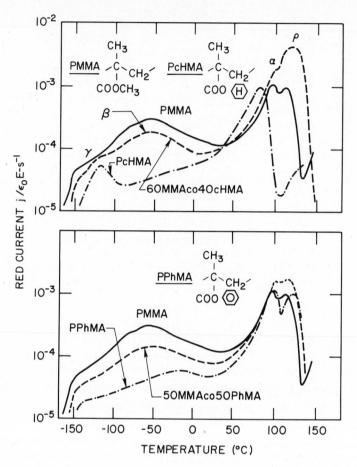

Figure 10.5:   TSDC thermogram of methacrylic homo- and copolymers with cyclic alkyl side groups (from Ref. 24).

also been studied (68) for elucidation of the mechanism of charge trapping in composites. Epoxy resins containing planar mica flakes as polymer composites were used in this study. Several inferences were drawn. The β -peaks for the resin and the composite and the γ-peak for the composite were considered to be due to dipole polarization, but the α-peak for the composite was attributed to space charge polarization occurring at the interface. The last conclusion was based on the fact that the current of the α-peak increased with the volume fraction of mica in the composite (Fig. 10.8). The plots for the dielectric dispersion of the α-peak for the resin (R) and composite (C) are shown in Fig. 10.9.

The authors observed that the rise in dielectric constant, ε', above 175°C for the composite indicated a dielectric relaxation in the high-temperature region, but the increase in the dielectric loss factor, ε'', above 175°C for the resin was likely to be due to DC or ohmic conduction, since no increase in ε' corresponding to the increase in ε'' occurred. The overall conclusion was that the region centered around 152°C

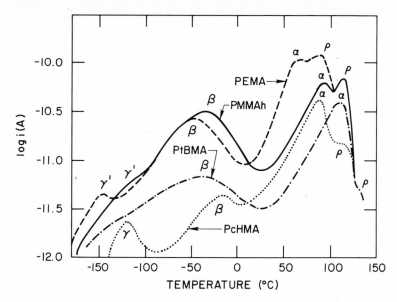

Figure 10.6:   TSDC spectra of some polymethacrylates. Polarization: E_p = 10 kV/cm, t_p = 30 min, T_0 = -196°C, T_p = 127°C (PMMA), 90°C (PEMA), 140°C (Pt BMA), and 120°C (PcHMA). Heating rate = 5°C/min (from Ref. 63).

was due to interfacial polarization from charge trapping of conductive ions in the resin matrix at the boundaries between the resin matrix and the mica flakes. In a recent study, it has been shown (69) that glass microsphere fillers in a radiation-cured unsaturated polyester composite had little effect on the intrinsic properties of the resin in the glass transition range, but interfacial polarization at the phase boundaries and interfaces was induced by the fillers. The latter effect was evident in the enhancement of the amount of measured depolarization current. It should be borne in mind, however, that ionizable radiation can create charged species in polymers, and some of these charges are capable of drifting at a suitably high temperature, contributing to the discharge current. Impurities in polymers can also produce charge enhancement. Impurity effects on the assessment of the compatibility of certain polymer blends have been studied (70).

The relaxation behavior of *n*-butylmethacrylate/methacrylic acid copolymers and their alkaline salts with various degrees of neutralization has been investigated (71) from 77 to 400 K by means of thermally stimulated depolarization and thermally stimulated polarization methods. Evidence of ionic clusters was found for higher than 50% neutralization. By studying the influence of polarization conditions, these authors have shown that the appearance and stability of the clusters are markedly dependent on thermal and electrical history of the samples. Additionally, clustering relaxation is closely related to the ionic conductivity via space-charge formation.

Development of preferred orientation in microstructure due to mechanical stress can be detected by TSDC measurement. Sacher (72) has shown that such orientation induced during manufacture of polyethylene terephthalate films can produce an elec-

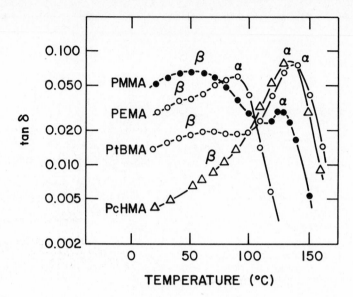

Figure 10.7:   Dielectric losses at 60 Hz of the polymethacrylates shown in Fig. 10.6 (from Ref. 64).

trical anisotropy that can be characterized by the depolarization technique. An important application-oriented analysis for the TSDC technique has been described by Sharp and Garn (73). It involves the use of TSDC measurement for evaluation of aging characteristics of electret materials for determination of their useful service life.

10.6 PRESENT AND POTENTIAL FUTURE APPLICATIONS OF THIN POLYMER ELECTRETS

Electrets formed by electrical polarization of thin polymer films have found a wide variety of practical applications. Projections have been made for future use of very thin electret films in active solid state and photovoltaic devices, electrical contacts and switches, solar photothermal adsorption devices, conductive coatings in optoelectronic devices, and as Josephson junctions in low-temperature devices.

Most of the applications of polymer electrets are based on their excellent charge-holding capabilities, the origin of which can be found in the dielectric characteristics of these polymers. The charge in electrets may be either real charge, such as that trapped in the polymer, or developed due to dipole orientation within the polymer. Displacement of real charges within domain structures gives rise to interfacial charge boundaries. Insulating (low dielectric) polymers, such as polytetrafluoroethylene (Teflon) can be made into very good charge storage devices, whereas use of polar polymers, such as polyvinylidenefluoride (PVDF), is based on their piezoelectric and pyroelectric properties.

The following summary of present and promising future applications of polymer electrets is largely from Sessler (74).

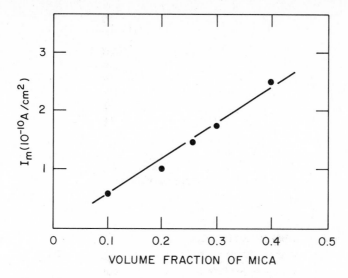

Figure 10.8:   Dependence of the maximum current, I_m , of the α-peak of the composite on volume fraction of mica (from Ref. 68).

10.6.1 Electret Transducers

The best-known example of a polymer electret acting as a transducer is the electret microphone developed by Sessler and West (58,59). The microphone is designed to consist of metallized acrylonitrile/butadiene/styrene backplate and a thin Teflon electret diaphragm stretched across it. The working principle of this microphone is based on the generation of an AC signal between the metallization of the backplate and the diaphragm, when sound waves impinging on the diaphragm causes it to vibrate. Such an arrangement has found application in telephone and speakerphone systems. Additionally, monocharge electret earphones based on nonmetallized electrets carrying only a single polarity charge are being used presently (75). Electret film contact sensor transducers hold promise of application in sophisticated instrumentation.

10.6.2 Electrostatic Imaging and Charge Transfer

The production of charge pattern on an appropriate carrier and its development with powders (electrophotography) was already studied (76). The breakthrough in this field came a few years later when investigations of photoconductive image formation led to the development of Xerographic reproduction methods (77). The polymer polyvinylcarbazole (PVC) in the form of a charge transfer complex with trinitrofluorenone (TNF) has been extensively reviewed (78,79).

Methods, such as electrostatic recording processes, are related to electrophotography, but these are not dependent on photographic methods. Recording on such carrier materials as polymers and paper by electrical discharge utilizes the electrostatic recording processes (80).

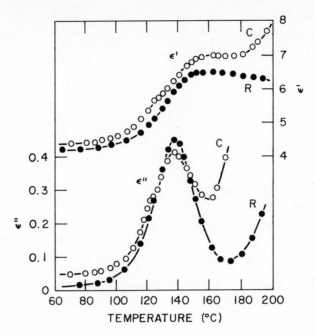

Figure 10.9:   Temperature dependence of ε' and ε'' of resin (R) and composite (C) measured at 1 kHz (from Ref. 68).

10.6.3 Piezoelectric and Pyroelectric Devices

A number of very recent electret applications are based on the piezoelectric and pyroelectric effects in polarized polymers, particularly PVDF. The piezoelectric effect of PVDF is utilized in microphones that have mechanical stability and a relatively large capacitance, simplifying amplification requirements.

As a pyroelectric material, PVDF has the advantages of excellent mechanical properties, availability as thin films and low thermal conductivity. The pyroelectric effect of PVDF has been used in fire detectors and burglar alarm systems. Further, suggestions for use of pyroelectric polymer electrets have been made by Quilliam (81).

10.6.4 Biomedical Applications

Potential future applications of polymer electrets are likely to be in the biological and medical fields. Attempts have been made to improve the blood compatibility of polymers for implants by negative charge deposition (82) and to detect electret properties of human bones and blood vessel walls (83). Electret bandages have been found to hasten the healing of wounds by improving the tensile strength of the wound over a given period of time (84).

10.6.5 Other Applications

A very useful application of polymer electrets is exemplified by gas filters made of polypropylene films (85). The film is corona charged and fibrillated to create a broad web because of the repulsive forces. The electret fibers capture charged particles by Coulombic forces and neutral particles by inductive forces. Filtration characteristics of electret filters with different fibers have been compared by VanTurnhout (85). An additional list of references on electret applications is appended.

10.7 CONCLUDING REMARKS

The essential features of thermally stimulated discharge current (TSDC) analysis techniques have been provided in this chapter. The treatment is by no means complete. Rather, the aim here has been to develop an essential understanding of the background and the basis for this powerful laboratory technique which has proven to be a very useful thermoelectric analytical tool for evaluation of polymer applications. These applications may range from determination of structure/property relationships of polymers to the practical utility of their charge storage capacity.

The effect of an impressed voltage field on the dielectric nature and the thermal properties of polymers in producing charge domains in the polymer is utilized in this technique. The experimental variables allow for specific charging of surface and interfaces which can be analyzed via thermospecific discharge.

Thermal relaxation of polarizations in polymer microdomains originating from the charging phenomenon can be resolved with sufficient sensitivity for the technique to be useful for very thin samples which can be created by one of the several available deposition techniques.

Although the majority of studies have been done typically with thin polymer films, the potential of the TSDC analysis technique for studying surface and interfacial characteristics of very thin films (<10 nm) has not been fully exploited.

The technique owes its origin and maturity to many contributors. There is no doubt that the current interest in characterization of very thin polymer films will provide the necessary impetus for further refinement of the technique.

10.8 REFERENCES

1.   S. Gray, *Phil. Trans. Roy. Soc., London Ser.*, I, 37, 285 (1732).

2.   M. Faraday, **Experimental Researches in Electricity**, Taylor, London (1839).

3.   O. Heaviside, **Electrical Papers**, Chelsea, New York, 488, (1892).

4.   M. Eguchi, *Proc. Phys. Math. Soc. Jpn.*, 1, 917 (1919).

5.   H. H. Wieder and S. Kaufman, *J. Appl. Phys.*, 24, 156 (1953).

6.   G. Wiseman, *Research and Development Studies on Electrets*, University of Kansas Press (1955).

7.   V. A. Johnson, *U.S. Government Research Reports TR-1045 and TR-1074*, Harry Diamond Laboratories, Washington, D.C. (1962).

8.   B. Gross and R. J. DeMoraes, "Polarization of the Electret," *J. Chem. Phys.*, 37, 710 (1962).

9.   P. V. Murphy and B. Gross, "Polarization of Dielectrics by Nuclear Radiation. II. Gamma-ray-induced Polarization," *J. Appl. Phys.*, 35, 171 (1964).

10. B. Gross, **Charge Storage in Solid Dielectrics**, Elsevier Publishers, Amsterdam (1964).

11. J. Euler, **Neue Wege zur Stromerzeugung**, Akad. Verlag, Frankfurt, Chap. 4 (1964).

12. M. L. Miller, "Persistent Polarization in Polymers. I. Relations Between the Structure of Polymers and Their Ability to Become Electrically Polarized," *J. Polym. Sci.*, A-2, 4, 685 (1966).

13. C. Bucci, R. Fieschi, and G. Guidi, "Ionic Thermocurrents in Dielectrics," *Phys. Rev.*, 148, 816 (1966).

14. M. L. Miller and J. R. Murphy, "Persistent Polarization in Polymers. II. Depolarization Currents," *J. Polym. Sci.*, A-2, 4, 697 (1966).

15. H. Frei and G. Groetzinger, "Uber Das Freiwerden Elektrischer Energie Beim Aufschmelzen Des Elektren," *Phys. Z.*, 37, 720 (1936)

16. B. Gross, "Time-Temperature Superposition Theory for Electrets," *J. Electrochem. Soc.*, 115, 376 (1968).

17. L. M. Baxt and M. M. Perlman, Eds., **Electrets and Related Electrostatic Discharge Phenomena**, Symposium Series, Electrochemical Society, New York (1968).

18. S. Mascarenhas, N. Januzzi, and C. Arguello, *J. Electrochem. Soc.*, 115, 382 (1968).

19. R. A. Creswell and M. M. Perlman, "Thermal Currents from Corona Charged Mylar," *J. Appl. Phys.*, 41, 2365 (1970).

20. T. Takamatsu and E. Fukada, *Polym. J.*, 1, 101 (1970).

21. A. C. Lilly, R. M. Henderson and P. S. Sharp, "Thermally Stimulated Currents in Mylar. High-field Low-temperature Case," *J. Appl. Phys.*, 41, 2001 (1970).

22. T. Nedetzka, M. Reichle, A. Mayer, and H. Vogel, "Thermally Stimulated Depolarization. Method for Measuring the Dielectric Properties of Solid Substances," *J. Phys. Chem.*, 74, 2652 (1970).

23. J. VanTurnhout, in **Advances in Static Electricity**, Vol. 1, W. F. deGeest, Ed., Auxilia, Brussels (1971).

24. J. VanTurnhout, "Thermally Stimulated Discharge of Polymer Electrets," *Polym. J.*, 2 (2), 173 (1971).

25. J. VanTurnhout, **Thermally Stimulated Discharge of Polymer Electrets**, Elsevier Scientific, New York (1975).

26. M. M. Perlman and S. Unger, "TSC Study of Traps in Electron-Irradiated Teflon and PE," *Intl. Conf. on Electrets, Charge Storage and Transport in Dielectrics*, Fall Meeting, Electrochemical Society, Miami Beach (1972).

27. J. Vanderschueren and J. Gasiot, in **Thermally Stimulated Relaxation in Solids**, P. Braunlich, Ed., Springer Verlag, Berlin, 138 (1979).

28. B. Gross, G. M. Sessler, and J. E. West, "TSC Studies of Carrier Trapping in Electron- and Gamma- Irradiated Teflon," *J. Appl. Phys.*, 47, 968 (1976).

29. J. C. Maxwell, **Electricity and Magnetism**, Vol. 1, Oxford University Press, London, 452 (1892).

30. K. W. Wagner, *Ann. Phys.*, 40, 817 (1913).

31. O. G. Altheim, "Properties of Amorphous and Crystalline Quartz in the Electrostatic Field," *Ann. Phys.*, 35, 417 (1939).

32. B. Chowdhury, "Thermally Stimulated Discharge Current for Epoxy Resins," *Am. Lab.*, 17 (1), 49 (1985).

33. M. L. Williams, R. F. Landel, and J. D. Ferry, "The Temperature Dependence of Relaxation Mechanisms in Amorphous Polymers and Other Glass-forming Liquids," *J. Amer. Chem. Soc.*, 77, 3701 (1955).

34. B. Gross, "On Permanent Charges in Solid Dielectrics II: Surface Charges and Transient Currents in Carnauba Wax," *J. Chem. Phys.*, 17, 866 (1949).

35. G. F. J. Garlick and A. F. Gibson, "The Electron Trap Mechanism of Luminescence in Sulphide and Silicate Phosphors," *Proc. Phys. Soc.*, 60, 574 (1948).

36. M. M. Perlman, "Thermal Currents and the Internal Polarization in Carnauba Wax Electrets," *J. Appl. Phys.*, 42, 2645 (1971).

37. B. Chowdhury, "Thermally Stimulated Conductivity at a Polymer Interface," *Am. Lab.*, 18(1), 28 (1986).

38. B. Chowdhury, "Direct Applicability of Time-Dependent Viscoelastic Response to Polymer Interfacial Depolarization," *Am. Lab.*, 19 (11) 35 (1987).

39. J. Vanderschueren, *Ph.D. Thesis*, University of Liege (1974).

40. N. M. Murayama and H. Hashizuma, *J. Polym. Sci., Polym. Phys. Ed.*, 14, 989 (1976).

41. C. Lacabanne and D. Chatain, *Makromol. Chem.*, 179, 2765 (1978).

42. V. K. Jain, C. L. Gupta, R. K. Jain, and R. C. Tyagi, "Thermally Stimulated Currents in Pure and Copper-doped Polyvinyl Acetate," *Thin Solid Films*, 48, 175 (1978).

43. J. Vanderschueren and A. L. Linkens, "Water-dependent Relaxation in Polymers. Study by the Thermally Stimulated Current Method," *Macromolecules*, 11, 1228 (1978).

44. See, for example, D. R. Paul, and S. Newman, Eds., **Polymer Blends**, Academic Press, New York (1978).

45. A. Joshi, L. C. Davis, and P. W. Palmburg, in **Methods of Surface Analysis**, A. W. Czanderna, Ed., Elsevier, New York, Chap. 5 (1975).

46. S. Schiller, O. Heisig, and K. Goedicke, in *Proc. 7th Intl. Vacuum Congress*, R. Dobrozemsky, Ed., Vienna, 1545 (1977).

47. C. Weissmantel, ibid., 1533.

48. R. F. Bunshah and D. M. Mattox, "Applications of Metallurgical Coatings," *Physics Today*, 33 (5), 50 (1980).

49. R. F. Bunshah, et al., **Deposition Technologies for Films and Coatings**, Noyes Publications, New Jersey (1982).

50. L. I. Meiksin and R. Glang, **Handbook of Thin Film Technology**, McGraw-Hill, New York (1970).

51.   M. C. Petty, P. J. Batey and G. G. Roberts, "A Comparison of the Photovoltaic and Electroluminescent Effects in GaP/Langmuir-Blodgett Film Diodes," *I.E.E. Proc. I.*, 132, 133 (1985).

52.   I. M. Dharmasada, G. G. Roberts, and M. C. Petty, *Elec. Lett.*, 16, 201 (1980).

53.   G. G. Roberts, M. C. Petty, and I. M. Dharmasada, *I.E.E. Proc. I.*, 128, 197 (1981).

54.   I. Langmuir, "Constitution and Fundamental Properties of Solids and Liquids. II. Liquids," *J. Amer. Chem. Soc.*, 39, 1848 (1917).

55.   K. B. Blodgett, "Monomolecular Films of Fatty Acids on Glass," *J. Amer. Chem. Soc.*, 56, 495 (1934).

56.   J. Batey, *Ph.D. Thesis*, University of Durham (1983).

57.   S. Nishikawa and D. Nukijama, *Proc. Imp. Acad. Tokyo*, 4, 290 (1928).

58.   G. M. Sessler and J. E. West, "Self-biased Condenser Microphone with High Capacitance," *J. Acoust. Soc. Amer.*, 34, 1787 (1962).

59.   G. M. Sessler and J. E. West, "Foil-electret Microphones," *J. Acoust. Soc. Amer.*, 40, 1433 (1966).

60.   J. Heijboer, *Brit. Polym.*, 1, 3 (1969).

61.   J. Heijboer, F. R. Schwarzl, and H. Thurn, *Kunststoffe*, Vol. 6, Nitsche and Wolf, Eds., Springer Verlag, Berlin, 633 (1962).

62.   J. Heijboer, "Molecular Significance of Secondary Damping Maxima," *Kolloid-Z.*, 148, 36 (1956); "The Movement of the Cyclohexyl Group in Glassy Polymers (a Dynamic Mechanical and Dielectric Investigations," ibid., 171, 7 (1960).

63.   J. Vanderschueren and A. Linkens, "Thermally Stimulated Depolarization and Dielectric Losses in Polymers. Some Problems Posed by the Comparison of Activation Energies," *J. Electrostatics*, 3, 155 (1977).

64.   J. Heijboer, *Macromol. Chem.*, 35A, 86 (1960).

65.   S. H. Carr, "Thermally Stimulated Discharge Current Analysis" in **Electrical Properties of Polymers**, D. A. Seanor, Ed., Academic Press, New York, 232 (1982).

66.   S. I. Stupp and S. H. Carr, "Spectroscopic Analysis of Electrically Polarized Polyacrylonitrile," *J. Polym. Sci., Polym. Phys. Ed.*, 16, 13 (1978).

67.   A. Bui, H. Carchans, J. Gustavino, D. Chatain, P. Gautier, and C. Lacabanne, *Thin Solid Films*, 21, 313 (1974).

68.   T. Tanaka, S. Hayashi, and K. Shibayama, "Thermal Depolarization-Current Study of Composites of Epoxy Resin Containing Mica Flakes," *J. Appl. Phys.*, 48, 3478 (1977).

69.   Z. Jelcic and F. Ranogajec, *Proc. Tihany Symp. Radiation Chem.*, 6th (2), 545 (1987).

70.   P. Alexandrovich, F. E. Karasz, and W. J. Mackmight, "Thermally Stimulated Discharge in Polymer Electrets: Compatibility and Impurity Effects," *J. Appl. Phys.*, 47 (10), 4251 (1976).

71.   J. Vanderschueren, L. Aras, C. Boonen, J. Niezette, and M. Corapci, "Evidence for Clustering in N-butyl Methacrylate Ionomers as Shown by Thermally Stimulated Currents," *J. Polym. Sci., Polym. Phys. Ed.*, 22, 2261 (1984).

72.   E. J. Sacher, *Macromol. Sci. Phys.*, 7, 231 (1973).

73.   E. J. Sharp and L. E. Garn, *Appl. Phys. Lett.*, 29, 480 (1976).

74.   G. M. Sessler, "Polymer Electrets", in **Electrical Properties of Polymers**, D. A. Seanor, Ed., Academic Press, New York, 274 (1982).

75.   H. J. Griese and G. Knock, *Funkschau*, 49, 1251 (1977).

76.   P. Selenyi, *U.S. Patent 1,818,760* (1931).

77.   C. F. Carlson, *U.S. Patent 2,221,776* (1940).

78.   W. D. Gill, **Photoconductivity and Related Phenomena**, Elsevier, Amsterdam, 303 (1976).

79.   J. Mort, "Polymers as Electronic Materials," *Adv. Phys. (GB.)*, 29, 367 (1980).

80.   U. Rothgordt, *Philips Tech. Rdsch.*, 36, 98 (1976/77).

81.   M. Quilliam, *Electron. Ind.*, 3, 23 (1977).

82.   P. V. Murphy and S. Merchant, in **Electrets, Charge Storage and Transport in Dielectrics**, M. M. Perlman, Ed., Electrochemical Society, Princeton, NJ, 627 (1973).

83.   S. Mascarenhas in **Electrets**, G. M. Sessler, Ed., Springer Verlag, Berlin, 321 (1980).

84.   J. J. Konikoff and J. E. West, *Annu. Rep. Conf. Electr. Insul. Dielectr. Phenomena*, National Academy of Sciences, Washington, D.C., 304 (1978).

85.   J. van Turnhout, J. W. C. Adamse, and W. J. Hoeneveld, "Electret Filters for High-Efficiency Air Cleaning," *J. Electrostat.*, 8, 369 (1980).

REFERENCES FOR ADDITIONAL READING

1.   G. R. Davis, **Physics of Dielectric Solids**, Institute of Physics Conf. Ser. No. 58, 50 (1981).

2.   R. Zahn, "Analysis of the Acoustic Response of Circular Electret Condenser Microphones," *J. Acoust. Soc. Am. (USA.)*, 69, 1200 (1981).

3.   G. M. Sessler, **Electrets**, Springer-Verlag, Berlin (1980).

4.   J. van Turnhout and G. M. Sessler, Eds., **Electrets**, Springer-Verlag, Berlin, Chap. 2 (1978).

5.   P. C. Mehendru, K. Jain, and P. Mehendru, "TSD Current Studies of Iodine-doped Polyvinyl-acetate Thin Films," *J. Phys. D*, 9, 83 (1976).

6.   M. Kryszevski, M. Zielinski, and S. Sapieha, *Br. Polym. J.*, 17, 212 (1976).

7.   T. Hashimoto, M. Shiraki, and T. Sakai, *J. Polym. Sci.*, 13, 2401 (1975).

8.   R. J. Gable, N. V. Vijayraghavan, and R. A. Wallace, *J. Polym. Sci., Polym. Chem. Ed.*, 11, 2387 (1973).

9.   D. A. Seanor, K. C. Frisch, and A. Patsis, Eds., **Electrical Properties of Polymers**, Technomic Publ., Westport, Conn. (1972).

10.   E. J. Sacher, *J. Macromol. Sci. Phys.*, 6, 449 (1970).

11

XPS/SIMS/AES FOR SURFACE AND INTERFACE CHARACTERIZATION OF THIN POLYMER FILMS

N. J. Chou

Thomas J. Watson Research Center
IBM Research Division
Yorktown Heights, New York

The advance in polymer science and technology has been so rapid in the last two and half decades that polymers now find pervasive use in practically every branch of engineering. Recent development of polyimides (1), a new class of processible, heat-resistant polymers, endowed with good mechanical stability and chemical inertness and a relatively low dielectric constant, further extended their use to advanced microelectronics, notably, the VLSI packaging technology. In the latter case, polyimides serve as insulating films in a multilayer metal/polymer structure built on top of alumina or glass/ceramic substrates (2,3). Since adhesion and "contaminant-based" degradation are a major concern in the performance and reliability of these composite structures, many investigators have focused their attention on the surface and interface chemistry, and its effect on the adhesion strength of thin polymer films in contact with metal or ceramic substrates as well as thin metal films on polymer or ceramic substrates. Various surface sensitive analytical techniques have been successfully used to characterize the polymer films and their interfaces with metals and ceramics.

Of these characterization techniques X-ray photoelectron spectroscopy (XPS), or commonly known as ESCA (electron spectroscopy for chemical analysis),* proves to be the most popular and powerful tool despite its limited spatial resolution. This is because the technique does not cause severe radiation damage in polymer films, and the sample-charging effect during XPS measurement can be easily handled with an electron flood gun. More importantly, XPS has the ability to determine chemical composition without having to resort to unduly complicated quantification procedures, and it provides the chemical state information from which changes in the molecular structure of polymer films can often be inferred.

---

* The acronym was first used by Siegbahn (4) to underscore the fact that both core-level and Auger electrons appear as peaks on the X-ray excited electron energy distribution curves. Surface scientists now prefer to use the term XPS in order to distinguish it from the electron-excited Auger electron spectroscopy. In this chapter, we will use these two terms interchangeably as the tradition dictates.

Auger electron spectroscopy (AES) is generally regarded as unsuitable for analyzing polymer films for two reasons: first, sample charging effect makes data acquisition very difficult, if not impossible, and second, the electron-induced radiation damage occurs at an unacceptably high rate. It is, however, an excellent tool for failure analysis in a high-performance multilayer microelectronic packaging structure. In this case, the sample charging effect can be usually circumvented by utilizing the metal lines and planes that are already present in the structure, and the high detection sensitivity and spatial resolution of the technique can be fully utilized. The interface exposed after adhesion failure, for example, can be examined by AES to determine the mode of failure and the presence, if any, of contaminants. It is known that one way to circumvent sample charging in a metal/polymer structure is to deposit a thin layer of gold on the structure, and depth profiling can be performed by ion etching through the overlayer.

Although XPS provides the multielement and chemical state information in the analysis of polymer films, it cannot always uniquely define their molecular structures since chemical shifts associated with similar bonds do not yield easily resolvable XPS peaks. When secondary ion mass spectroscopy (SIMS) is used at very low primary ion beam currents (<10 nA/cm^2) it becomes what is usually known as static SIMS (SSIMS). It is a surface sensitive technique since the removal of one monolayer of the analyzed material at such beam dosages can last several hours. The fragmentation patterns of the cluster ions, and molecular and monatomic ions emitted in the process provide the information regarding the chemical structure of the surface. Applied to polymer films, SSIMS requires some means of eliminating sample charging, and low-energy electron flooding is frequently used. This often gives rise to uncertainties resulting from electron-stimulated desorption (ESD) or decomposition (5). A modified version of SIMS is known as fast atom bombardment mass spectrometry (FABMS) (6), in which neutral particles are used to bombard the target, thus offering an alternate means of analyzing polymer surface and interface. It should be noted that while SIMS-type measurements are orders of magnitudes more sensitive than XPS and AES, their quantification requires a careful and tedious procedure because secondary ion emission is very sensitive to the electronic state of the atoms to be ionized and to the matrix from which ions are emitted (7).

Other spectroscopic methods such as attenuated total reflection or multiple internal reflection Fourier transform infrared spectroscopy (FTIR-ATR or FTIR-MIR) (8), ion scattering spectroscopy (ISS) (9), high-resolution electron energy loss spectroscopy (HREELS) (10), and Raman spectroscopy (11) have also been used in polymeric studies. They each have their advantages and limitations. It has become increasingly clear that in many occasions, several properly chosen surface science techniques must be used either jointly or in tandem in order to unravel the complex nature of polymer surface and interface (12).

This chapter will be devoted to a discussion of three most popular and routinely used techniques, namely, XPS, SIMS, and AES. Since all these techniques are surface science related, many similarities exist among them in terms of sample preparation and handling, vacuum requirement and system organization, instrumentation, placement of accessaries, and so on. To avoid repetition, similar subjects will be covered only once under one of the three techniques. For example, the procedures for sample preparation and handling, which will be covered in some detail under XPS, will not be discussed under SIMS or AES.

11.1 X-RAY PHOTOELECTRON SPECTROSCOPY (XPS)

XPS has gradually evolved into a powerful surface sensitive analytical tool in the last two decades since Siegbahn and coworkers (4) first established in 1967 that the energy-analyzed electrons, photoemitted during irradiation of a solid sample by monochromatic X-rays, exhibited sharp peaks which corresponded to the binding energies (BE's) of core-level electrons in the sample. Carefully catalogued, these BE peaks provide a quick means of identifying the chemical constituents in the specimen. Since the energy of photoemitted electrons decay exponentially with the distance they travel in the specimen, only those electrons that escape from the outermost few monolayers of the specimen can retain their identity in terms of energy. XPS has therefore a sampling depth of < 50 Å in metals and polymers.

Since BE's are sensitive to local charge environment, the XPS peaks are expected to shift significantly as the chemical environment is changed. As much as 10 eV or more in chemical shift has indeed been observed, for example, between core levels (C 1s) of carbon atoms in different chemical environments. On other occasions, however, chemical shifts can be disappointingly small even when changes involving significant charge redistribution have occurred. This is because relaxation and secondary excitation always occur during core-hole production (i.e., removal of a core electron) and XPS peaks measure the energy difference between the ground state of the sample and the final state of the ionized sample. In this case, it is very difficult to predict the correspondence between XPS peaks and the ground state properties. Experimentally determined values of BE's for known substances are therefore frequently used to extract chemical state information from the measured XPS spectra.

A substantial body of literature has been created dealing with the subject of binding energies, both theoretically and experimentally. Particularly valuable to the study of polymers is the work of Clark and coworkers (13), who have over the past 10 years systematically determined and compiled the BE's of polymers and relevant model compounds.

In this section, we shall show how XPS can be advantageously used to characterize polymer surfaces and interfaces. To do this, we shall describe the instrumentation and experimental techniques of XPS in some detail, and discuss the interpretation of XPS data in terms of primary and satellite spectral features. We shall examine various problems associated with XPS measurement, and use some recent illustrative examples to explore various approaches to their solution.

11.1.1 Instrumentation

An electron spectrometer consists of three principal parts: an excitation (ionization) source, an electron energy analyzer, and a detector. They are housed in an ultrahigh vacuum (UHV) chamber, equipped with turbomolecular and ion pumps. Where spatial resolution is required, the system should be isolated from the vibration of a turbopump. In an XPS system, soft X-rays of characteristic energies are usually used to excite core-level electrons from a specimen.

Other components and accessories available in the spectrometer system may include various fixtures used: (1) to transport a specimen from an air-lock or a preparation chamber into the spectrometer, (2) to heat or cool the sample at the analysis position, (3) to expose the sample to various gases, (4) to deposit materials onto the sample, or (5) to "clean" the sample surface with ion sputtering in the preparation chamber. UHV compatible modular accessories such as specimen manipulator, crystal cleaver, fracture stage, and so on, are now commercially available from UHV equipment manufacturers.

X-Ray Sources: Since the spread in kinetic energy of the X-ray excited electrons depends directly on the natural width of the characteristic X-ray lines, Mg and Al anodes are universally used in the X-ray source. Mg K_α and Al K_α lines have a natural width of 0.7 and 0.85 eV, respectively. This limits the overall energy resolution achievable in practice to a value ranging from 0.85 to 1.2 eV, since line broadening also occurs in other parts of the spectrometer. Improvement in energy resolution has been made by monochromatizing the X-rays.

Monochromatization of X-rays is accomplished by diffraction in a crystal of appropriate lattice spacing. For Al K_α radiation with a wavelength $\lambda = 0.83$ nm, quartz crystal is a natural choice, since the spacing of its $10\overline{1}0$ planes is 0.475 nm. The method of monochromatization used in the majority of commercial instruments is fine focusing. As shown in Fig. 11.1, the quartz crystal is placed on the circumference of a Rowland or focusing circle, its shape ground or bent to approximate that of a sphere. The Al anode, lying also on the circle, is irradiated with a tightly focused electron beam to give off Al K_α X-rays, which are dispersed and refocused by the quartz crystal at another point on the circle where the specimen is mounted. With a proper choice of the diameter of the Rowland circle and the electron beam spot size, a line width of 0.2 eV for Al K_α lines has been obtained on commercially available XPS instruments. Instrumental broadening and sensitivity (signal-to-noise) considerations, however, reduce the overall resolution, and the best energy resolution achieved to date on most of the spectrometers with monochromatized sources is 0.6 eV.

Monochromatization removes satellite interference and improves signal-to-noise ratio by eliminating the bremsstrahlung continuum in the X-radiation spectrum. However, it reduces the photon flux by a factor of about 40 when comparison is made between a standard nonmonochromatized source and the source with a 1-m-diameter monochromator.

Monochromatized soft X-ray sources commercially available are line sources. As such, they have inherent limitations. The photoionization cross section of core level electrons, for example, varies with primary photon energy. A fixed line source may be near the cross section maximum for one group of core levels while near minimum for another. With a continuous source one would be able to scan through the photon spectrum to pick out the features that would have been missed by the line source.

A continuously tunable, intense soft X-ray source is provided by synchrotron radiation facilities. Accelerating electrons in the torus (or "doughnut") of a synchrotron emit photons in a continuous spectrum having an intensity maximum proportional to the curvature of the electron path and inversely proportional to the cube of the electron energy (14). When the acceleration is nearly relativistic, the radiation is concentrated in a very small solid angle tangential to the electron orbit, and can be readily tapped to pass through a monochromator for energy selection or tuning.

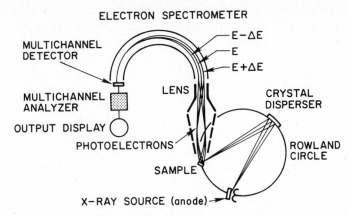

Figure 11.1: Principle of monochromatization of X-rays. Selection of a particular wavelength is achieved by diffraction in a crystal of suitable lattice spacing and quartz is a convenient crystal to use for the Al Kα lines. Placed on the surface of a Rowland circle, the quartz crystal is accurately ground to match the curvature. The X-ray source is situated at another point on the circle. Photons of the required wavelength will be focused at a third point on the circle where the sample is placed. For the Al K α_1 line, the angle between the incident and diffracted beams is 23°. A typical circle diameter is 0.5 m. From Kelly and Tyler (17).

Although "pocket-size" synchrotrons are now commercially available, their large sizes and prohibitive costs make it unlikely for an investigator to install such a device in a surface science laboratory. XPS users must either take their own spectrometers to the national synchrotron facilities or use those already available on a beam line. In addition, synchrotron radiation is not continuously tunable in the full range of electron energy. Limited by monochromator capability, the tunable range is 20 to 500 eV at the present time. However, advances in monochromator design and fabrication are expected to extend the energy range to those of discrete line sources very soon. Meanwhile synchrotron radiation, despite its present limitations, still stands out as an attractive excitation source for XPS because its intensity is about two orders of magnitude stronger than the standard line sources.

One of the drawbacks of XPS is its lack of spatial resolution. In contrast to advances in the field of ion and Auger electron microscopies, efforts to reduce the XPS probing area have so far managed to reduce the X-ray beam spot size to only ~ 200 μm in diameter, and XPS systems with such capability are appropriately called "small spot" ESCA machine. There are two basic approaches to improving the spatial resolution: (1) to use a tightly focused electron beam for X-ray production, and (2) to employ an electron lens system to collect photoelectrons emitted from a very small area. Both approaches call for increased X-ray intensity, which is limited by the practical problem of anode cooling. The latest version (15) of small spot ESCA uses a rotating anode and an imaging optics, and has reportedly a spatial resolution of 25 μm and an energy resolution of 0.3 eV (determined by Fermi level measurement).

Electron Energy Analyzers: There are two basic designs of electron analyzers used in commercially available ESCA spectrometers. In a double-pass or two-stage cylindrical mirror analyzer (CMA), a reflecting potential V is applied between two coaxial cylinders of radii r_o and r_i. As shown in Fig. 11.2, photoejected electrons entering the analyzer through the first entrance slit at an angle α are deflected by the applied potential V, and only those of a particular energy E_0 satisfying the relationship $E_0/eV = k/\ell n(r_i/r_o)$, can pass through two successive stages of the analyzer to reach the detector. It can be shown (16) that for $\alpha = 42°18'$, the analyzer becomes a second-order focusing instrument, and k assumes the limiting value of 1.3098. For a given r_i/r_o, the ratio of the pass energy to applied potential is then fixed. The popular commercial CMA's have a ratio of 1.7.

Under the second-order focusing conditions, $\Delta\alpha$ and, consequently, the entrance and exit slits in the CMA can be made large. Since the entire azimuthal angle of 2π can be used, CMA is therefore a device of very high luminosity.

The XPS spectrum or the energy distribution of photoelectrons is obtained by ramping the negative potential applied to the outer cylinder of the CMA. The energy resolution of a CMA can be improved by retarding the photoelectrons from an initial kinetic energy E_0 to a fixed pass energy E_p. This is done by tying the potential of the two spherical retarding grids at the front end of the CMA to the potential of the inner cylinder of the CMA (Fig. 11.2). As a result, the resolution is improved by a factor of E_p/E_0, whereas the sensitivity is reduced by a factor of $(E_p/E_0)^{0.5}$. As the pass energy is decreased to achieve an absolute resolution of better than 1.2 eV, the signal intensity drops off more precipitously than predicted due to perhaps the onset of the adverse effect of pre-retardation. In some CMAs, a rotary slit is provided to enable angle resolved measurement without changing the area of the irradiated spot. However, the signal intensity in this case is reduced significantly.

Another more popular version of electron energy analyzer for XPS measurement is the concentric hemispherical analyzer (CHA), developed by Siegbahn in the sixties (4). Shown schematically as part of an XPS spectrometer (Fig. 11.1), it is constructed of two concentric hemispherical surfaces of radii r_i and r_o. A potential V is applied between the two surfaces such that the outer is reflecting. A median equipotential surface of radius R_0 exists between the hemispheres, and if the concentric hemispheres are geometrically perfect, $R_0 = (r_i + r_o)/2$. The entrance and exit slits of the CHA are both centered on R_0. In general, electrons entering the analyzer at an angle α with respect to the normal at the center of the slit will travel in elliptical orbits in the CHA. If an electron of kinetic energy E enters the analyzer at the center of the slit at $\alpha = 0°$, and travels in a circular orbit of R_0, then the following relationship must hold

$$e \times V = E\left(\frac{r_o}{r_i} - \frac{r_i}{r_o}\right)$$

where e is the electron charge and V the reflecting potential in volts. This gives the ideal pass energy. During the measurement, the entire CHA is isolated from ground and floats at a varying potential which scans through the desired range of energy, while V is maintained at a fixed value or at a fixed ratio with respect to the floating

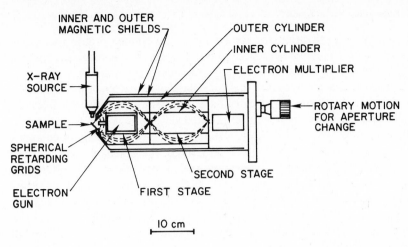

Figure 11.2:   Schematic representation of a double-pass CMA, used for both AES and XPS. The exit aperture from the first stage is the entrance aperture to the second stage. At the front end of the analyzer are two hemispherical retarding grids centered on the source area of the sample which retard photoelectrons to a constant-pass energy for XPS. For AES, the grids are at ground potential as is the inner cylinder. An externally operated rotary motion mechanism allows the entrance and exit apertures of the second stage to be changed from large sizes for XPS to small sizes for AES. The electron gun is situated internally on the axis of the CMA, but the X-ray source is external and positioned as close to the sample as the geometry will allow (32,91).

potential.   One of the advantages CHA has over CMA is that a multielement dispersion compensated electrostatic lens can be conveniently placed in front of CHA to improve the energy resolution, and to effect position-sensitive multichannel detection (17).

Detection Schemes: In principle, multiple detection schemes can be used in both CMA and CHA (18).   However, since the focal surface in the CHA is planar whereas it is conical in the CMA, multielement detection schemes are exclusively used in the former for practical engineering reasons.   In a multichannel detection scheme, the conventional single element high-gain electron multiplier, usually a channeltron or a spiralttron, is replaced by a microchannel plate assembly, which consists of an array of electron multipliers, often 256 × 60, and performs position-sensitive detection.   The output of the channel plate can be either directed to a phosphor screen where light pulses are created and picked up by a scanning video camera, or connected to a resistive anode where the magnitude and position of the deposited charge are computed electronically (19).

When single element electron multipliers are used, the multiplier output is directly connected to the pulse counting electronics via passive DC-decoupling elements.

11.1.2 Experimental Techniques

Static Charge Neutralization and Referencing: A common problem encountered in the XPS studies of polymer surfaces and interfaces is sample charging, since polymer films are usually electrical insulators. With nonmonochromatized Mg or Al Kα sources, sample charging reaches a steady state in a matter of a few seconds after X-ray irradiation (20). A steady-state charging cannot be achieved for a monochromatized source unless an electron flood gun is used. In this case, a controlled beam of low-energy ($<$ 10 eV) electrons impinge on the analyzed area to neutralize the positive charge buildup. Care should be taken, however, to minimize electron-bombardment-induced desorption or decomposition. This often means the energy with which the electrons reach the sample surface should be below some threshold value (5), and it has been found that the best results can be achieved with low-energy ($<$ 1 eV) electrons from a source very close to the sample (21). The establishment of a steady-state charging is important in that charging contributes to the change in the kinetic energy of photoelectrons and, in most cases, the broadening of spectral peaks. In a quantitative analysis the magnitude of the static charge and extraneous broadening must be determined before the true values of BE can be obtained.

Since it is rather difficult without an elaborate arrangement to compensate the surface charge exactly so that the surface potential will be zero, two static charge referencing methods are frequently used in the study of polymer films. One method is to refer all BE's, particularly the resolvable components of C 1s spectra, to that of hydrocarbon, that is, the BE of 1s electrons from a carbon atom bonded to either a hydrogen or another carbon, which has a generally accepted value of 285.0 eV. Another convenient way of monitoring chemical shift of an element is to determine its Auger parameter, α, defined as the difference between the BE and the Auger kinetic energy measured from the same XPS spectrum. It can be shown (22) that the magnitude of α is not affected by surface charging, and thus provides a quick means of checking if a change in chemical environment has occurred. This method is particularly useful for comparison between two spectra obtained under different surface conditions.

Sample Preparation: In the early days of XPS study of polymer films, no special care was taken in sample preparation and handling, and the measurements were not conducted under truly UHV conditions. Polymer samples (spun-cast films, fibers, or in powder form) were introduced into the spectrometer in "as-received" conditions with a specimen holder to which the samples were attached, frequently by a double-sided adhesive tape. Since polymer surfaces usually have a low sticking coefficient for ambient gases and are stable in vacuum at ambient temperatures, surface contamination and outgassing did not present a serious problem.

However, as one went beyond composition and impurity monitoring, and tried to study surface structures and interface effects, one began to realize that the buildup of hydrocarbon-like surface contaminants within the time span of a measurement can affect the C 1s profile, thus obscuring the structure information it carries. Furthermore, interface studies involving highly reactive metals would require extremely low residual gas pressures in the spectrometer lest reactions with the system ambient should occur, which might interfere or obscure the interfacial interaction under investigation (23). In recent years, therefore, XPS study of polymer surfaces and interfaces is conducted under UHV conditions (24a). Whenever possible, the specimens are synthesized, prepared, or processed in a chamber directly connected to the spectrometer (24b), and XPS measurements are made without exposing the specimen to room ambient. Materials used for synthesis must be as pure as possible, and well-proven procedures should be followed to ensure sample uniformity, stoichiometry, and completeness of curing. Surface cleaning techniques developed in semiconductor industries can be appropriately adapted to polymer studies where specimens cannot be prepared in situ. To minimize static charging, polymers, or their precursors that are soluble in organic solvents are preferably spun-cast to form very thin films on conductive substrates, and subsequently heated to remove the solvents and complete the curing process.

Methods of Measurement: Most XPS studies begin with multiple qualitative survey scans, which serve to determine the regions of interest, and to ascertain if the X-ray-induced desorption or damage occurs within the time span of the measurement by monitoring the time dependence of the XPS spectra. The latter is particularly important in studying surface modification effects. Numerous studies have, for example, reported X-ray-induced desorption of fluorine and fluorine-containing species from the plasma-modified or fluorinated polymer surfaces under irradiation (25). Survey scans are also useful in monitoring the amount of overlayer deposited in an experiment to study interface reactions. As Fig. 11.3 shows, a series of spectra obtained from a polyimide surface on which the Cr overlayers were successively deposited under UHV conditions (26). The thickness of the overlayer and possible hydrocarbon or other ambient contamination were monitored by the increasing Cr and diminishing substrate (PI) signals. It should be noted that the thickness of the overlayer prepared by vacuum deposition is usually determined indirectly by means of a quartz crystal monitor, or by calculated dosage (or exposure) in terms of langmuirs (1×10^{-6} torr.s), the latter being even less accurate since the sticking coefficient is usually not determined in the experiment. Given the inelastic mean free path of the electrons in the overlayer material (27), the overlayer attenuation of XPS signals from the substrate can be used to check the inferred thicknesses quantitatively (26,28).

In the regions of interest, XPS peaks are measured in a high-resolution mode. For a CHA or a double-pass CMA, the pass energy can be varied to achieve an optimized trade-off between the data collection time and the acceptable resolution. As indicated in Section 11.2.1, adverse effects of the retarding grid in a CMA make it impractical to achieve a better than 1.2-eV resolution, and this is why a CHA is preferred in the high-resolution work.

There are two ways of setting the pass energy for XPS measurement using a CHA: (1) the pass energy is varied to maintain a fixed ratio with respect to the electron energy being scanned; and (2) the pass energy is fixed at a constant value during the entire energy scan. In the former case, the CHA operates in the mode of constant relative resolution (CRR), while in the latter a constant absolute resolution is obtained at a fixed analyzer transmission (FAT). The CRR mode is harder to quantify, but imparts better detectivity to small peaks at low kinetic energies than the FAT. However, the FAT mode is more popular because of the ease of quantification.

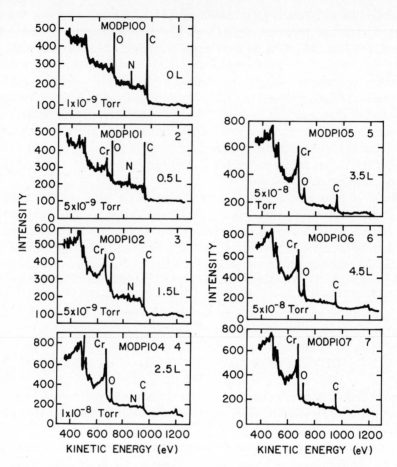

Figure 11.3: X-ray photoelectron spectra of a bare PI surface with successively increased amount of Cr overlayer: (1) freshly cured PI; (2) 1 monolayer (L) coverage; (3) 1.5 L; (4) 2.5 L; (5) 3.5 L; (6) 4.5 L and (7) same, but at a different spot of the surface. The deposition pressure is given for each deposition run in the plot.

In interface studies where sequential deposition is used, it is always advisable to make XPS measurement at different takeoff angles to check the extent of inter-action. It is highly desirable if other techniques such as scanning electron microscopy (SEM) and transmission electron microscopy (TEM) can be used either in parallel or in tandem to check the uniformity of deposition since the probing area of XPS is rela-tively large.

11.1.3 Interpretation of Data

To derive information from the XPS data, one must be always remindful of the fact that the sampling depth of the technique is only a few tens of angstroms. What is determined of this thin surface layer is not necessarily representative of the bulk film.

Other techniques should be used to establish the correspondence between the surface layer and the bulk film.   Among them, Fourier transform IR and Raman spectroscopies are probably the most suitable for polymer films.   The depth-profiling technique which works well for AES or SIMS is not recommended for XPS polymer studies, since the chemical state information will be lost due to sputtering damage. Furthermore, XPS is less sensitive than SIMS or AES, and XPS measurements cannot be made simultaneously with sputtering because of the secondary electrons generated in the process.   However, the depth information obtained by angle-resolved (or variable takeoff angle) measurement, though limited to the thin overlayer, can be very important in investigating surface modification and plasma-etching effects.

Core-Level Binding Energies and Shake-up Satellites: The binding energy (BE) of a core-level electron is related to its measured kinetic energy by the equation

$$E_k = h\nu - E_b - e\Phi$$

where E_k is the measured kinetic energy of the electron, $h\nu$ the energy of exciting radiation (1253.6 and 1486.5 eV for Mg K_α and Al K_α, respectively), and Φ the work function, its precise value being dependent on the sample and spectrometer.

The major constituents in polymeric materials are light elements such as H, C, N, O, S, Si, and halogens.   Their core-level binding energies have been, by and large, determined and compiled by Clark and his coworkers (13) using pure polymers and model compounds.   A large volume of pertinent data is also contained in the NIST XPS database recently compiled by the US Institute of Standards and Technology (formerly National Bureau of Standards) from a critical review of the published literature up to 1985.

Of the enumerated elements, the BE's of carbon 1s electrons are the most informative in terms of bond structures.   When carbon is bound to hydrogen or another carbon, the BE of its 1s electron is 285.0 eV (29), which is often used as BE reference.   Halogens induce the largest shifts in the BE of C 1s, the magnitude increasing with the electronegativity of the substituent.   The shift to the higher BE per each substituent ranges from 1 eV for Br, 1.5 eV for Cl to about 2.9 eV for F.   As much as 8.5-eV shift has been observed for the C 1s in CF_3 functional of a hexafluorodianhydride-oxydianiline polyimide (HFDA-ODA PI) film (30).   The chemical shift of carbonyl carbon in a cured pyromellitic dianhydride-oxydianiline polyimide (PMDA-ODA PI) film amounts to 3.8 eV (31), which is similar to the BE shift of C 1s in a O-C-O structure.   This is just one example of ambiguities in BE shift data.   The secondary effect of X in a C-O-X is small ($\pm$ 0.4 eV) except for nitrate ester, in which an additional 0.9-eV shift is observed.   Metal carbides, on the other hand, induce a shift of about 1 eV to the lower BE side (32).

The primary substituent effect of nitrogen functionalities varies with the substituent, and C 1s shifts of 0.2, 0.6, 1.8, and 1.8 eV are obtained for $-N(CH_3)_2$, $-NH_2$, $-NCO$, and $-NO_2$, respectively.   C 1s in $-C \equiv N$ exhibits a 1.4-eV shift.

The O 1s and N 1s both vary in a narrow range of about 2 eV at about 533 and 399 eV, respectively.   Oxidized nitrogen functionals, however, exhibit much higher N 1s BE's, which vary from 405 to 408 eV for a change from $-ONO$ to $-ONO_2$.

Secondary spectral features such as shake-up satellites are frequently observed for polymeric materials containing unsaturated hydrocarbons.   Theoretical studies by Clark et al. (33) showed that these satellites are associated with $\pi \rightarrow \pi^*$ transitions.

Attempts to use the satellites as additional piece of information to resolve the ambiguity that the core-level signals may have regarding different surface structures of polymers have not been very successful (34). However, they can be conveniently used as a means of monitoring surface damage of polymer films irradiated by energetic ions, as was observed during SIMS measurement of polystyrene films (35).

11.1.4 Surface and Interface Studies

XPS studies of thin polymer films fall roughly into two categories: (1) surface structure studies, including those dealing with surface modification and degradation in different ambients; and, (2) interface studies, which examine the interaction of polymer films with materials which come in contact with the films. It is not our purpose in this chapter to give a comprehensive review of these studies. This has been done in several excellent papers and monographs (36). Instead, we shall put our emphasis on recent developments in the field.

Surface Analysis: One of the reasons for the rapid development of XPS in the past has been its applicability to the study of polymer surfaces. Both surface contamination and surface segregation (depletion) of additives in polymers and plastics can have significant effects on their properties as commercial products, including adhesive bondability, static chargeability, fire retardance, oxidation resistance, and machinability. These were practical problems which XPS tried to solve in its initial "trouble-shooting" role in the plastic and polymer industries. An interesting example was given by Briggs (36d), who investigated surface cracking of low-density polyethylene (LDPE) moldings. Comparison of the XPS survey spectra between a normal workpiece and one which exhibited surface cracks (Fig. 11.4) showed that the cracked specimen had been surface contaminated by a silicone (used as a mold release agent), which was believed to be responsible for stress corrosion in LDPE.

XPS was soon engaged in the structural study of polymers, which began with the fluorinated systems (37) and was then extended to other classes of polymers (polyethylene, polypropylene, and copolymers). This was followed by the study of chemically or plasma-modified surfaces of these polymers. The complexity of C 1s spectral profile of fluorinated polymers depends heavily on curve fitting for detailed peak assignments. From the studies of simple fluoropolymers, Clark et al. (37) were able to verify the curve fitting results and identify individual species resulting from plasma fluorination of high density polyethylene (HDPE) surfaces. An elegant example of using model polymer in XPS study of surface reaction kinetics is due to Clark and Dilks (38), who took advantage of a random alternating copolymer of ethylene and tetrafluoroethylene which has a simple C 1s profile consisting of only two peaks associated with the $-CH_2-$ and $-CF_2-$ groups and studied its crosslinking process in an Ar plasma. The crosslinking was tracked through the loss of fluorine and formation of CF groups. More recently, XPS was used to study polymerization in plasma or glow discharge (39).

In the last few years, PMDA-ODA PI (3), with previously mentioned desirable properties, has found wide use in microelectronics as insulating and pattern-delineating materials. XPS was thus employed to monitor the completeness of curing processes and to determine moisture uptake by cured PI (32a). With increased use of reactive ion etching (RIE) as a technique to delineate fine patterns in PI, it is not surprising to see that many investigators have tried to use XPS to elucidate the

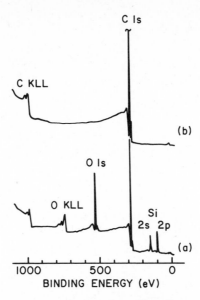

Figure 11.4:   Survey-scan spectra from the surface of low-density polyethylene (LDPE) moldings: (a) specimen exhibiting unacceptable surface cracking; (b) normal specimen. The cracked specimen has become contaminated on the surface with a silicone, probably of the type used as a molding release agent, which is also a stress corrosion agent for LDPE (36d).

kinetics and mechanism of RIE (40). Comparison of the results of various studies shows that ambiguities in BE shifts have led to different postulated species on similarly treated surfaces. Apparently, other techniques must be used to resolve the difference.

An organic chemistry approach to identifying functional groups of similar BE's is to label them by derivatization. The procedure, analogous to isotope tagging in mass spectrometry, involves using specific derivatizing agents to react with one particular functional group and label it with a distinctive element (41). The method has the advantage of not requiring other highly specialized instrumentation or spectroscopic equipment. In some cases, it may even have the additional advantage of increased sensitivity because the labeling element can have a larger cross section than those of O, C, and N. It is important, however, to bear in mind that the specific agents are difficult to find and derivatization often gives rise to side effects such as surface reorganization and migration of functional groups, which can defeat the purpose of labeling.

Interface Studies: Concerns over the adhesion and reliability of thin PI films in a microelectronic device or packaging structure have prompted a number of XPS studies, which exemplify how XPS can be advantageously used in the interface studies.

The BE's of the constituents of a cured PI and its precursor polyamic acid film have been determined in a number of carefully executed experiments (31,42). Assignment of these BE's has been worked out semiempirically (31) using appro-

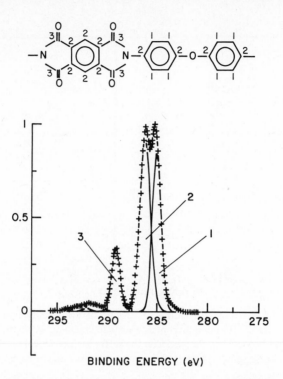

Figure 11.5: Carbon 1s spectra of PMDA-ODA polyimide. Assignment of peaks has been worked out semiempirically and carbon atoms contributing to the observed peaks are identified in the plot (30).

priate model monomer compounds, and is shown for a cured PI film in Fig. 11.5. Silverman et al. (43) have later made molecular orbital calculations to synthesize the C 1s core-level spectra, which turned out to be in very good agreement with, and thus appeared to have theoretically verified, the resolved components of the experimental C 1s spectra.

These baseline data have set the stage for the metal/polyimide interface studies to follow. Adapting Burkstrand's approach (44) to the PMDA-ODA systems, Chou and Tang (45) investigated the effect of thermodynamic properties on the interaction between various metals and PI, and identified the differences in interfacial reaction among various metals with PI, which can be correlated to their adhesive strength to PI (46). Their experiments showed that during deposition of Cr or Ni, the carbonyl component of the C 1s spectra diminished at a much faster rate than what was expected from simple overlayer attenuation. Systems with Ag and Cu, on the hand, exhibited normal attenuation for the carbonyl component. In addition, the 2p core-level BE's of these metals behaved in a way consistent with whether interfacial oxidation had occurred or not. A thermodynamic model involving interfacial oxidation of metals by carbonyl oxygen was proposed, with the prediction that Al, Sn, Ti would be interfacially oxidized at room temperature. It was later pointed out (47) that there are two possible reaction pathways following the electron transfer from metal to carbonyl oxygen, and that the scheme of the electron transfer varies with the increasing metal coverage and is indistinguishable to XPS. Their observation inspired a number of investigations, which shifted from their macroscopic approach to detailed studies of interaction mechanisms on a microscopic scale (23,48-62). Using synchrotron radiation and monitoring C 1s spectra as a function of Cr coverage for PI and model polymers containing structural fragments of PI, Jordan et al. (48) not only confirmed that carbonyl groups are primary sites of interaction for Cr during the initial stage of metallization, but also observed that the initial reaction stage was followed by gradual formation of electron rich, carbidelike carbon species. Molecular orbital calculations, on the other hand, showed that the observed evolution of C 1s spectra with initial Cr coverage could be replicated theoretically if formation of π-arene complexes between Cr and PMDA-ODA was assumed (58). Clabes (23), however, argued that in such calculations the initial core level changes were more sensitive to the amount of charge transfer into the delocalized π system of the PMDA unit than to any local specific transfer. To remove the ambiguity of different structures having no unique spectral signatures, it was suggested that additional experimental evaluation should be made in other metal/polymer systems. Information regarding the reaction pathways following the electron transfer was sought in systems involving Ce and Cs (23,59). Complimentary information was also sought in a comparison between the XPS spectra of electrochemically reduced and Cr-covered PMDA-ODA. Of particular interest to XPS users is the information that can be extracted from the core-level spectra of PMDA-ODA recorded at different takeoff angles for the clean surface and with an intermediate Ce coverage (Fig. 11.6). Note the enhanced feature of C 1s and O 1s spectra at the shallower takeoff angle because of the shallow interaction zone. Furthermore, the N 1s spectra exhibit no new discrete state but a long tail to the lower BE, attributable perhaps to a distribution of nitrogen atoms in different coordinations.

For reactive metal (Ti, Al, Ni, or Cr) and PI systems studied by XPS, accumulated experimental evidence seems to favor a stepwise coverage-dependent mechanism in which the interfacial reaction is initiated by electron transfer from metal to the carbonyl oxygen of the electroactive PMDA subunit of PI. Further reaction with the deposited material gives rise to polymer-bound metal oxide and metal nitride intermediates. With continuing deposition bond scission occurs, leading to formation of metal carbide, nitride and oxide (23,48,59,60).

A different scheme of spectral evolution with increasing metal coverage was proposed by Weaver's group (61,62) for the Co/PI system. Based on the instability of Co nitride at room temperature they suggested that formation of a Co-O-C complex (Fig. 11.7) accompanied by a bond breakage was responsible for the development of "carbidelike" and N=C "nitridelike" peaks, which were also observed in the Al and Cr on PI systems. This scheme of evolution seems to favor the π-arene complexing model, which assumes the hexagonal centers of the PMDA subunits as the preferential sites of accommodation for metal atoms in the initial stage of deposition.

It is worth noting that in all these XPS studies, the attempt to obtain a complete picture of the interface formation on an atomistic scale was met with ambiguities in BE shifts. Various approaches were employed to resolve the problem, which ranged from simple thermodynamic arguments (60,61) to quantum mechanical calculations (58), from comparison of XPS data of analogous model systems (50,54) to extensive

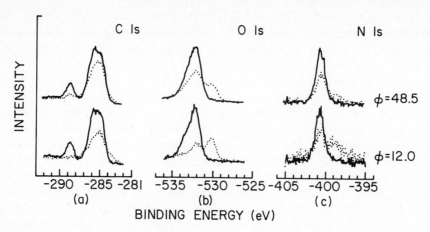

Figure 11.6:   XPS core level spectra of PMDA-ODA as a function of takeoff angle for the clean surface and with intermediate Ce coverage (23).

experimental evaluation of similar but chemically better defined systems (23,59). While it is generally agreed that carbonyl groups of the PMDA subunits are the primary sites of interaction in the initial stage of metallization, the detail schemes of electron transfer and subsequent reaction pathways remain an open question. Obviously, additional molecular structure specific information obtained by other techniques from one and the same system are needed to remove the uncertainties. Secondary ion mass spectroscopy is one of the promising surface sensitive techniques which we will discuss in the following section.

11.2 SECONDARY ION MASS SPECTROSCOPY (SIMS)

Application of SIMS to material characterization can be divided into four categories: the bulk analysis, the depth composition profiling, chemical imaging or mapping, and surface characterization. The technique consists of bombarding a specimen surface with ions of KeV energies, and mass-analyzing the secondary ions, namely, the positively and negatively charged atomic and molecular particles, which are emitted as a result of the ion/solid surface interactions. SIMS has been until recently considered as primarily a concentration depth-profiling tool. Because of its high sensitivity, the technique has become one of the important analytical tools in semiconductor industry. To achieve the capability of detecting elements at ppb levels, the primary ion fluxes have to be high so that several monolayers of material can be removed in a second. This is what is now referred to as "dynamic" SIMS or DSIMS. Obviously, DSIMS is not appropriate for surface studies. At high ion doses, several perturbing processes can occur in the surface zone, including amorphization, generation, or destruction of surface compounds, preferential sputtering, knock-on implantation, and bombardment-induced adsorption, which cause the surface to change its structure and composition continuously until a steady-state sputtering or material removal is achieved. It was also recognized in the very early stage of SIMS's development that SIMS data cannot be readily quantified, since secondary ion emission is very

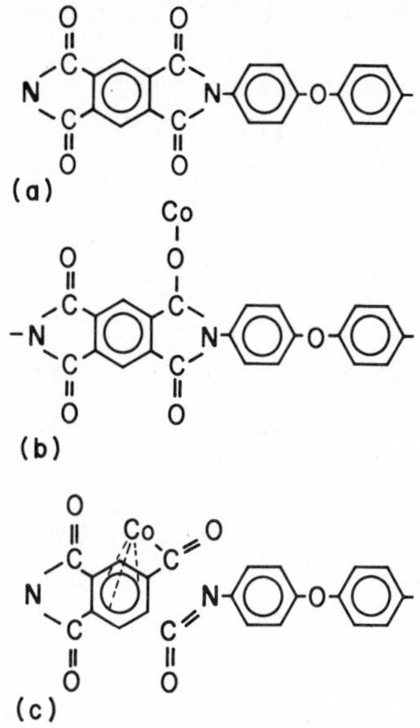

Figure 11.7:   Possible room-temperature Co/PI bonding configurations: (a) shows PI before chemical attack; (b) shows Co directly attached to carbonyl oxygen groups, this does not produce a new N-bonding environment. The configuration favored by Weaver et al. (61) is shown in (c) where Co leads to disruption of PI, forming the reaction products.

sensitive to the electronic state of the atoms or molecules to be ionized, and to the substrate from which the secondary ions are emitted (63,64). Laborious, system-dependent calibration had to be done in order to relate the SIMS signal intensity to the composition in the bulk or thin films.

When the primary ion current is reduced to < 10 nA/cm^2 and, more preferably, to ≤ 1 nA/cm^2, the time it takes to remove a monolayer of material can be several hours or days. Equipped with a high-transmission mass filter, a high-gain detector, and pulse-counting electronics, SIMS becomes a surface sensitive technique, usually referred to as static SIMS or SSIMS. Although SSIMS has proved to be useful for characterization of the surfaces of inorganic materials (65), its applicability to polymer surfaces was not fully appreciated until around 1980 when it was demonstrated that SSMIS gave different fragmentation patterns for polymers with different structures (66). The possibility of obtaining structural information on polymer surfaces not provided by XPS stimulated a great deal of research activities in this area. In the last few years systematic studies were carried out by Briggs and his coworkers (35,67), by Brown and Vickerman (68), by van Ooij and Brinkhuis (69), and by Benninghoven and his group (70), using mass sprectrometers with higher mass range and better sensitivity and resolution. In these studies optimum conditions for polymer surface analysis were determined in terms of minimizing beam damage, expanding the mass range of spectra and improving the quality of negative ion spectra. The fingerprinting capability of SSIMS for a variety of polymers was demonstrated, and the fragmentation patterns were found to be predominantly determined by the molecular structure of a sample and not strongly dependent on the primary ion species used.

Particularly noteworthy is the very recent development of the time-of-flight SIMS (TOF-SIMS) (70) which has greatly enhanced SIMS's capability by extending into the study of molecular weight distribution of polymers and by minimizing the nagging problem of peak assignment in polymer surface structure studies. The development of the reflectron-type TOF mass spectrometer has now improved the mass resolution to the extent that the chemical composition of the secondary ions can be determined directly from the spectra and peaks associated with oxygen- and nonoxygen-containing ions easily distinguished in the 200 Dalton range (70b,70d).

Despite the aforementioned rapid development, neither DSIMS nor SSIMS has matured into a routine characterization technique for polymer surfaces and interfaces. In the following sections, we shall discuss the instrumentation for SIMS in general, and then direct our attention to the individual modes of measurement. We shall place more emphasis on SSIMS than DSIMS since the former is less problematic in terms of ion damage and hence yields easier-to-interpret data.

11.2.1 Instrumentation

SIMS is in essence a special version of mass spectrometry, where ion species to be analyzed are created by bombarding the specimem surface with a beam of energetic ions. A rudimentary SIMS system consists of an ion source, a mass analyzer, and a detector with associated electronics. In addition, the three components are to be housed in a UHV chamber where the SIMS experiment is performed. A computerized data acquisition system is certainly a necessity for efficient data management.

Frequently, SIMS is installed in a UHV system which also has AES and XPS capabilities. In this case all facilities such as sample manipulator, heating, and cooling jigs, and so forth, which are usually available in an AES/XPS system will be available to SIMS, too.

Ion Sources: The most popular ion guns used in SIMS measurements are those which use either Penning or high voltage discharge to initiate a sustained gas discharge that produces gas ions (71). Positive ions are extracted from the discharge chamber and focused on the sample surface with an ion lens. A pair of X- and Y-deflection plates are added after the ion lens so that the ion beam can be rastered over an area of the surface. The beam spot size varies typically between 0.1 and 3.0 mm in diameter depending on the ion optics used. The ion current density varies with voltage and pressure and can reach as much as 200 μA/cm^2 under optimized operating conditions. Most of these ion guns are differentially pumped. Ar and Xe gases are usually used for ion beam production.

Recent development of liquid metal ion sources (72) has opened up the possibility of SIMS imaging or chemical mapping under SSIMS conditions. In these sources one of the liquid metals such as Ga, In, Sn, Au, or Cs is drawn up to the surface of a tungsten needle through the surrounding capillary channel, and forms a fine conical tip on the needle under the influence of a strong electrostatic field (Fig. 11.8). The tip, formed by the balance between the electrostatic force and surface tension, emits an intense beam of positively charged metal ions, which are extracted through a circular aperture. Prewett and Jefferies showed that a 0.5 $-$ μm beam spot can be obtained with Ga as the liquid metal source (73). They were able to obtain a SIMS

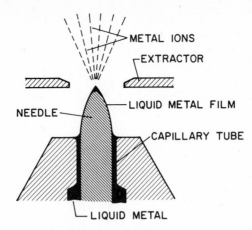

Figure 11.8:   Schematic of the principle of operation of a liquid-metal field emission ion source. A needle with a tip of radius 1 to 10 μm dips into a reservoir of liquid metal through a close-fitting capillary tube. The liquid metal is drawn along the needle over its tip by a capillary action, provided the liquid wets the material of the needle. A high voltage of between 4 and 10 kV is then applied to an extractor electrode placed a short distance in front of the tip, where-upon the liquid metal is drawn out into a cusplike protrusion from which positive metal ions are extracted by field emission. Since high currents, of the order of 100 μA, can be drawn from the very small volume of the cusp, the source has a high brightness and the ion beam can thus be focused into a very small spot while retaining usefully high currents (73).

image of a microelectronic circuit with the Ga ion source. At a low beam current ($\leq$ 10 nA) when the space charge crowding effect is minimized, the lateral resolution can be improved to ~ 1000 Å (74).

In a TOF-SIMS, the mass resolution in the low mass range is limited by the overall time resolution, determined by the pulse width of the primary ion beam, the response of the detector, and the time resolution of the registration electronics. In the high mass range it is determined by the dispersion of the analyzer. To obtain a mass-separated, monoenergetic pulsed beam, the ion sources in the TOF-SIMS are commoly equipped with a deflector for beam chopping and mass separation and an axial bunching device to reduce the pulsewidth (70b).

Mass Analyzer: There are two basic types of mass analyzers in the commonly used SIMS systems. A double focusing mass analyzer consists of an electric and a mag-netic sector (Fig. 11.9). For certain chosen angles of the sectors, higher-order focusing can be achieved (75), and the analyzer will have very good sensitivity and a good transmission characteristics, and tolerate large velocity spread of ions. Mass analyzers of this type with electric sector in the reverse geometry allow position sensi-tive multichannel detection, a feature which is highly desirable in SSIMS operations (Fig. 11.10) (76).

The quadrupole mass analyzer is a more popular type of instrument in SIMS systems probably because of its adaptability to UHV systems. It is much easier to install a quadrupole filter in a UHV chamber than to provide a pumping system for a magnetic analyzer. The quadrupole analyzer employs a combination of an RF and a

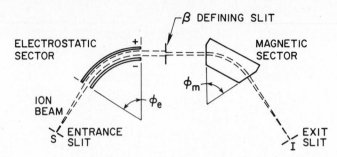

Figure 11.9:   Components of a double focusing mass analyzer (not to scale).

DC traverse electrostatic hyperbolic fields to achieve path stability for ions with a specific value of Daltons (Fig. 11.11). Mass scanning is usually achieved by varying the DC potential. It is a high transmission device, with a low mass discrimination characteristics. Its accessible mass range is 1 to 1200 Daltons.

For velocity dispersive SIMS measurements, the quadrupole filter can be front-fitted with a CMA-type velocity filter (77).

Completely different electrostatic devices are used in the TOF spectrometers as mass analyzers, which combine a sector field with linear drift spaces to achieve an effective flight path as long as practical (70a,70b).

Detection and Data Acquisition: Channeltron and counting electronics are widely used with quadrupole mass filters although the needs for multichannel detection are mounting with further development of SSIMS (76). A computerized data acquisition system is an essential part of SIMS for efficient manipulation and management of the enormous amount of data that can be generated in an experiment.

While the TOF-SIMS has demonstrated its superiority in mass resolution and unlimited mass range, it requires not only sophisticated hardware for the primary beam system and mass analyzer, but also a high-efficiency detection system and an electronic registration system, capable of accumulating a single scan spectrum in a total flight time of the order of $(150 \sim 200)$ μs with a time resolution of a few nanoseconds. It will certainly take some time for such a TOF-SIMS to become commercially available.

11.2.2 Experimental Technique and Interpretation of SIMS Data

DSIMS: Depth Profiling of Polymer Films: While SSIMS studies of polymer surfaces have been actively pursued by several groups of scientists, very few studies on depth profiling have been reported in the literature so far (78). The reason for the lack of interest is several-fold. Since heavy radiation damage by ion bombardment in polymeric materials have been observed at ion doses much below 10^{14} ions/cm², structural information is not likely to be gained in depth profiling where ion fluences are typically 10^{17} to 10^{19} ions/cm². Development of microscopic erosion textures during sputtering (79-81) frequently makes it difficult, if not impossible, to calibrate the sputtering rate. These microscopic defects also give rise to spurious time-dependent changes in SIMS spectra. Sample charging presents another problem in

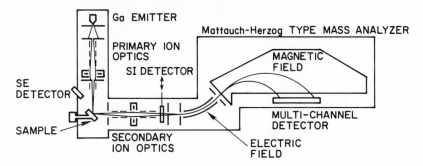

Figure 11.10:   Schematic diagram of the multichannel detection in a submicrometer secondary ion mass spectrometer (76).

SIMS analysis of polymers, causing loss or instability of spectra. Simultaneous electron bombardment or Au coating of sample surfaces has been used to achieve charge compensation in depth profiling (56). Electron bombardment gives rise to electron-induced secondary ion emission, whose contribution to the SIMS spectra must be determined separately. For SIMS using negative secondary ions, the charge compensation procedure is even more delicate (59).

Previous studies on depth profiling of polymer films have been therefore understandably confined to atomic secondary ions. Nevertheless, these studies have shown that SIMS profiling can be a powerful tool for more than just monitoring impurities in polymer films and interfaces. For example, isotope monitoring during SIMS profiling has been applied to the study of moisture absorption by PI films (83). Figure 11.12 shows the relative intensity of O^{18}, normalized with respect to O^{16}, as a function of sputter time after the PI film was exposed to H_2O enriched with O^{18}. The uptake of moisture was clearly demonstrated.

SSIMS: Characterization of Polymer Surfaces: Systematic studies conducted by various groups of scientists have shown that in SSIMS characterization of polymer surfaces, the primary ion current density should be in the range of 1 to 4 nA/cm^2 and the total ion fluences used during spectral acquisition should be kept below 10^{13} ions/cm^2. Defocused low-energy (< 700 eV) electron beams with current densities several times the ion current density can be used to achieve charge compensation. The positive and negative ion yields can be separately optimized by adjusting the sample bias and electron beam parameters. Under these conditions, the contribution of electron-stimulated secondary ion emission to the SIMS spectra will be negligible, and no noticeable ion damage will occur in most cases. As a rule of thumb, 1 to 4 KeV Ar^+ or Xe^+ ions are used in these SSIMS measurements.

While XPS and AES have matured in the last two decades, the static SIMS is still a developing technique as far as polymer studies are concerned. Earlier efforts exploring the applicability of SSIMS to polymers have demonstrated the potential advantages it will have over XPS, namely, greater molecular specificity, greater sensitivity and shallower sampling depth (84). Although experimental conditions in these studies have not been optimized in terms of charge neutralization and damage monitoring, their results indicated that cluster patterns of fragments C_mH_n from dif-

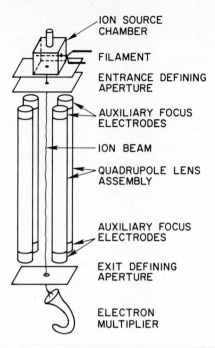

Figure 11.11:   Relative positions of the components in a quadrupole mass analyzer.

ferent polymers exhibit different characteristic relative intensities within the clusters and between the clusters. The ability of SSIMS to provide molecular information was demonstrated by a comparison between the SSIMS and XPS spectra of *n*-butyl, *sec*-butyl, *iso*-butyl, and *tert*-butyl polymers (66b). While XPS data could not distinguish structural isomers, SSIMS spectra exhibited two features which can be used to differentiate the isomers clearly. First of all, relative intensities of the 57-dalton peak (i.e., $C_4H_3^+$) varied in an inverse order according to the expected stability of primary (*n*-), secondary (*sec*-, *iso*-), and tertiary (*tert*-) carbonium ions. Secondly, bond breaking events within the butyl group provide the clues for distinguishing isomers. Consistent with the process depicted in Fig. 11.13, the SSIMS spectra exhibited three highest peaks of 27, 29, and 41 Dalton ions for poly(*sec*-butyl methacrylate), and two highest peaks at 27 and 39 for poly(*iso*-butyl methacrylate).

Systematic SSIMS studies of various classes of polymers have been carried out in last few years (67-69,70). Their results have confirmed the general validity of previous observations, which were initially regarded as too optimistic because many issues such as ion damage, ESD, and charge neutralization had not been addressed in the earlier investigations.

Several important observations can be made from these studies.

1. SSIMS fragmentation patterns are consistent with the generally accepted radiative degradation mechanism, which consists in main chain scission induced by side chain elimination, followed by radical rearrangements which lead to backbone breakage (67f).

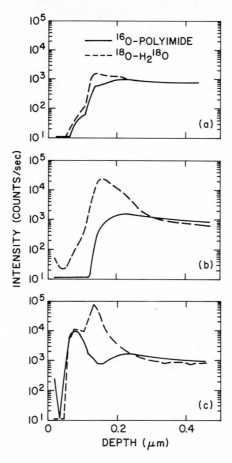

Figure 11.12: SIMS profile of Ti/thin Cu films on polyimide: (a) as-deposited, showing O^{16} and O^{18} from ambient humidity exposure (normalized for isotopic abundance); (b) exposed to H_2O^{18} and (c) exposed to H_2O^{18} and annealed in nitrogen at $350°C$ (83).

2. Manipulation of surface potential by adjusting specimen bias and charge-neutralizing electron current density allows optimization of negative ion yields (66). Optimized negative SIMS spectra, which were previously regarded as uninformative (68a), actually provide better fingerprinting capability than positive SIMS because (a) many ions in the negative SIMS retain their side chains, giving rise to higher mass spectral patterns, and (b) the negative spectra are less populated with hydrocarbon fragments (67f) as in the case of positive SIMS.

3. In contrast to negative SIMS, the positive SIMS spectra exhibit low-mass C_mH_n patterns characteristic of polymer backbone and sidechain units (67).

4. Secondary ion formation is independent of the charge state of the bombarding primary particles, and SSIMS and FABMS are equivalent techniques when charge neutralization in the former is optimized.

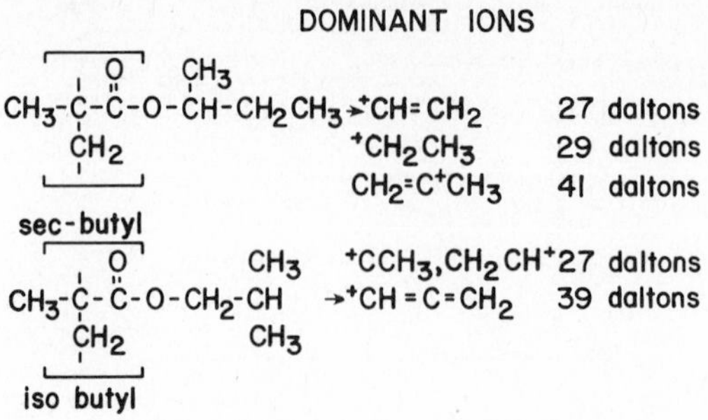

DOMINANT IONS

CH₃-C-C-O-CH-CH₂CH₃ →⁺CH=CH₂	27 daltons
⁺CH₂CH₃	29 daltons
CH₂=C⁺CH₃	41 daltons

sec-butyl

CH₃-C-C-O-CH₂-CH   ⁺CCH₃,CH₂CH⁺ 27 daltons
→⁺CH=C=CH₂   39 daltons

iso butyl

Figure 11.13:   Bond breaking events in isomeric carboniums (66b).

5. It is noteworthy that while the importance of optimizing charge neutralization in SSIMS measurement is widely recognized, very few quantitative studies on this subject have been reported in the literature. Wittmaack (81a) observed that charge compensation involves direct recombination between the charge induced by incident ions and the injected electrons. Only very recently has it been reported (81b) that the problem of surface charging in a thick insulating sample can be successfully solved by a low-energy electron flood gun.

6. Successful exploitation of the TOF-SIMS technique has clearly demonstrated that with improvement in mass resolution, the peak assignment, that is, the positive identification of secondary ions, can be made directly from the spectra. Until recently, the same task has to be performed in several indirect ways: by varying the molecular structure of the material, by measuring the spectra of related materials or constituents, and by deuteration.

Despite these promising observations, it is premature to conclude that SSIMS will soon become the only routine surface analysis technique needed for characterization of polymers. Currently, the technique has several limitations. First of all, most of the structural information is reflected in the changes in peak ratios (69) rather than in the appearance of new characteristic fragments. Since the characteristic peaks involved are limited in number, changes in their ratios will be difficult to interpret if several fragmentation processes occur simultaneously. Furthermore, these peak ratios are strongly dependent on the types of instrument used and on the experimental conditions under which the SSIMS measurement is made. Optimization of experimental conditions is again system-dependent. Comparison of the spectra obtained on different systems is therefore very complicated. Due to the lack of understanding of the charge neutralization process, the capability of negative SSIMS has not been sufficiently explored. In addition, insufficient public data base makes it necessary for an investigator to work out the fragmentation patterns for the polymer he or she studies. Even with the ease of peak assignment now afforded by the TOF-SIMS, it is still no trivial task to translate the spectral information into the original surface structure since intra- and intermolecular rearrangement processes can take place during the SSIMS measurement.

11.3 AUGER ELECTRON SPECTROSCOPY (AES)

When a beam of energetic electrons or X-ray photons impinges on a solid surface, the atoms in the surface layer will be ionized. If a core-level electron is removed from an atom, the excited atom may relax to a lower energy state by electronic rearrangement. In a radiationless Auger process, the excess energy between the initial and the final relaxed state will be spent to eject another electron, leaving the atom doubly ionized. The ejected electron, named Auger electron after its discoverer (85), will therefore have a kinetic energy characteristic of the transition, and hence the parent atom.

In this section, we shall discuss the traditional AES, which uses <u>electrons</u> of up to 10 KeV energies as the excitation source to bombard a sample, and energy-analyzes the secondary electrons emitted from the surface. When Auger transitions occur a few tens of angstroms within the surface, the Auger electrons will emerge from the surface with little or no loss of energy, giving rise to peaks on the energy distribution curve of the secondary electrons. These Auger peaks bear the "fingerprints" of the elements which are present in the shallow surface layer of the sample.

11.3.1 Instrumentation

An AES system is constructed of a stainless steel UHV chamber, which houses at least an electron gun, an energy-dispersive or velocity-selecting analyzer (usually a single stage or double stage CMA) with an electron multiplier assembly, and a sample holder. The system is preferably equipped with an ion gun, which can be used for surface cleaning and depth profiling. Both the principles and construction of CMA and ion guns have been discussed in the previous sections dealing with XPS and SIMS.

Since Auger peaks are small peaks superimposed on the large background of secondary electrons, they are easier to detect by differentiating the energy distribution with respect to energy. In the early stage of AES development, the Auger spectra were obtained almost without exception in a derivative form, $d\,[EN(E)]/dE$, by superimposing a small AC modulating signal on the velocity-selecting voltage and detecting the amplitude of the first harmonic component of the multiplier output. With the development of scanning Auger microprobes (SAM), where data acquisition and parameter control are computerized, the Auger data are frequently collected in the $EN(E)$ mode, and the spectra recorded and presented in the $dEN(E)/dE$ form with the differentiation performed by the computer.

As an analytical tool, AES not only has the fingerprinting capability, but also enjoys a detection sensitivity of 1 to 2% of a monolayer (about ten times more sensitive than XPS), and the ability to perform efficient analysis of composition distribution in thin films by depth profiling. With the lateral spatial resolution and the secondary electron-imaging capability provided by SAM, AES has become an analytical technique routinely used in microelectronics, in metallurgical and petroleum industries, and in many other production practices (86).

At present the spatial resolution of SAM which can be achieved routinely is about 1000 Å although the electron beam spot can be as small as 500 Å or less in diameter. Electron-scattering effects in a solid and sample surface roughness account for most of the difference. Depth profiling is best performed by rastering an Ar^+ beam over an area around the spot to be analyzed and measuring the Auger spectra simultaneously or intermittently, the latter mode being necessary when the film is very thin, and there is a likelihood that it may take longer to acquire a spectrum than removing the film. The probing electron beam should be placed during profiling in the center of sputtering crater to avoid the edge effect (87), particularly when the ion beam is stationary.

11.3.2 Interpretation and Quantification of Data

Auger Energies: An Auger process involves three electrons. When an electron in the K shell [atomic level in j-j coupling convention (88)] of an atom is ionized by the incoming radiation (electrons or photons), the hole created can be filled by an electron from the L_1 shell. The excess energy is given to a third electron, say, in the $L_{2,3}$ shell, which is subsequently ejected. The transition and ejected electron are then labeled as $KL_1L_{2,3}$ transition and electron, respectively. The Auger electrons can also come from the valence band, and one can then have, for example, a $KL_{2,3}V$, KVV, or $L_{2,3}VV$ electron, depending on the origin of the other two electrons. The kinetic energies of Auger electrons are element specific. The kinetic energy of a $KL_1L_{2,3}$ electron from an atom with atomic number z, for example, is given to the first approximation by an empirical equation (89)

$$E_{KL_1L_{2,3}}(z) = E_K(z) - \frac{1}{2}[E_{L_1}(z) + E_{L_1}(z+1)] - \frac{1}{2}[E_{L_{2,3}}(z) + E_{L_{2,3}}(z+1)]$$

where the first term on the right refers to the ionization energy of the K shell, and the second and third terms are average ionization energies for the appropriate shells with z+1 referring to the element next higher up in the periodic table. The rationale for averaging the ionization energies is that after the loss of a core electron, the nuclear charge is effectively increased by 1.

A more plausible semiempirical approach to computation of Auger energies is to make use of relaxation and interaction energy terms as follows

$$E_{KL_1L_{2,3}} = E_K - E_{L_1} - E_{L_{2,3}} - I - R^{in} - R^{ex}$$

where E's on the right are BE of various shells indicated by the subscripts, I is the interaction energy between the two-hole state and the final state, and R's are intra- and extraatomic relaxation energies associated with the inward relaxation (or collapse) of the outer orbitals and readjustment of charge from other atoms, respectively. The calculation is semiempirical because BE's in the expression are determined experimentally.

Compilation of Auger electron energies based on the above equations is available in reference books as well as in chart forms (90). However, for identification of chemical elements, the published atlases of recorded spectra (91) are more convenient to use. Auger spectra are traditionally presented in a derivative form, and Auger energies are identified by negative-going peaks. This is, of course, a choice of convenience because the true peak positions correspond to the crossover points between the positive- and negative-going peaks in the differentiated spectra, and are not easily definable without the knowledge of the background.

Auger Intensities and Fine Structures: Efforts to quantify Auger data have not been as successful as XPS in the past. While Auger energies can be computed with an acceptable accuracy, the Auger intensities cannot. Unlike photoejection processes in XPS, the Auger intensities involve not only the ionization cross section but also the transition probability. There is no simple empirical or semiempirical formula which can provide a set of useful Auger intensity data. Until recently (92), the problem of spectral background subtraction, necessary to estimate the peak height and area and to observe the line shapes, has not been fully explored. Quantification of the relationship between the intensity and the concentration at present relies on a simplistic but useful parameter "relative sensitivity" or "sensitivity factor" defined by (91)

$$S_{A,Ag} = \frac{I_A^\infty}{I_{Ag}^\infty}$$

which normalizes the Auger peak-to-peak height of pure element A to that of pure Ag. The concentration of element i in a multicomponent system can then be determined by

$$C_i = \frac{\dfrac{I_i}{S_{i,Ag}}}{\displaystyle\sum^n \dfrac{I_i}{I_{i,Ag}}}$$

A lucid and concise discussion of various factors affecting the Auger signal intensity has been given by Seah (93), who also expounded on the measurement problems involved in using the peak-to-peak height as a measure of Auger intensity. We will merely point out here that in practice the accuracy of this approach to quantification is probably no better than $\pm$ 10 %.

Effects that give rise to BE changes in XPS can also produce Auger chemical shifts. The difference between the Auger and XPS chemical shifts, as shown by Wagner (22), is characteristic of a chemical state and results from the difference in final state relaxation energies. Due to the difficulty in obtaining AES spectra from polymers, the Auger chemical shift has not been utilized in the field of polymer studies.

Except for a few cases such as C, S, and Si, Auger fine structures resulting from chemical effects have not been exploited in practical analysis, even though carbon KLL fine structures had been used to identify carbides from graphitic carbon in the early days of AES application. This is because the line shapes and fine structures of Auger spectra are more complicated than those of XPS.

11.3.3 Depth Profiling

In AES depth profiling Auger intensities are monitored as a function of time during sputtering, commonly known as sputtering profile. To obtain a true distribution of concentration with depth from the sputtering profile, it is necessary to determine the sputter or erosion rate and to convert the Auger intensities into local concentrations. A true concentration profile can be reconstructed only if the sample surface recedes in parallel to itself during sputtering and various instrumental, material-related and, radiation-induced effects on the instantaneous composition of the receding surface can be corrected for. The material-related and radiation-induced effects on surface composition during sputtering were mentioned briefly in Section 11.2. A detailed discussion of the computational techniques for correcting these effects and instrumental broadening will not be presented here since AES depth profiling, if applicable, is qualitative at best in the study of thin polymer films. Readers interested in the quantification of AES depth-profiling data are referred to the monograph by Briggs and Seah (94).

11.3.4 Application of AES to Polymer Surface and Interface

Normally AES is not a technique suited either for surface analysis or for composition profiling of polymer films. Electron-stimulated desorption or decomposition alters the surface composition while ion- and electron-induced bond breaking and cross-linking cause changes in the properties of the material. These changes inevitably lead to morphological changes such as surface roughening, texture, and hillock formation (79-81), and so on, during sputtering, and render it impossible to calibrate the sputtering rate. Perhaps, most serious of all is the problem of sample charging which can prevent acquisition of useful Auger spectra. However, AES can be advantageously employed in some areas of thin polymer film studies.

Surface and Interface Analysis in Microelectronic Structures: Increased use of polymer films in microelectronic device and packaging structures not only stimulated theoretical and experimental studies of adhesion involving polymers, metals, and ceramics, but also necessitated adhesion failure analysis at the device level. AES can be used to determine the mode of failure involving metal-to-polymer adhesion, and to identify the possible cause of failure. An interesting example was recently provided by Furman et al. (83), who studied the effect of ambient exposure on the reliability of adhesive bondage between Cu/Ti and PI in an accelerated test where the sample was exposed to moisture and heated in forming gas at 350°C for 30 min. The initial peel strength of Cu/Ti on PI was 50 g/mm, but dropped to nil after heating. AES examination of the peeled-off strips showed convincingly that the failure mode was cohesive (i.e., it occurs in the PI film) before exposure, but changed to adhesive (i.e., at the interface) after heating (Fig. 11.14a). Auger depth profiling through the Cu/Ti overlayer indicated that Ti at the interface was oxidized during heating (Fig. 11.14b).

To prove that the oxygen supply for the interfacial oxidation did not come from the annealing ambient, the heating experiment was repeated on samples with Cu/Ti deposited on Si substrates. Auger profiling showed no evidence of oxygen penetration through the overlayer. Finally, a SIMS isotope experiment was used to show the absorption of moisture by the PI film (Fig. 11.12), and it was concluded that in the case of Cu/Ti on PI, Ti was oxidized by the oxygen or moisture in the PI film.

As the cited example shows, applicability of AES profiling and surface analysis to polymer films depends largely on whether the static charging can be reduced to a manageable degree. In metal/polymer adhesion studies, the peeled strips consist of very thin polymer films on metal substrates, and charging is minimized. In the case of AES profiling of metal on PI, static charging has become a problem as the sput-

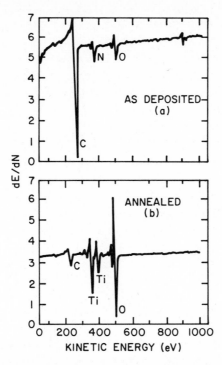

Figure 11.14:   Auger spectra of strip surfaces peeled from polyimide: (a) as-deposited and (b) after 350°C forming gas annealing (83).

tering front moved into the PI film as evidenced by the diminishing oxygen signal in Fig. 11.15.

11.4 CONCLUDING REMARKS

In this chapter we have covered three characterization techniques, namely, XPS, SIMS, and AES. Our object is to provide a basic understanding of these techniques for a material scientist, polymer engineer, or someone who is searching for an appropriate characterization technique to help resolve problems encountered in surface-related study of polymer films. The sections dealing with instrumentation, experimental procedure, and interpretation of data under the topic of each technique were, therefore, written with the aim that the general information from this chapter will enable the reader to make a proper choice of the equipment and experimental approach once he or she has decided which technique will be used. Due to the similarity that exists among the three techniques, the presentation of material has been so organized as to avoid repetition. To keep these techniques in proper perspective, we have examined at some length the advantages and limitations of each technique, and showed various approaches to overcoming these limitations.

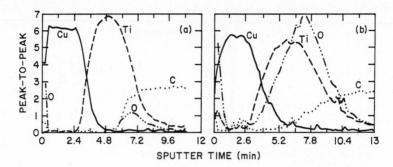

Figure 11.15:   Auger depth profiles of Ti/thin Cu films on polyimide: (a)as-deposited and (b) after 350°C forming gas annealing (83).

11.5 REFERENCES

1.   (a) C. E. Sroog, "Polyimides," *J. Polym. Sci.* C16(2), 1191 (1967); (b)L. B. Rothman, "Properties of Thin Polyimide Films," *J. Electrochem. Soc.*, 127(10), 2216 (1980).

2.   K. Niwa. N. Kamehara, K. Yokouchi, and Y. Imanaka, "Multilayer Ceramic Circuit Board with a Copper Conductor," *Adv. Ceram. Mater.*, 2, 382 (1987).

3.   A. J. Blodgett, "Multilayer Ceramic, Multi-chip Module," *Proc. Electr. Comp. Conf.*, 30. 263 (1980).

4.   K. Siegbahn, C, Nordling, A. Fahlman, K. Hamrin, J. Hedman, G. Johanson, T. Bergmark, S. Karlson, I. Lindgren, and B. Lindberg, **ESCA, Atomic, Molecular and Solid State Structure Studies by Means of Electron Spectroscopy**, Almqvist and Wiksells Boktryckeri AB, Uppsula (1967).

5.   See, for example, J. A. Kelber, and M. L. Knotek, "Electron-stimulated Desorption from Partially Fluorinated Hydrocarbon Thin Films: Molecules with Common Versus Separate Hydrogen and Fluorine Bonding Sites," *Phys. Rev. B*, 30, 400 (1984) and references cited therein.

6.   See, for example, (a) D. J. Surman and J. C. Vickerman, "Surface Analysis of Glasses by Fast Atom Bombardment Mass Spectrometry," *Appl. Surf. Sci.*, 9, 108 (1981); (b) D. H. Williams, "Fast Atom Bombardment Mass Spectrometry: a Powerful Technique for the Study of Polar Molecules," *J. Am. Chem. Soc.*, 103, 5700 (1981); (c) C. C. Fenselau, "Fast Atom Bombardment," in **Ion Formation from Organic Solids**, A. Benninghoven, Ed., Springer, New York, 90 (1983); (d) D. Briggs, A. Brown, J. A. Van der Berg, and J. C. Vickerman, "A Comparative Study of Organic Polymers by SIMS and FABMS," in **Ion Formation from Organic Solids**, A. Benninghoven, Ed., Springer-Verlag, 162 (1983); (e) M. Doerr, I. Luderwald, and H. R. Schulten, "Characterization of Polymers by Field Desorption and Fast Atom Bombardment Mass Spectroscopy," *Z. Anal. Chem.*, 318, 339 (1984).

7. See, for example, T. R. Lundquist, "Energy Dependence of the Ionization Probability of Sputtered Cu and Ni," *Surf. Sci.*, 90, 548 (1979) and references cited therein.

8. J. R. Ferraro and L. J. Basile, **Fourier Transform IR Spectroscopy**, Academic Press, New York (1979).

9. See, for example, review paper by T. M. Buck "Low-energy Ion Scattering Spectrometry," in **Methods of Surface Analysis**, A. W. Czanderna, Ed., Elsevier, Amsterdam, 75 (1975).

10. H. Ibach, **Electron Energy Loss Spectroscopy and Surface Vibrations**, Academic Press, New York (1982).

11. R. K. Chang, **Surface Enhanced Raman Spectroscopy**, Plenum Press, New York (1982).

12. See, for example, (a) N. J. Chou, C. H. Tang, J. Paraszsczak, and E. Babich, "Mechanism of Oxygen Plasma Etching of Polydimethyl Siloxane Films," *Appl. Phys. Lett.*, 46, 31 (1985), where multiple internal reflectance IR spectroscopy was used to resolve the ambiguity in XPS data; (b) T. J. Hook, J. A. Gardella, Jr., and L. Salvati, Jr., "Multitechnique Surface Spectroscopic Studies of Plasma-modified Polymers," *J. Mater. Res.*, 2, 117 (1987) where XPS, ISS and FTIR were used to study plasma modified polymer surface.

13. D. T. Clark and A. Harrison, "ESCA Applied to Polymers. XXXI. A Theoretical Investigation of Molecular Core Binding and Relaxation Energies in a Series of Prototype Systems for Nitrogen and Oxygen Functionalities in Polymers," *J. Polym. Sci., Polym. Chem. Ed.*, 17, 957 (1981) and previous papers cited therein.

14. Ernst-eckhard Koch, **Handbook on Synchrotron Radiation**, North Holland, New York (1983).

15. See "ESCA-300", product brochure of Scienta Instrument AB, Uppsala, Sweden (1989).

16. V. V. Zashkvara, M. I. Kosunskii, and O. S. Kosmachev, "Focussing Properties of Electrostatic Mirror with a Cylindrical Field," *Sov. Tech. Phys.*, 11, 96 (1966).

17. M. A. Kelly and C. E. Tyler, "A Second Generation ESCA Spectrometer," *Hewlett-Packard J.*, 24, 2 (1973).

18. B. Wannberg, U. Gelius, and K. Siegbahn, "Design Principles in Electron Spectroscopy," *J. Phys. E*, 7, 149 (1974).

19. S. S. L. Product Brochure for Imaging Detector, *Surface Science Laboratories*, Kervex Corp. (1987).

20. D. T. Clark, "Structure, Bonding and Reactivity of Polymer Surfaces Studied by means of ESCA," *C.R.C. Critical Review in Solid State and Material Sciences*, 1 (1978).

21. See, for example, (a) G. Johansson, J. Hedman, A. Berndtsson, M. Klasson, and R. Nilsson, "Calibration of Electron Spectra," *J. Electron Spectr. Relat. Phenom.*, 2, 295 (1973); (b) C. P. Hunt, C. T. H. Stoddart, and M. T. Seah, "The Surface Analysis of Insulators by SIMS: Charge Neutralization and Stabilization of the Surface Potential," *Surf. Interface Anal.*, 3, 157 (1981).

22. (a) C. D. Wagner, "Auger Parameter in Electron Spectroscopy for the Identification of Chemical Species," *Anal. Chem.* 47, 1201 (1975); (b) C. D. Wagner, "Chemical Shifts of Auger Lines and the Auger Parameter," *Faraday Disc. Chem. Soc.*, 60, 291 (1975).

23. See, for example, J. G. Clabes, "Correlation of Interface Chemistry and Growth mode of Ce on Polyimide (Pyromellitic Dianhydride-oxydianiline)," *J. Vac. Sci. Technol.*, A 6, 2887 (1988).

24. (a) Brundle and Roberts were probably the first to conduct XPS study under UHV conditions. See, for instance, C. R. Brundle and M. W. Roberts, "Some Observations on the Surface Sensitivity of Photoelectron Spectroscopy," *Proc. Roy. Soc.*, A331, 383 (1972); (b) S. Kowalczyk, Y. Kim, G. Walker, and J. Kim, "Polyimide-copper Interface: Growth Sequence Dependence and Solvent Effects," *J. Vac. Sci. Technol.*, A 6, 1377 (1988).

25. See, for example, (a) D. R. Wheeler and S. V. Pepper, "Effect of X-ray Flux on Polytetrafluorethylene in X-ray Photoelectron Spectroscopy," *J. Vac. Sci. Technol.*, 20, 226 (1982); (b) N. J. Chou, J. Paraszczak, E. Babich, J. Heidenreich, Y. S. Chaug, and R. D. Goldblatt, "X-ray Photoelectron and Infrared Spectroscopy of Microwave Plasma Etched Polyimide Surfaces," *J. Vac. Sci. Technol.*, A 5, 1321 (1987).

26. N. J. Chou, D. W. Dong, J. Kim, and A. C. Liu, "An XPS and TEM Study of Intrinsic Adhesion Between Polyimide and Cr Films," *J. Electrochem. Soc.*, 131, 2335 (1984).

27. M. P. Seah and W. Dench, "Quantitative Electron Spectroscopy of Surfaces: A Standard Data Base for Electron Inelastic Mean Free Paths in Solids," *Surf. Interface Anal.*, 1, 2 (1979).

28. J. L. Jordan, P. N. Sanda, J. F. Morar, C. A. Kovac, F. J. Himpsel, and R. A. Pollack, "Synchrotron-radiation Excited Carbon 1s Photoemission Study of Cr/organic Polymer Interfaces," *J. Vac. Sci. Technol.*, A 4, 1046 (1986).

29. Actually a spread of values from 284.6 to 285.2 eV was reported in the literature. A value of 284 8 eV was also used as see, for instance, P. Swift, "Adventitious Carbon - The Panacea for Energy Referencing?," *Surf. Interface Anal.*, 4, 47 (1982).

30. P. L. Buchwalter, B. D. Silverman, L. Witt, and A. R. Rossi, "X-ray Photoelectron Spectroscopy Analysis of Hexafluorodianhydride-oxydianiline Polyimide: Substantiation for Substituent Effects on Aromatic Carbon 1s Binding Energies," *J. Vac. Sci. Technol.*, A 6, 226 (1987).

31. (a) H. J. Leary, Jr. and D. S. Campbell, "ESCA Studies of Polyimide and Modified Polyimide Surfaces," *A.C.S. Symp. Ser. 162*, 419 (1981); (b) P. L. Buchwalter and A. I. Biase, "ESCA Analysis of PMDA-ODA Polyimide," in **Polyimides: Synthesis, Characterization and Applications**, K. L. Mittal, Ed., Plenum Press, New York, Vol. 1, 537 (1984).

32. C. D. Wagner, E. Davies, N. C. MacDonald, P. W. Palmberg, G. E. Riach, and R. E. Weber, in **Handbook of X-ray Photoelectron Spectroscopy**, G. E. Mullenburg, Ed., Physical Electronics Division, Perkin-Elmer Corp. (1978).

33. (a) D. T. Clark, D. B. Adams, A. Dilks, J. Peeling, and H. R. Thomas, "Some Aspects of Shake-up Phenomena in some Simple Polymer Systems," *J.*

Electron Spectrosc. & Relat. Phenom., 8, 51 (1976); (b) D. T. Clark and A. Dilks, "ESCA Studies of Polymers. XIII. Shake-up Phenomena in Substituted Polystyrenes," *J. Polym. Sci., Polym. Chem. Ed.*, 15, 15 (1977).

34.   J. J. O'Malley, H. R. Thomas, and G. Lee, "Surface Studies on Multicomponent Polymer Systems by X-ray Photoelectron Spectroscopy," *Macromolecules*, 12, 496 (1979).

35.   D. Briggs and A, B. Wootton, "Analysis of Polymer Surfaces by SIMS. I. An Investigation of Practical Problems," *Surf. Interface Anal.*, 4, 109 (1982).

36.   See, for example, (a) B. D. Ratner, "Characterization of Graft Polymers for Biomedical Applications," *J. Biomed. Mater. Res.*, 14(5), 665 (1980); (b) D. Briggs, "X-ray Photoelectron Spectroscopy for the Investigation of Polymeric Materials," in **Electron Spectroscopy: Theory, Techniques and Applications**, C. R. Brundle and A. D. Baker, Eds., Vol. 3, Academic Press, London (1979); (c) D. Briggs, "Industrial Applications of XPS: Recent Developments in Catalysis and Polymer Adhesion Studies," *Applied Surf. Sci.*, 6, 188 (1980); (d) D. Briggs, "Application of XPS in Polymer Technology," in **Practical Surface Analysis by Auger and X-ray Photoelectron Spectroscopy**, D. Briggs and M. P. Seah, Eds., Wiley, New York, 359 (1983).

37.   D. T. Clark, W. J. Feast, W. K. R. Musgrave, and J. Ritchie, "Application of ESCA (Electron Spectroscopy for Chemical Analysis) to Polymer Chemistry. VI. Surface Fluorination of Polyethylene," *J. Polym. Sci., Polym. Chem. Ed.*, 13, 857 (1975).

38.   D. T. Clark and A. Dilks, "ESCA Applied to Polymers (XVIII): RF Glow Discharge Modification of Polymers in Helium, Neon, Argon and Krypton," *J. Polym. Sci., Polym, Chem. Ed. 16,*, 911 (1978).

39.   A. Dilks and E.Kay, "Plasma Polymerization of Ethylene and the Series of Fluoroethylenes: Effluent Mass Spectrometry and ESCA Studies," *Macromolecules*, 14(3), 855 (1981).

40.   (a) F. Egitto, V. Vukanovic, R. Horwarth, and F. Emmi, "Significance of the Oxygen to Fluorine Atomic Concentration Ratio in rf Plasma for Etching Organic Materials," in **Proc. 7th Symp. Plasma Chem.**, C. J. Timmermand, Ed., Vol. 3, 983 (1985); (b) M. Kogoma and G. Turban, "Mechanism of Etching and of Surface Modification of Polyimide in RF and LF SF_2 -O_2 Discharges," *Plasma Chem. Plasma Proc.*, 6, 349 (1986); (c) N. J. Chou, J. Parazsczak, E. Babich, Y. S. Chaug, and R. Goldblatt, "Mechanism of Microwave Etching of Polyimide in Oxygen and CF_4 gas Mixtures," *Microelectronic Eng.*, 5, 375 (1986).

41.   W. M. Riggs and D. W. Dwight, "Characterization of Fluoropolymer Surfaces," *J. Electron. Spectrosc.*, 5, 447 (1974).

42.   J. Russat, "Characterization of Polyamic Acid/Polyimide Films in the Nanometric Thickness Range from Spin-deposited Polyamic Acid," *Surf. Interface Anal.*, 11, 414 (1988).

43.   B. D. Silverman, J. W. Bartha, J. G. Clabes, and P. S. Ho, "Molecular Orbital Analysis of the XPS Spectra of PMDA-ODA Polyimide and its Polyamic Acid Precursor," *J. Polym. Sci., Polym. Chem. Ed.*, 24, 3325 (1986).

44. (a) J. M. Burkstrand, "Formation of a Copper-Oxygen-Polymer Complex of Polystyrene," *Appl. Phys. Lett.*, 33, 387 (1978); (b) J. M. Burkstrand, "Substrate Effects on the Electronic Structure of Metal Overlayers - an XPS Study of Polymer-Metal Interfaces," *Phys. Rev. B*, 20, 4853 (1979); (c) J. M. Burkstrand, "Metal-polymer Interfaces: Adhesion and X-ray Photoenission Studies," *J. Appl. Phys.*, 52, 4795 (1981); (d) J. M. Burkstrand, "Copper-Polyvinyl Alcohol Interface: A Study with XPS," *J. Vac. Sci. Technol.*, 20, 440 (1982).

45. N. J. Chou and C. H. Tang, "Interfacial Reaction During Metallization of Cured Polyimide: An XPS Study," *J. Vac. Sci. Technol.*, A 2, 751 (1984).

46.   K. S. Kim and J. Kim, "Elasto-Plastic Analysis of the Peel Test for Thin Film Adhesion," *Trans. Amer. Soc. Mech. Eng., J. Eng. Mat. and Tech.*, 110, 266 (1988).

47. N. J. Chou and C. H. Tang, "A Thermodynamic Model for Predicting Formation of Chemical Bonds Between Metals and Cured Polyimides During Metallization," in **Colloidal and Surface Chemistry in Computer Technology**, K. L. Mittal, Ed., Plenum Press, New York, 287 (1987).

48. J. L. Jordan, J. F. Morar, G. Hughes, R. A. Pollack, and F. J. Himpsel, "Polyimide and the Polyimide-Metal Interaction," *N.S.L.S. Annual Report*, 125 (1984).

49. P. S. Ho, P. O. Hahn, J. W. Bartha, G. W. Rubloff, F. K. LeGoues, and B. D. Silverman, "Chemical Bonding and Reaction at Metal/polymer Interfaces," *J. Vac. Sci. Technol.*, A 3, 739 (1985).

50. P. N. Sanda, J. W. Bartha, B. D. Silverman, P. S. Ho, and A. R. Rossi, "Model Compound Approach for Polymer-Metal Interfaces: ESCA Studies," *Mater. Res. Symp. Proc.*, 40, 238 (1985).

51. P. O. Hahn, G. W. Rubloff, J. W. Bartha, F. LeGoues, R. Tromp, and P. S. Ho, "Chemical Interactions at Metal-Polymer Interfaces," *Mater. Res. Soc. Symp. Proc.*, 40, 251 (1985).

52. J. W. Bartha. P. O. Hahn, F. Legoues, and P. S. Ho, "Photoemission Spectroscopy Study of Aluminum-Polyimide Interface," *J. Vac. Sci. Technol.*, A 3, 1390 (1985).

53. N. J. DiNardo, J. E. Demuth, and T. C. Clark, "Interaction of Thin Metal Films with the Polyimide Surface: Electron Energy Loss Spectroscopy of Surface Vibrations and UV Photoemission of Electronic States," *Chem. Phys. Lett.*, 121, 239 (1985).

54. P. N. Sanda, J. W. Bartha, J. G. Clabes, J. L. Jordan, C. Feger, B. D. Silverman, and P. S. Ho, "Interaction of Metals with Model Polymer Surfaces: Core Level Photoemission Studies," *J. Vac. Sci. Technol.*, A 4, 1035 (1986).

55. J. L. Jordan, C. A. Kovac, J. F. Morar, and R. A. Pollack, "High-resolution Photoemission Study of the Interfacial Reaction of Cr with Polyimide and Model Polymers," *Phys. Rev. B*, 36, 1369 (1987).

56. C. A. Kovac, J. L. Jordan, and R. A. Pollack, "Metal Atom Reactions with Polymer Films," *Mater. Res. Symp. Proc.*, 72, 247 (1986).

57. F. S. Ohuchi and S. C. Freilich, "Metal Polyimide Interface: A Titanium Reaction Mechanism," *J. Vac. Sci. Technol.*, A 4, 1039 (1986).

58. A. R. Rossi, P. N. Sanda, B. D. Silverman, and P. S. Ho, "A Theoretical Study of the Bonding and XPS Spectra of Cr Interacting with a Polyimide Model Compounds," *Organomet.*, 6, 580 (1987).

59. J. G. Clabes, M. L. Goldberg, A. Viehbeck, and C. A. Kovac, "Metal-Polymer Chemistry. I. Charge-transfer Related Modifications of Polyimide (Pyromellitic Dianhydride-4,4'-oxydianiline)," *J. Vac. Technol.*, A 6, 985 (1988).

60. M. L. Goldberg, J. G. Clabes, and C. A. Kovac, "Metal-Polymer Chemistry. II. Chromium-Polyimide Interface Reactions and Related Organometallic Chemistry," *J. Vac. Sci. Technol.*, A 6, 991 (1988).

61. S. G. Anderson, H. M. Meyer, III, and J. H. Weaver, "Temperature-dependent X-ray Photoemission Studies of Metastable Co/polyimide Interface Formation," *J. Vac. Sci. Technol.*, A 6, 2205 (1988).

62. H. M. Meyer III, S. G. Anderson, L. J. Atanasoska, and J. H. Weaver, "X-ray Photoemission Investigation of Clustering and Electron Emission and Trapping at the Gold/Polyimide Interface," *J. Vac. Sci. Technol.*, A 6, 30 (1988).

63. V. R. Deline, "Factors Influencing Secondary Ion Yields," in **Secondary Ion Mass Spectrometry SIMS II**, A. Benninghoven, C. A. Evans, R. A. Powell, R. Shimizu, and H. A. Storms, Eds., Springer Series in Chemical Physics, 34, 48 (1979).

64. T. R. Lundquist, "Correlation Between the Spectral Ionization Probability of the Sputtered Atoms and the Electron Density of States," in **Secondary Ion Mass Spectrometry SIMS II,** in A. Benninghoven, C. A. Evans, R. A. Powell, R. Shimizu, and H.A. Storms, Eds., Springer Series in Chemical Physics, 34, 44 (1979).

65. (a) A. Benninghoven, "Surface Investigation of Solids by the Statical Method of Secondary Ion Mass Spectroscopy (SIMS)," *Surface Sci.*, 35, 437 (1973); (b) W. J. Van Ooij and A. Kleinhesselink, "SIMS for the Study of Polymer Surface," *Appl. Surf. Sci.*, 4, 324 (1980);

66. (a) J. A. Gardella, Jr., and H. M. Hercules, "Static Secondary Ion Spectrometry," *Anal. Chem.*, 52, 226 (1980); (b) J. A. Gardella, Jr., and H. M. Hercules, "Comparison of Static Secondary Ion Mass Spectrometry, Ion Scattering Spectroscopy, and X-ray Photoelectron Spectroscopy for Surface Analysis of Acrylic Polymers," *Anal. Chem.*, 53, 1879 (1981); (c) J. E. Campana, J. J. DeCorpo, and R. J. Colton, "Characterization of Polymeric Thin Films by Low-damage Secondary Ion Mass Spectrometry," *Appl. Surf. Sci.*, 8, 337 (1981).

67. (a) D. Briggs, "Analysis of Polymer Surfaces by SIMS. III. Preliminary Results from Molecular Imaging and Microanalysis Experiments," *Surf. Interface Anal.*, 5, 113 (1983); (b) D. Briggs, M. J. Hearn, and B. D. Ratner, "Analysis of Polymer Surfaces by SIMS. IV. A Study of Some Acrylic Homo- and Co-polymers," *Surf. Interface Anal.*, 6, 184 (1984); (c) D. Briggs and M. J. Hearn, "Analysis of Polymer Surfaces by SIMS. V. The Effects of Primary Ion Mass and Energy on Secondary Ion Relative Intensities," *Int. J. Mass*

Spectrometry Ion Processes, 67, 47 (1985); (d) D. Briggs, "SIMS for the Study of Polymer Surfaces," *Surface Sci.*, 9, 391 (1986);. (e) M. J. Hearn and D. Briggs, "Analysis of Polymer Surfaces by SIMS. XII. On the Fragmentation of Acrylic and Methacrylic Homopolymers and the Interpretation of Their Positive and Negative Ion Spectra," *Surf. Interface Anal.*, 11, 198 (1988).

68. (a) A. Brown and J. C. Vickerman, "Static SIMS for Applied Surface Analysis," *Surf. Interface Anal.*, 6, 1 (1984); (b) A. Brown and J. C. Vickerman, "Positive and Negative Ion Static SIMS Spectra of Polymers," *Surf. Interface Anal.*, 8, 75 (1986).

69. (a) W. J. van Ooij and R. H. G. Brinkhuis, "Interpretation of the Fragmentation Patterns in Static SIMS Analysis of Polymers. I. Simple Aliphatic Hydrocarbons," *Surf. Interface Anal.*, 11, 430 (1988); (b) R. H. G. Brinkhuis and W. J. van Ooij, "Identification of Positive Secondary Ions in Static SIMS Spectra of Poly(methylmethacrylate) using the Deuterated Polymer," *Surf. Interface Anal.*, 11, 214 (1988);

70. see, for example, (a) P. Steffens, E. Niehuis, T. Friese Greifendorf, and A. Benninghoven, "A Time-of-flight Mass Spectrometer for Static SIMS Application," *J. Vac. Sci. Technol.*, A 3, 1322 (1985); (b) E. Niehuis, T. Heller, H. Feld, and A. Benninghoven, "Design and Performance of a Reflectron Based Time-of-flight Secondary Ion Mass Spectrometer with Electrodynamic Primary Ion Mass Separation," *J. Vac. Sci. Technol.*, A 5(4), 1243 (1987); (c) J. Lub, P. N. T. van Velzen, D. van Leyen, B. Hagenhoff, and A. Benninghoven, "TOF-SIMS Analysis of the Surface of Insulators. Examples of Chemically Modified Polymers and Glass," *Surf. Interface Anal.*, 12, 53 (1988); (d) E. Niehuis, P. N. T. van Velzen, J. Lub, T. Heller, and A. Benninghoven, "High Mass Resolution Time-of-flight Secondary Ion Mass Spectrometry: Application to Peak Assignment," *Surf. Interface Anal.*, 14, 135 (1989) and the references cited therein.

71. For a brief discussion of the design of various ion sources, see, e.g., G. Sidenius, "Gas and Vapor Ion Sources for Low-energy Accelerators," in **Low-energy Ion Beams, 1977**, *Inst. Phys. Conf. Series No. 38*, K. G. Stephens, J. H. Wilson, and J. L. Moriozzi, Eds., Institute of Physics, Bristol and London (1978).

72. G. Taylor and M. D. van Dyke, "Electrically Driven Jets," *Proc. Royal Soc. (London)*, A313, 453 (1969).

73. P. D. Prewett and D. K. Jefferies, "Liquid Metal Field-emission Ion Sources and Their Applications," *Int. Conf. Low Energy Ion Beams*, Bath (1980).

74. S. Tomita, K. Okuno, F. Soeda, and A. Ishitani, "Application of SIMS Technique to Industrially Used Organic Materials," in **Secondary Ion Mass Spectrometry SIMS V**, A. Benninghoven, et al., Eds., Springer Series in Chemical Physics, 44, 545 (1985).

75. J. D. Morrison, "Ion Focusing, Mass Analysis and Detection," in **Gaseous Ion Chemistry and Mass Spectrometry**, J. H. Furtrell, Ed., Wiley, New York, 107 (1986).

76. See, for example, Y. Nihei, H, Satoh, S. Tatsuzawa, and M. Owari, "High Spatial Resolution Secondary Ion Mass Spectrometry with Parallel Detection System," *J. Vac. Sci. Technol.*, A 5, 1254 (1987).

77. See, for example, "Monitoring of Plasma Process," *VG Quadrupoles Product Brochure*, VG Quadrupoles Ltd, (1986).

78. K. Okuno, S. Tomita, and A. Ishitani, "Application of SIMS Technique to Organic Polymers," in **Secondary Ion Mass Spectrometry SIMS IV**, A. Benninghoven et al., Eds., Springer Series in Chemical Physics, 36, 392 (1984).

79. R. Michael and D. Stulik, "Depth Profile Measurement in PMMA-PTFE Sandwich Structure Using Dynamic FAB-MS," *Appl. Surf. Sci.*, 28, 367 (1987).

80. C. A. Chang, J. E. E. Baglin, A. G. Schrott, and K. C. Lin, "Enhanced Cu-teflon Adhesion by Presputtering Prior to Deposition," *Appl. Phys. Lett.*, 51, 103 (1987).

81. (a) K. Wittmaack, "Charge Compensation in SIMS Analysis of Polymer Foils Using Negative Secondary Ions," *Surf. Interface Anal.*, 10, 311 (1987). (b) B. Hagenhoff, D. van Leyen, E. Niehuis, and A. Benninghoven, "TOF-SIMS of Insulators by Pulsed Charge Compensation with Low Energy Electrons," in **Proc. SIMS VI**, A. Benninghoven et al., Eds., Wiley, 235 (1988).

82. R. Chujo, T. Nishi, Y. Sumi, T. Adachi, H. Naitoh, and H. Franzel, "Depth Profiling of Polymer Blends and Optical Fiber with aid of SIMS," in **Secondary Ion Mass Spectrometry SIMS IV**, A. Benninghoven et al., Eds., Springer Series in Chemical Physics, 36, 389 (1984).

83. B. K. Furman, S. Purushothaman, E. Castellani, and S. Renick, "Mechanicl and Surface Analytical Studies of Ti/Polyimide Adhesion," Proc. Symp. on Multilevel Metallization, Interconnection, and Contact Technologies, L. B. Rothman and T. Herndon, Eds., *Electrochem. Soc. Meeting*, Oct. 21-22, San Diego, CA (1986).

84. M. J. Hearn, D. Briggs, S. C. Yoon, and B. D. Ratner, "SIMS and XPS Studies of Polyurethane Surfaces. II. Polyurethanes with Fluorinated Chain Extenders," *Surf. Interface Anal.*, 10, 384 (1987).

85. P. Auger, "Theoretical Considerations of the Directions of Photoelectron Emission," *Compt. Rend.*, 180, 65 (1925).

86. M. P. Seah, "AES in Metallurgy," in **Practical Surface Analysis by Auger and X-ray Photoelectron Spectroscopy**, D. Briggs and M. P. Seah, Eds., 247, Wiley, New York (1983).

87. H. J. Mathieu and D. Landolt, "Influence of Gases in UHV Systems on Auger Electron Spectroscopy," *J. Microsc. Spectr. Electron.*, 3, 113 (1978).

88. For a discussion of coupling schemes, see, for instance, J. C. Slater, **Quantum Theory of Matter**, Chapter 10, McGraw-Hill, New York (1968).

89. M. F. Chung and L. H. Jenkins, "Auger Electron Energies of the Outer Shell Electrons," *Surface Sci.*, 21, 253 (1970).

90. (a) W. A. Coghlan and R. E. Clausing, "Auger Catalog. Calculated Transition Energies Listed Energy and Element," *Atomic Data*, 5, 317 (1973); (b) F. P. Larkins, "Semiempirical Auger-electron Energies for Elements $10 \leq Z \leq 100$," *Atomic Data and Nuclear Table*, 20, 311 (1977).

91. L. E. Davies, N. C. MacDonald, P. W. Palmber, G. E. Riach, and R. E. Weber, *Handbook of Auger Electron Spectroscopy*, Physical Electronics Division, Perkin-Elmer Corp. (1978).

92. See, for example, J. A. D. Matthew, M. Prutton, M. M. El Gomati and D. C. Peacock, "The Spectral Background in Electron Excited Auger Electron Spectroscopy," *Surf. Interface Anal.*, 11, 173 (1988).

93. M. P. Seah, "Quantification of AES and XPS," in **Practical Surface Analysis by Auger and X-ray Photoelectron Spectroscopy**, in D. Briggs and M. P. Seah, Eds., Wiley, New York, 189 (1983).

94. S. Hofman, "Depth Profiling," in **Practical Surface Analysis by Auger and X-ray Photoelectron Spectroscopy**, D. Briggs and M. P. Seah, Eds., Wiley, New York, 141 (1983).

12

PROSPECTS FOR EXAMINATION OF POLYMER MOLECULES WITH SCANNING TUNNELING MICROSCOPE AND ATOMIC FORCE MICROSCOPE

D. H. Reneker

Polymers Division
National Bureau of Standards
Gaithersburg, Maryland

Gerd K. Binnig and Heinrich Rohrer received the Nobel Prize in 1986 for the invention of the scanning tunneling microscope (STM). Quate (1) reviewed the early development of this instrument, along with some early applications and descriptions of its promise. Hansma and Tersoff (2) described both the experimental and theoretical principles upon which the microscope is based. They showed an image of the biopolymer molecule, DNA, which was obtained by Binnig and coworkers before 1985. Early applications to conducting solids yielded incisive information about their surfaces, in vacuum, air, and in liquids, both insulating and conducting. This paper is a review of some observations of small molecules and polymer samples that are leading toward useful techniques for imaging polymer molecules.

Related instruments and experimental techniques described below have also been developed. These include atomic force microscopy (AFM), inelastic tunneling scanning microscopy, and other tip-scanning microscopies. It is the purpose of this paper to review the application, both demonstrated and potential, of all these methods to the elucidation of interesting aspects of the structure of polymer molecules.

Many descriptions of STM and AFM instrument designs can be found in the bibliographies of the papers cited in this paper. The designs are evolving rapidly at the time of this writing. Several commercial models of STMs are available. The equipment and techniques for processing of data into useful images are also improving rapidly, and will not be discussed further in this paper. The early work of Young (3), on scanning microscopes for the examination of surfaces, anticipated many of these developments.

12.1 METHODS

Brief descriptions of the principles of scanning tunneling microscopy, atomic force microscopy, related techniques, and some experimental methods are provided in the sections that follow.

12.1.1 Scanning Tunneling Microscopy

The essential element of a scanning tunneling microscope (STM) is a conducting tip separated by a nonconducting gap from the surface of a conductor. The gap is small enough that electrons move across it by quantum mechanical tunneling. The probability of an electron tunneling between the tip and the conductor decreases exponentially with gap length, which is usually around 1 nm. As a result, most of the tunneling current flows through the atom on the tip that has the shortest perpendicular distance to the conductor. For an atomically smooth sample surface, and the tip materials commonly used, this atom is unique. Multiple tunneling sites on the tip are active when more complicated surfaces are examined. This complicates the interpretation of some measurements.

When the tunneling current flows through the single atom that is closest to the conducting surface, the tunneling current is very sensitive to the lateral position of the tip with respect to atom scale features of the conducting surface. Features parallel to the surface and smaller than 0.1 nm are often resolved.

Even though molecules are generally not conductors, the presence of a molecule in the tunneling gap affects the tunneling probability in the experiments described below. Whether the probability increases or decreases in the presence of the molecule is not yet established in a general way.

Figure 12.1 shows some of the features of a contemporary scanning tunneling microscope. The elimination of low-frequency vibrational modes that move the tip relative to the sample is a challenge dealt with successfully in this design. The tip is mounted on a piezoelectric cylinder which has electrodes positioned so that the cylinder can be caused to bend right or left, to bend front to back, or to extend and contract along the axis, by the application of appropriate voltages. One end of the piezoelectric cylinder is mounted in a metal shell which, in operation, is held on three screws which serve to adjust the separation between the tip and the sample. The three screws are adjusted by hand until the separation between the tip and the sample is a few hundred micrometers. Then the fine advance screw is turned by a stepping motor that tilts the metal shell about the tops of the coarse advance screws and causes the tip to approach the sample in very small steps. The tunneling current is measured after each step, and the motion stopped when tunneling is detected. This permits the tip to be brought to within a few nanometers of the surface without mechanical contact. The entire assembly, along with electronic circuitry that supplies a bias voltage of around 0.1 V to the tip and serves as a preamplifier for the tunneling current, are enclosed in a metal shield, and supported on a vibration-free mount.

A servomechanism adjusts the length of the piezoelectric cylinder to increase or decrease the gap and maintain a constant tunneling current of around 1 nA. The tunneling current is sensitive to changes in the tip to sample distance as small as 0.001 nm. The changes in the tip position are recorded as the tip is carried in a raster pattern over the sample by voltages applied to the electrodes that bend the piezoelectric cylinder. The resulting data may be presented in many ways, the most common of which is an image in which the tip position is represented by a gray scale at each position of the sample explored by the tip. A typical image contains from 40,000 to 160,000 picture elements.

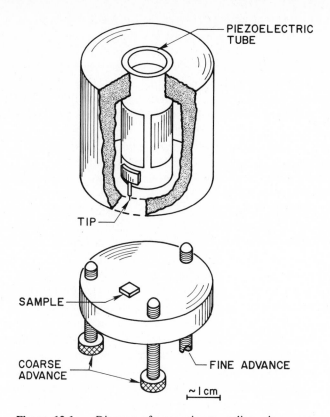

PIEZOELECTRIC
TUBE

TIP

SAMPLE

COARSE
ADVANCE

FINE ADVANCE

~1 cm

Figure 12.1:   Diagram of a scanning tunneling microscope.

12.1.2 Replica Techniques with the STM

The surface morphology of insulating structures can be examined by coating the insulator with a thin conducting layer. Conducting layers only a few nanometers thick can be applied to most solid surfaces by evaporation of a platinum and carbon mixture, using methods developed to shadow samples for the transmission electron microscope. This thin layer allows the larger-scale morphological features to be observed. Conducting coatings were used as described below to examine biological and molecular structures.

Shen et al. (4) and Jaklevic et al. (5) examined insulating surfaces coated with an ultrathin smooth gold film applied by vacuum evaporation onto liquid-nitrogen-cooled samples. Lateral resolutions of 1 nm were achieved.

12.1.3 Atomic Force Microscopy

To examine the surface of nonconducting solids at the atomic scale, Binnig et al. (6) developed an instrument called the atomic force microscope (AFM). In the AFM, a fragment of diamond or other hard material is attached to a spring and brought close enough to a surface to engage the force of the atoms on the surface. The force is measured by detecting the deflection of the spring with a tunneling gap, with a laser interferometer (7), or with an optical lever (8). A feedback system is used to adjust the vertical position of the AFM tip above the sample surface to keep the deflection of the spring, and therefore the force, constant as the tip is scanned over the sample. The AFM can be thought of as a refinement of the stylus profilometer in which the forces between an atom on a scanning tip and the surface of a solid are measured. The AFM explores the interatomic potential functions with no transfer of electrons between the tip and the sample, while the STM requires that the sample be conducting since electrons must pass through it. Atomic force microscopy is rapidly developing as a valuable technique for the examination of the surfaces of nonconducting objects, be they polymeric, molecular, or mineral.

Figure 12.2 is a diagram of an AFM that uses an optical lever to detect bending of a small cantilever that carries a diamond fragment to within the range of the surface forces of the sample. The diagram is based on a design used by Drake et al. (9) in the laboratory of Paul K. Hansma. Light from a laser is carried to the instrument by an optical fiber which isolates the laser from the microscope, as was first done in this type of microscope by R. Barrett at Stanford University. The light reflects from the cantilever onto a photodiode which compares the amount of light falling on each of two electrodes, and is thereby very sensitive to small deflections of the light beam.

The microfabricated cantilever has a length of only 100 μm. The optical lever ratio used was around 800. The force with which the diamond pushes against the sample is as low as 2×10^{-9} N. In this instrument, the entire sample is carried in a raster pattern by a single tube piezoelectric translator similar to that which carries the tip in the STM shown in Fig. 12.1. The length of the tube is controlled by a servomechanism that keeps the deflection of the cantilever near a constant value. The tip position data are used to form an image in the same way as for the STM.

The mechanical approach mechanism that brings the diamond point to the sample is the same as that described for the STM. This microscope can operate in air, or with the tip immersed in a liquid such as water. Operation in a liquid allows the applied force to be more precisely known and controlled, and provides an interesting environment for the study of biological or chemically reactive surfaces.

12.1.4 Inelastic Scanning Tunneling Microscopy

As an electron tunnels through an insulator it undergoes strong interactions with the bound electrons in the insulator. Energy may be transferred between the tunneling electron and the vibrating molecule. This phenomenon is the basis for the active experimental area called inelastic tunneling spectroscopy.

Jaklevic and Lambe (10) observed that the vibrational spectra of molecules in oxide tunnel junctions could be measured by the second derivative of the voltage/current curve of the tunneling electrons. Inelastic tunneling spectroscopy was recently reviewed in a book edited by Hansma (11).

Persson and Demuth (12) examined the possibility of measuring the vibrations of an individual molecule with the tunneling microscope. Molecules of sorbic acid absorbed onto graphite were examined by Smith (13) and Smith et al. (14) with an STM at liquid helium temperature. Large changes in the first derivative of the voltage/current curve were observed at energies corresponding to the vibrational modes of the sorbic acid molecule. The tunneling conductivity measured with the STM increased by as much as a factor of 10 at the energies of the molecular

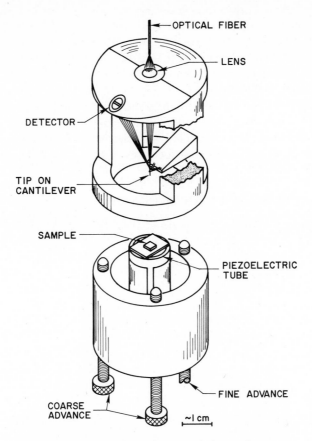

Figure 12.2:   Diagram of an atomic force microscope.

vibrations. The changes in the voltage/current curves seen in conventional scanning tunneling spectroscopy are much smaller. Gata and Antoniewicz (15) have proposed several theoretical models to describe inelastic tunneling through molecular adsorbates.

When the tip moved 0.5 nm to another part of a sorbic acid molecule, the voltages at which the changes in the first derivative occurred were different. This indicates that the vibrational modes observed were localized in the part of the molecule directly under the tip. These experiments suggest that it may become possible to form maps showing the sites within a molecule where particular vibrational resonances occur.

12.1.5 Other Tip Scanning Microscopies

Wickramasinghe (16) observed that the scanning tunneling microscope has taught us how to position a tip with subatomic accuracy over a sample surface and scan it to form an image. He suggests the use of this ability to examine other physical properties of the surface in addition to the electronic states and surface morphology.

Included in his list is charge imaging, observed by oscillating the tip. This could become an interesting technique for polymers and other insulators where questions about charge trapping sites or ferroelectric domain boundaries exist.

Teague (17) also reviewed this subject. He included scanned tip microscopies that utilize field emission, capacitance, and near-field optical microscopy, in addition to those mentioned already in this paper.

Hansma (18) described the construction and use of a scanning ion-conductance microscope. The probe is an electrolyte-filled micropipette. Images of the topography of nonconducting surfaces covered with electolytes were obtained. Images of 0.8 μm pores in a membrane filter demonstrated its use for imaging ion currents through channels in membranes.

12.2 MOLECULAR SUBSTANCES INVESTIGATED

This section provides examples of the use of the STM and AFM to examine molecules. The results illustrate the atomic, or near-atomic detail which can be observed. Although most of the molecules examined are somewhat smaller than a typical polymer, a segment of a polymer molecule is similar in many ways to the molecules in the experiments summarized below. These examples serve to illustrate the sort of information that can be expected from the examination of polymer molecules. Direct observation of branches, copolymer groups, and crosslinks appears to be possible now, with details of atomic structure likely future prospects.

A typical polymer molecule, contracted into its densest conformation, typically occupies a volume a few tens of nanometers in diameter. The dense conformation may be either a crystal or an amorphous solid. Access to the segments of the polymer molecule can be gained by unraveling it from the solid and attaching it to a surface, or by "winding" it from one particle to another. Experimental techniques to gain access to segments are now being sought.

12.2.1 Small Molecules

Foster et al. (19) reported the purposeful attachment of single molecules of dimethyl phthalate and other compounds to a graphite surface by applying a short voltage pulse to the tip of an STM. The tip was immersed in a liquid layer of the compound covering the cleavage face of highly oriented pyrolytic graphite. Pulses of around 3.5 V and 100-ns duration attached irregular structures to the surface. Shapes varied from one experiment to the next and sizes were commensurate with that of the molecules. No dependence on the polarity of the tip was observed.

The attached molecule can be removed from the surface with a similar voltage pulse applied as the tip passes. Alternatively, the second pulse may reduce or enlarge the size of the structure attached to the graphite. The authors suggest that enlargement results from the attachment of a second molecule adjacent to the first, and that reduction in size is due to removal of only part of the molecule. Reduction in size is suggested as evidence for a change in the chemical structure of the molecule. Fragments as small as a benzene ring were left attached to the cleavage surface. The authors noted that this is the smallest spatially localized change in the structure of a molecule yet observed.

Possible artifacts, such as the imaging of structure on the tip by a sharp protrusion on the graphite, or the transfer of atoms between the tip and surface without immobilization of the molecules from the liquid were discussed. While such possibilities cannot be completely ruled out, the molecular attachment phenomena is the explanation favored by the authors.

Lippel and Wilson (20) observed isolated copper phthalocyanine molecules sublimed onto clean silicon or gold surfaces. The molecules were about 1.5 nm in diameter. Some displayed a four-fold symmetric structure while others in the same images appeared as disks against well-resolved backgrounds. Atomic resolution of copper phthalocyanine is reported in work from the same laboratory (21). Gimzewski et al. (22,23) observed rows of copper phthalocyanine molecules adsorbed on a silver surface. The rows showed clearly resolved periods of 1.4 and 2.6 nm.

Wilson et al. (24) observed coherent molecular arrays, molecular registration at step edges, boundaries of rotational and translational domains, the internal symmetry of benzene, and an example of molecular diffusion of benzene in coadsorbed arrays of benzene and carbon monoxide on rhenium (111) planes. An image of benzene from a similar experiment on a rhodium surface is given in a recent news article (25). Ohtani et al. (26) observed STM images of the (3 × 3) superlattice of benzene and carbon monoxide coadsorbed on rhenium (111) surfaces. A well-ordered array of ringlike features associated with individual benzene molecules was observed. They suggest that examination of carefully chosen metal adsorbate surfaces with the STM is a promising way to observe surface chemical processes, such as molecular diffusion, nucleation phenomena, and step or defect-related reactivity. Hallmark et al. (27) investigated short chain organic molecules on the (111) surface of gold in a high vacuum.

Hubacek et al. (28) have observed the adsorption and ordering of catalytic species such as phosphotungstic acid and rhenium carbonyl complexes on graphite. Adsorption of solvent molecules such as acetone was also observed. Hubacek et al. (29) observed cloudy patches of acetone drifting across images of atoms at the surface of cooled graphite as the surface was exposed to acetone vapor. Ono et al. (30) investigated polyphosphoric acid on graphite, and made current versus tip voltage measurements at room temperature.

12.2.2 Polymers

Dovek et al. (31) examined submonolayer films of poly(octadecylacrylate) on graphite, molybdenum disulfide, and gold substrates. Images obtained by STM and AFM showed good agreement on each substrate. Fibrils of width around 0.8 nm and groups of such fibrils were observed and interpreted as individual polymer chains or small bundles of parallel chains. The fibrils could be cut in the STM by applying a pulse of about 4 V for a duration of 100 ns.

Polymer structures produced by the precipitation of ultrahigh molecular weight polyethylene onto mica from a solution containing an entangled network of molecules were shadowed with a platinum/carbon mixture and examined by Reneker et al. (32) with both the STM and the transmission electron microscope. As the entangled molecules precipitated into crystals, small fibers were formed which served as nuclei for further crystallization. These processes produced irregular branched structures with 20-nm-diameter branches. Adjacent areas examined with the STM and the transmission electron microscope are compared in Fig. 12.3. This work shows that

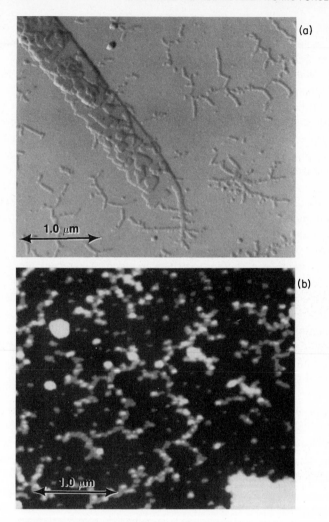

Figure 12.3:   Polymer structures produced by the precipitation of ultrahigh molecular weight polyethylene onto mica from a solution containing an entangled network of molecules. (a) As observed with the transmission electron microscope. A 1-μm bar is marked on the figure. (b) As observed with the STM, at the same magnification. The light (high) area in the lower right corner of (a) is the tip of a folded chain lamellar crystal similar to that shown extending toward the center of (a) from the upper left corner.

the surfaces of polymer objects coated with a thin conducting layer can be imaged with the STM in air at a lateral resolution commensurate with that obtained from a replica examined in the transmission electron microscope, and with much better resolution in the thickness direction. The STM avoids the difficulties associated with the removal of the replica, a fact likely to be of considerable practical importance.

Very small fibers of polytetrafluoroethylene drawn by mechanically detaching tiny particles of the polymer from evaporated gold surfaces were examined with the STM by Reneker and Howell (33). Some other experiments aimed at the production of isolated segments of a polymer molecule on a conducting surface were described in that paper and its references.

Rows of metal atoms on a graphite surface can be considered as a special kind of polymer, although the metal atoms are bonded to the substrate about as strongly as they are bonded to each other. Stroscio (34), in the course of observing single atom adsorbates, noted that cesium forms linear chains on GaAs. Long chains are not observed because, unlike more conventional polymers, atoms can easily attach to the sides of the metal chains and convert them to patches of metal. Adsorbed oxygen atoms on the (110) surface of GaAs were found by Stroscio et al. (35) to produce both positive and negative contours, depending upon the bias voltage of the tip.

12.2.3 Polymeric Conductors

Bonnell et al. (36) made measurements of the electronic structure by tunneling spectroscopy of fully and partially protonated polyaniline films, leading to implications about proposed models for the insulator-to-metal transition.

Elings and Wudl (37) examined the surface of pyrolyzed organic compounds, graphite and glassy carbon. Complex structures were found, including rows of unresolved atoms separated by 0.15 to 0.5 nm.

Bando et al. (38) examined two-dimensional networks of the organic superconducting molecule called (BEDT-TTF)2X.

12.2.4 Liquid Crystals

Foster and Frommer (39) reported on the observation of liquid crystals on graphite. Near atomic resolution was achieved. Two-dimensional periodicity comparable to that observed with X-rays was observed.

Several molecules that form liquid crystal systems were imaged at a resolution that produced atomic scale information by Foster and Frommer (40). Each crystal system had a distinct and different packing over fields of 20 × 20 nm. The image from one kind of molecule that contained both aromatic and aliphatic parts showed a pattern of light and dark regions that suggest that the tunneling probability was greatly different in the two regions.

12.2.5 Biological Molecules

Beebe et al. (41) observed uncoated calf thymus DNA with the STM. The DNA molecules were dried from a salt water solution onto graphite. The major and minor grooves of the double helix were observed, as shown in Fig. 12.4.

Drake et al. (9) obtained images of an ordered array of polyalanine molecules formed on a glass surface by a simple process involving deposition from a solution. The polyalanine molecules were imaged with an atomic force microscope with a resolution of less than 0.5 nm, as demonstrated by the observation of the molecular features of the cleavage plane of mica. Images were obtained both with the protein immersed in water, and dry, as shown in Fig. 12.5. The force exerted on the molecule by the tip of the atomic force microscope was around 2×10^{-9} N. This paper also contains a series of ten images of the polymerization of fibrin, the molecular fabric of blood clots. These images illustrate the potential of the AFM for revealing biological processes as they occur in real time in a liquid environment.

Lindsay (42-44) used a scanning tunneling microscope for the examination of biopolymers in an aqueous buffer solution. An additional electrode inserted into the cell was used to "electroplate" biopolymer molecules such as cytochrome-c, tRNA, and DNA onto a gold substrate. The DNA formed aggregates in which the rodlike molecules were packed side by side. The approximately 3.5-nm twist period of the DNA was evident. The measured dependence of tunneling current on the tip to substrate distance in these samples suggests that the contrast mechanism depends upon an elastic interaction between the tip and the sample.

Travaglini (45) evaporated platinum carbon mixtures of the sort used for electron microscopy to provide a conducting coating for biological molecules. RecA-DNA complexes and DNA molecules were imaged in interesting detail in this way.

Amino acids adsorbed on highly oriented pyrolytic graphite were observed with the STM operating in air by Feng et al. (46). Individual molecules were found, but clusters, dimers or monomolecular arrays were more often observed.

12.2.6 Molecular Membranes

Imaging of a lipid bilayer (cadmium arachidate) at molecular resolution by scanning tunneling microscopy was reported by Smith et al. (47). A bilayer of cadmium icosanoate (arachidate) was deposited onto a graphite substrate by the Langmuir-Blodgett technique. The STM images of a surface of the film containing the aliphatic ends showed a partially ordered triclinic unit cell. The unit cell had an area, viewed along the long axis of the molecules, of 0.194 nm (2). The unit cell attributed to the lipid bilayer was distinctly different from that of the graphite, which was observed on nearby areas of the same sample. The lipid bilayer was more than 5.0 nm thick so it appears, somewhat surprisingly, that the tunneling electrons were transferred through an "insulating" layer of at least that thickness. The resistivity along the long axis of the lipid molecule was estimated to be around 103 ohm-cm, much less than that of a typical organic solid.

About 10 frames/s were recorded at a high-scan rate and presented as a movie. The ordered bilayer was consistently observed to flow across the viewing field of the STM at a rate typically near 0.4 nm/s. The nearby areas of the graphite had much lower drift rates, leading to the conclusion that the bilayer was diffusing across the graphite. Although patches of bilayer and less well-resolved material were observed, no edges separating the bilayer from the graphite were found. The authors suggest that many biological membranes should be accessible to study by scanning tunneling microscopy. Similar results on cadmium arachidate and di-myristoyl-phosphatidic acid were obtained by Lang et al. (48).

Sleator (49) used gold/alkyl thiol chemistry with a covalent monolayer of 18 carbon atom alkyl chains to mimic the outer part of a membrane bilayer. Although atomic steps on the gold were observed, the alkyl chains were not observed. He suggested that the reason might be lack of conduction, lack of horizontal resolution, or that motion of the alkyl chains might have made it impossible to see them.

Voelker et al. (50) described plans to work with calf skin collagen fibrils which are 1.5 nm in diameter and 300 nm long, and with lipid membranes.

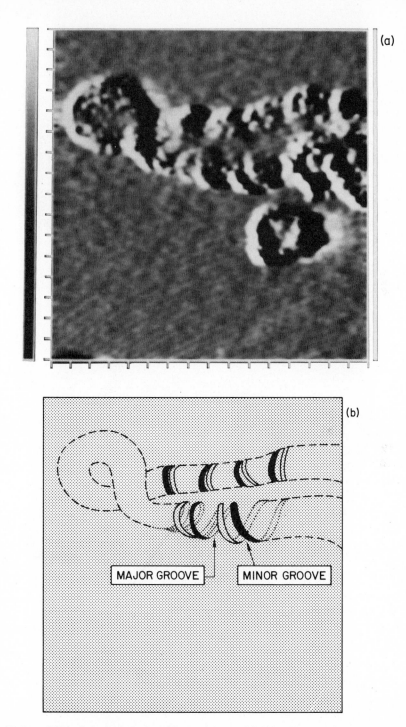

Figure 12.4:   DNA deposited on graphite and imaged in air with the STM. (a)Processed image of a DNA molecule. The field is 40 nm on each edge. (b)Schematic of the DNA structure in (a). See Ref. 41.

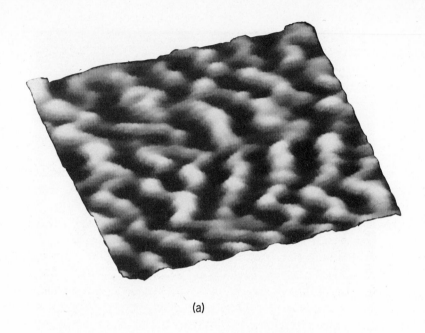

(a)

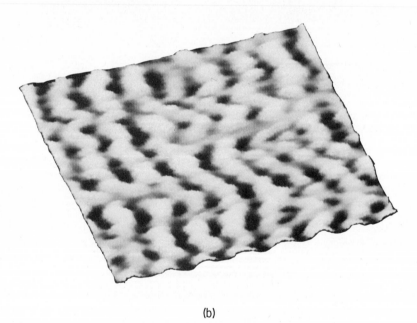

(b)

Figure 12.5:   AFM images of an array of polyalanine molecules on a glass surface. (a) Covered with water, (b) dry. The field is 3.4 nm on each edge. See Ref. 9.

Use of the atomic force microscope to obtain images of organic molecules polymerized in monolayers produced by Langmuir-Blodgett methods was described

by Marti et al. (51). The monolayers were made from
n-(2-aminoethyl)-10,12-tricosadiynamide (AE-TDA) molecules. The monolayers
were supported on glass with the hydrophobic ends of the AE-TDA molecules facing
the diamond tip of the AFM. The rows of molecules in the monolayer are known to
have a side-by-side spacing of about 0.5 nm. Images from the AFM show parallel
rows with a side-by-side spacing of 0.55 ± 0.05 nm. The observation of rows rather
than the ends of individual molecules is attributed by the authors to the effects of the
direction of the polymer chains that connect the AE-TDA molecules together. Some
structure was visible along the rows.

Arrays of molecules at the surface of an amino acid crystal (DL-leucine) were
imaged by Gould et al. (52) with an AFM. Comparison of the observed positions of
the molecules at the surface with the results of X-ray diffraction analysis of the position of the molecules inside the crystal indicates that the spacings between the molecules are not significantly affected by the surface. The separation between
molecules, which were clearly resolved in filtered images, was 0.539 nm along the
crystallographic a-axis and 0.519 nm along the c-axis.

Rabe et al. (53) examined submonomolecular layers of substituted
poly(phthalocyaninatosiloxane) and polymerized amphilic diacetylene on a graphite
surface, and on carbon fibers. Fuchs and Schrepp (54) observed Langmuir-Blodgett
films prepared from monomeric and polymeric materials.

12.2.7 Biological Structures

Zsasadzinski et al. (55) demonstrated the use of the STM to observe
platinum/carbon replicas of insulating molecules at a resolution of better than 5 nm.
They used electron beam evaporation to deposit a thin, conducting layer of platinum
and carbon onto the fracture surface of frozen dimyristoyl-phosphatidyl-choline
(DMPC). The conducting layer was then supported so that the DMPC could be
removed and the side originally in contact with the DMPC examined in the scanning
tunneling microscope. DMPC is known from X-ray diffraction and electron
microscopy to be "rippled" with a period of 12 or 13 nm. The images made with the
STM showed that the fracture surface had ripples with a spacing of 13 nm and an
amplitude of about 4.5 nm. They observed that metal freeze-fracture replicas of
biomembranes can greatly enlarge the type and number of surfaces that can be
examined with the STM. The STM measures the vertical dimension of the replicas
more conveniently than the transmission electron microscope.

Amrein et al. (56) described images of unstained air-dried purple membranes and
DNA-recA protein complexes. Tunneling into the top of at least five layers of membrane was observed. Internal details on the DNA complex on a scale of around 1 nm
were observed.

Blackford et al. (57) imaged the proteinaceious sheath of a cell which has a wall
thickness of about 12.5 nm, a diameter of about 500 nm, and a length of several
micrometers. Images of the entire cell sheath deposited on graphite, as well as
higher-resolution images of the sheath surface after coating with evaporated gold or
platinum were obtained.

12.3 SUMMARY

The application of the STM and the AFM to molecular substances is growing rapidly. A wide range of capabilities have been demonstrated. The solution of many kinds of problems that rest on the structure, electronic properties, or vibrational properties of molecules may be found by measurements with these versatile new techniques.

Small molecules have been imaged on conducting surfaces in air, in liquids, and in vacuum. Interesting images of large molecules and molecular aggregates have been made with replica techniques. Langmuir-Blodgett films and biological membranes thicker than the normal tunneling distances used have produced images with molecular scale structure resolved. Polymer molecules, both synthetic and biological, have been successfully imaged. Conducting molecular solids created by synthetic chemistry or by pyrolysis of organic solids have been examined with atomic scale resolution. Chemical changes in the structure of both small molecules and polymers that are caused by an electrical pulse applied to the molecule through the tip of the STM have been observed. Singularities in the derivative of the volt-ampere curve of a tip over a molecule at liquid helium temperature were associated with expected vibrational modes of molecules.

Better models of the influence of the stationary electronic states of a molecule on the tunneling probability are needed. The interactions between the tunneling electrons and the vibrational states of the molecule are poorly understood. Tunneling through structures at least 5 nm thick made of organic molecules requires more detailed explanation. Does the tip plough through such samples? Is an unfamiliar electron transport mechanism activated? The tunneling mechanism for DNA in solution needs clarification.

Of the many other effects of tunneling electrons at the molecular scale which are expected we will mention one. Passage of a tunnneling electron may produce controlled changes in molecular conformation. The resulting control of the motion of polymer molecules could lead to transport mechanisms that could unwind and carry long polymer molecules past the tip. Such transport mechanisms could speed the systematic investigation of the structure of typical polymer molecules, which contain hundreds of thousands or even millions of repeat units, each of which is a potential site of a side branch, or other variation. Direct reading of the genetic code of a DNA molecule with the STM has not yielded to the first investigators, but still seems a possibility. The ability to manipulate long DNA molecules at the molecular scale could help make this possibility a practical process.

A host of information storage schemes can be foreseen. Thomson (58) observed that if a bit of information could be stored in the area of an atom, 10^{15} bits/cm^2 could be stored, and conluded that an STM storage device with useful read and write rates and a large storage capacity should be achievable. Polymer molecules offer the potential of an atomic scale storage medium that would have some of the access and storage density characteristics of magnetic tape. The STM may give experimental access to man-made computing devices at the molecular level of the sort described by Hopfield et al. (59).

While we have sought to mention examples of progress in observation of polymer molecules and other related substances, we have certainly not been able to keep pace with the rapid developments in the field. We apologize in advance for the important papers that we have missed.

12.4 ACKNOWLEDGMENTS

The author is grateful to Howard H. Harary and Freddy A. Khoury for many helpful comments.

12.5 REFERENCES

1.   C. J. Quate, "Vacuum Tunneling: a New Technique for Microscopy," *Physics Today*, 26, Aug. (1986).
2.   P. K. Hansma and J. Tersoff, "Scanning Tunneling Microscopy," *J. Appl. Phys.*, 61, R1 (1988).
3.   R. D. Young, "Surface Microtopography," *Physics Today*, 42, Nov. (1971).
4.   W. Shen, J. T. Chen, R. C. Jaklevic, L. Elie, and G. V. Puskorius, "Topographic Images of Insulating Surfaces with the Scanning Tunneling Microscope," *Bull. Amer. Phys. Soc.*, 33, 431 (1988).
5.   R. C. Jaklevic, L. Elie, W. Shen, and J. T. Chen, "Application of the Scanning Tunneling Microscope to Insulating Surfaces," *J. Vac. Sci. Technol.*, A6, 448 (1988).
6.   G. Binnig, C. F. Quate, and C. Gerber, "Atomic Force Microscope," *Phys. Rev. Lett.*, 56, 930 (1986).
7.   G. M. McClelland, R. Erlandsson, and S. Chiang, "Atomic Force Microscopy: General Principles and a New Implementation," in *Review of Progress in Quantitative Non-Destructive Evaluation*, Vol. 6, Plenum, New York (1987).
8.   N. M. Amer and M. Meyer, "A Simple Optical Method for the Remote Sensing of Stylus Deflection in Atomic Force Microscopy," *Proc. Roy. Microscopical Soc.*, 23, Part 3, Supplement, S20 (1988).
9.   B. Drake, C. B. Prater, A. L. Weisenhorn, S. A. C. Gould, T. R. Albrecht, C. F. Quate, D. S. Cannell, H. G. Hansma, and P. K. Hansma, "Imaging Crystals, Polymers, and Processes in Water with the Atomic Force Microscope," to be published in *Science* (1989).
10.   R. C. Jaklevic and J. Lambe, "Molecular Vibration Spectra by Electron Tunneling," *Phys. Rev. Lett.*, 17, 1139 (1966).
11.   P. K. Hansma, Ed., **Tunneling Spectroscopy: Capabilities, Applications, and New Techniques**, Plenum, New York (1982).
12.   B. N. J. Persson and J. E. Demuth, "Inelastic Electron Tunneling from a Metal Tip," *Solid State Commun.*, 57, 769 (1986).
13.   D. P. E. Smith, "New Applications of Scanning Tunneling Microscopy," *Thesis*, Ginzton Laboratory No. 4217, W. W. Hansen Laboratories of Physics, Stanford University, Stanford, CA (1987).
14.   D. P. E. Smith, M. D. Kirk, and C. F. Quate, "Molecular Images and Vibrational Spectroscopy of Sorbic Acid with the Scanning Tunneling Microscope," *J. Chem. Phys.*, 86, 6034 (1987).

15. M. A. Gata and P. R. Antoniewicz, "Model for Inelastic Tunneling Through Chemisorbed Species," *Proc. Roy. Microscop. Soc.*, 23, Part 3, Supplement, S23 (1988).

16. H. K. Wickramasinghe, "Novel Scanned Tip Microscopies," *Bull. Amer. Phys. Soc.*, 33, 413 (1988).

17. E. C. Teague, "Scanning Tip Microscopies: an Overview and some History," *Proc. 46th Annual Meeting of the Electron Microscopy Society of America*, San Francisco Press Inc., 1004 (1988).

18. P. K. Hansma, B. Drake, O. Marti, S. A. C. Gould, and C. B. Prater, "The scanning Ion-conductance Microscope," *Science*, 243 , 641 (1989).

19. J. S. Foster, J. E. Frommer, and P. C. Arnett, "Molecular Manipulation Using a Tunnelling Microscope," *Nature*, 331, 324 (1988).

20. P. H. Lippel and R. J. Wilson, "Adsorbate Deformation as a Contrast Mechanism in STM Imaging of Biopolymers in an Aqueous Environment," *Proc. Roy. Microscop. Soc.*, 23, Part 3, Supplement, S37 (1988).

21. R. Baum, "Molecules Imaged at Atomic Resolution," *Chem. Eng. News*, pp. 27-8, Feb. 6, 1989. See also P. H. Lippel, R. J. Wilson, M. D. Miller, C Woll, and S. Chiang, "High Resolution Imaging of Copper-Phthalocyanine by Scanning Tunneling Microscopy," *Phys. Rev. Abstr.*, 20, 171 (1989).

22. J. K. Gimzewski, E. Stoll, and R. R. Schlittler, "Scanning Tunneling Microscopy of Individual Molecules of Copper-Phthalocyanine Adsorbed on Polycrystalline Silver Surfaces," *Surf. Sci.*, 181, 267 (1987).

23. J. Gimzewski, J. H. Coombs, R. Moller, and R. A. Schlittler, "Scanning Tunneling Microscopy of Molecular Clusters of Copper Phthalocyanine Adsorbed on Silver Surfaces," *Proc. Roy. Microscop. Soc.*, 23, Part 3, Supplement, S35 (1988).

24. R. J. Wilson, H. Ohtani, S. Chiang, and C. M. Mate, "Direct Observations of Coadsorbed Arrays of Benzene and Carbon Monoxide on Rh(111) by Scanning Tunneling Microscopy," *Proc. Roy. Microscop. Soc.*, 23, Part 3, Supplement, S9 (1988).

25. Editor, "Benzene Snapshot Captures Images of Individual Rings," *Chem. Eng. News*, August 1, 5, 1988.

26. H. Ohtani, R. J. Wilson, S. Chiang, and C. M. Mate, "Scanning Tunneling Microscopy Observations of Benzene Molecules on Rh(111) Surfaces," *Phys. Rev. Lett.*, 60, 2398 (1988).

27. V. M. Hallmark, S. Chiang, R. J. Wilson, and J. F. Rabolt, "Studies of Organized Organic Monolayers on Thin Metal Films by Scanning Tunneling Microscopy," *Proc. Roy. Microscop. Soc.*, 23, Part 3, Supplement, S3 (1988).

28. J. S. Hubacek, R. Brockenbrough, S. Skala, G. Gammie, M. P. Keyes, J. R. Shapley, and J. W. Lyding, "Scanning Tunneling Microscopy of Graphite Adsorbed Molecular Species," *Bull. Amer. Phys. Soc.*, 33, 527 (1988).

29. J. S. Hubacek, G. Gammie, R. Brockenbrough, S. Skala, F. R. Shapley, and J. W. Lyding, "Scanning Tunneling Microscopy of Graphite Adsorbed Molecular Species," *Proc. Roy. Microscop. Soc.*, 23, Part 3, Supplement, S11 (1988).

30.   M. Ono, W. Mizutani, M. Shigeno, K. Saito, N. Morita, T. Yoshioka, and K. Kajimura, "Measurements of Polyphosphoric Acid on HOPG," *Proc. Roy. Microscop. Soc.*, 23, Part 3, Supplement, S40 (1988).

31.   M. M. Dovek, T. R. Albrecht, W. J. Kuan, C. A. Lang, R. Emch, P. Grutter, C. W. Frank, R. F. W. Pease, and C. F. Quate, "Observation and Manipulation of Polymers by Scanning Tunneling and Atomic Force Microscopy," *Proc. Roy. Microscop. Soc.*, 23, Part 3, Supplement, S6 (1988).

32.   D. H. Reneker, J. Schneir, H. Harary, and B. Howell, "Examination of Precipitated Ultra-high Molecular Weight Polyethylene with a Scanning Tunneling Microscope," *Bull. Amer. Phys. Soc.*, 34, 752 (1989).

33.   D. H. Reneker and B. F. Howell, "Preparation of Polymer Molecules for Examination by Scanning Tunneling Microscopy," *J. Vac. Sci. Technol.*, A6, 553 (1988).

34.   J. A. Stroscio, "Observation of Single Atom Adsorbates with the STM," *Bull. Amer. Phys. Soc.*, 33, 338 (1988).

35.   J. A. Stroscio, R. M. Feenstra, D. M. Newns, and A. P. Fein, "Voltage-dependent Scanning Tunneling Microscopy Imaging of Semiconductor Surfaces," *J. Vac. Sci. Technol.*, A6, 499 (1988).

36.   D. A. Bonnell, M. Angelopoulos, D. R. Clarke, and J. Shaw, "Spatial Variation of Electronic Structure in Conducting Polymers (Polyaniline) by Scanning Tunneling Spectroscopy," *Proc. Roy. Microscop. Soc.*, 23, Part 3, Supplement, S32 (1988).

37.   V. Elings and F. Wudl, "Tunneling Microscopy on Various Carbon Materials," *J. Vac. Sci. Technol.*, A6, 412 (1988).

38.   H. Bando, N. Morita, H. Tokumoto, W. Mizutani, H. Anzai, M. Tokumoto, K. Murata, N. Kinoshita, T. Ishiguro, and K. Kajimura, "STM Measurements on Organic Superconductors," *Proc. Roy. Microscop. Soc.*, 23, Part 3, Supplement, S3 (1988).

39.   J. S. Foster and J. E. Frommer, "Scanning Tunneling Microscopy on Surface-adsorbed Organics: Liquid Crystals on Graphite," *Bull. Amer. Phys. Soc.*, 33, 526 (1988).

40.   J. S. Foster and J. E. Frommer, "STM Imaging of Liquid Crystals on Graphite," *Proc. Roy. Microscop. Soc.*, 23, Part 3, Supplement, S33 (1988).

41.   T. P. Beebe, Jr., T. E. Wilson, D. F. Ogletree, J. E. Katz, R. Balhorn, M. B. Salmeron, and W. J. Siekhaus, "Direct Observation of Native DNA Structures with the Scanning Tunneling Microscope," *Science*, 243, 370 (1989).

42.   S. M. Lindsay, "Scanning Tunneling Microscopy of Biopolymers in an Aqueous Environment," *Bull. Amer. Phys. Soc.*, 33, 257 (1988).

43.   S. M. Lindsay, T. Thundat, and L. Nagahara, "Imaging Biopolymers under Water by Scanning Tunneling Microscopy," in **Biological and Artificial Intelligence Systems**, E. Clementi and S. Chin, Eds., ESCOM Science Publishers, p.125 (1988).

44.   S. M. Lindsay, T. Thundat, and L. Nagahara, "Adsorbate Deformation as a Contrast Mechanism in STM Images of Biopolymers in an Aqueous Environ-

ment: Images of the Unstained, Hydrated DNA Double Helix," *J. Microscopy*, 152, 213 (1988).

45. M. Amrein, A. Stasiak, H. Gross, E. Stoll, and G. Travaglini, "Scanning Tunneling Microscopy of RecA-DNA Complexes Coated with a Conducting Film," *Science*, 240, 514 (1988).

46. L. Feng, C. Z. Hu, and J. D. Andrade, "Scanning Tunneling Microscopic Images of Amino Acids," *Proc. Roy. Microscop. Soc.*, 23, Part 3, Supplement, S33 (1988).

47. D. P. E. Smith, A. Bryant, C. F. Quate, J. P. Rabe, C. Gerber, and J. D. Swalen, "Images of a Lipid Bilayer at Molecular Resolution by Scanning Tunneling Microscopy," *Proc. Nat. Acad. Sci.* (USA), 84, 969 (1987).

48. C. A. Lang, J. K. H. Horber, T. W. Hansch, W. M. Heckl, and H. Mohwald, "Scanning Tunneling Microscopy of Langmuir-Blodgett Films on Graphite," *J. Vac. Sci. Technol.*, A6, 368 (1988).

49. T. Sleator, "Scanning Tunneling Microscopy of Model Biological Systems," *Bull. Amer. Phys. Soc.*, 33, 257 (1988).

50. M. A. Voelker, S. R. Hameroff, J. D. He, E. L. Dereniak, R. McCuskey, C. W. Schneiker, L. S. Bell, T. Chvapil and L. B. Weiss, "STM Imaging of Collagen and Lipid Membranes," *Proc. Roy. Microscop. Soc.*, 23, Part 3, Supplement, S42 (1988).

51. O. Marti, H. O. Ribi, B. Drake, R. R. Albrecht, C. F. Quate, and P. K. Hansma, "Atomic Force Microscopy of an Organic Monolayer," *Science*, 239, 50 (1988).

52. S. Gould, O. Marti, B. Drake, L. Hellemans, C. E. Bracker, P. K. Hansma, N. L. Keder, M. M. Eddy, and G. D. Stucky, "Molecular Resolution Images of Amino Acid Crystals with the Atomic Force Microscope," *Nature*, 332, 332 (1988).

53. J. P. Rabe, M. Sano, D. Batchelder, and A. A. Kalatchev, "Polymers on Graphite: Molecular Images and Perturbations of the Substrate," *Proc. Roy. Microscop. Soc.*, 23, Part 3, Supplement, S40 (1988).

54. H. Fuchs and W. Schrepp, "Surface Characterization and Modification of Langmuir-Blodgett Films by Scanning Tunneling Microscopy," *Proc. Roy. Microscop. Soc.*, 23, Part 3, Supplement, S46 (1988).

55. J. A. N. Zasadzinski, J. Schneir, J. Gurley, V. Elings, and P. K. Hansma, "Scanning Tunneling Microscopy of Freeze-fracture Replicas of Biomembranes," *Science*, 239, 1013 (1988).

56. M. Amrein, G. Travaglini, R. Durr, A. Stasiak, H. Gross and H. Rohrer, "Scanning Tunneling Microscopy of Unstained Purple Membranes and RecA-DNA Complexes," *Proc. Roy. Microscop. Soc.*, 23, Part 3, Supplement, S3 (1988). See also M. Amrein, R. Durr, A. Stasiak, H. Gross, and G. Travaglini, "Scanning Tunneling Microscopy of Uncoated RecA-DNA Complexes," *Science*, 243, 1708 (1989).

57. B. Blackford, M. Watanabe, D. Dahn, and M. H. Jericho, "STM Imaging of the Complete Bacterial Cell Sheath of Methanospirillium Hungatei," *Proc. Roy. Microscop. Soc.*, 23, Part 3, Supplement, S32 (1988).

58. D. J. Thomson, "The STM as an Information Storage Device," *Proc. Roy. Microscop. Soc.*, 23, Part 3, Supplement, S29 (1988).

59. J. J. Hopfield, J. N. Onuchic, and D. N. Beratan, "A Molecular Shift Register Based on Electron Transfer," *Science*, 241, 817 (1988).

13
ADHESION MEASUREMENT OF THIN
POLYMER FILMS BY INDENTATION

M. J. Matthewson*

Almaden Research Center
IBM Research Division
San Jose, California

J. E. Ritter

Mechanical Engineering Department
University of Massachusetts at Amherst
Amherst, Massachusetts

13.1 INTRODUCTION

The adhesion of thin polymer films is usually crucial to their successful performance. Despite its importance, adhesion is difficult to quantify in a physically meaningful way. While there is an extensive literature on thin film adhesion (for example, Ref. 1 presents a selected bibliography of over 300 references), no entirely satisfactory test method has yet emerged. This chapter describes the indentation technique which has recently been developed for thin film adhesion measurement. This technique is unique in being able to determine in one experiment the two adhesion parameters required to uniquely specify interfacial strength. The technique, while applied here to adhesion of polymer films, is generally applicable to other film materials.

Any adhesion test requires that the interface between the thin film and the substrate be subjected to stress (usually shear but sometimes normal tension or a combination of the two) sufficiently large to cause failure. The indentation technique, shown schematically in Fig. 13.1 for the case of a ball indenter, is a particularly simple and convenient method for applying an in-plane stress to the film. The indenter is loaded normally onto the film, typically using an Instron machine, which deforms and displaces laterally. This lateral motion of the film results in a shear stress across the interface which, at sufficiently high indenter loads, causes an interfacial crack to initiate and subsequently propagate. The crack is readily observed if a transparent substrate is used. If the crack is not visible by direct observation it may be detected by ultrasonic imaging or acoustic emission. While as yet no measurements have been made on films thinner than about 10 μm there is no reason in principal why very much thinner films can not be used, providing that the debond crack is still visible.

---

* Present address: Fiber Optic Materials Research Program, Rutgers University, P.O. Box 909, Piscataway, New Jersey 08855.

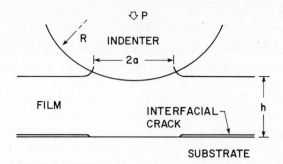

Figure 13.1: Schematic diagram of the indentation test for adhesion measurement. The indenter (a sphere here) is loaded onto the film until a crack forms in the interface.

It is frequently overlooked that interfacial strength needs to be specified by two independent parameters. The first, usually called the interfacial fracture resistance, is analogous to the fracture toughness of a bulk solid and is a measure of the energy required to create a unit area of interfacial crack. The interfacial fracture resistance is normally measured by determining the stress or load required to extend an interfacial crack of known geometry. The resistance is then calculated from energy balance considerations. The second parameter is a strength parameter which depends on the fracture resistance and strength-controlling defects as well as on the nature of the strength measurement technique (shear, tension, etc.) and on any residual deposition stresses in the film. The importance of knowing both adhesion parameters may be illustrated with the following example. If an interface is found to be unacceptably weak then this could be caused by either having an inherently weak interface because of poor bonding (i.e., low fracture resistance), or by having large interfacial defects. These two possible problems would be solved by different techniques so the exact cause of the weakness needs to be known, but can not be determined by either a strength or fracture resistance measurement alone.

The indentation technique is unique in its ability to measure both adhesion parameters in one experiment. The indenter is loaded onto the film until a critical load is reached where fracture initiates; further loading causes the interfacial crack to grow stably. Thus, this test is able to examine both the initiation and propagation stages of fracture. In contrast, other test techniques look only at one stage. For example, the "pin pull" test (in which a pin is glued to the thin film and then pulled off) examines initiation only since pull-off represents the onset of catastrophic failure. On the other hand, the "peel" test examines the stable propagation of an interfacial crack. Thus, these two tests measure quite different parameters since the initiation stage is determined by the strength of the interface while the propagation stage is determined by the interfacial fracture resistance.

Although the indentation technique is in its infancy and much work remains to be done before it is fully understood, particularly in the analysis of the propagation stage, substantial understanding of the failure mechanisms has been achieved and we will now describe how the indentation test can be used to measure interfacial failure.

13.2 THE INITIATION STAGE OF FAILURE

Matthewson (2) and Ritter et al. (3) studied the adhesion of a range of thin polymeric films on rigid substrates using the indentation technique and found that, even though the details of the film deformation were quite variable, the adhesive failure initiated in essentially the same way for all systems studied. For example, Fig. 13.2 [from Ritter et al. (3)] shows a typical sequence of micrographs of the indenter/film contact zone viewed through the substrate for a 4-mm-diameter sphere indenting a 98-μm-thick epoxy film on a glass substrate. As the indenter load is increased, a debond crack initiates close to the contact edge at some quite well-defined load (at about 40 N in this case, Fig. 13.2c) and grows around the edge of the contact and extends radially outwards; it does not extend far into the contact region because of the high compressive stress across the film/substrate interface in this region. Further loading extends the debond crack, though in the case shown in Fig. 13.2, it is not exactly circular due to local fluctuations in adhesive strength. The debonded zone does not extend further on unloading and is still visible under zero indentation load (Fig. 13.2f).

The stress field acting across the film/substrate interface at the debond crack initiation position, at or near the contact edge, is approximately pure radial shear since there is negligible normal stress across the interface at this position. Matthewson (2) proposed a shear stress failure criterion for the polyester/abraded glass system he studied, which predicts that the failure initiates when the interfacial shear stress at the contact edge exceeds some critical value, τ_c. Ritter et al. (3) showed that this criterion was applicable to a range of film materials with different mechanical properties and adhesions. It is necessary to calculate the interfacial shear strength, τ_c, from measurements of the critical load to initiate debonding, P_c, and the contact geometry at debonding but the exact relationship depends on the type of deformation of the film beneath the indenter which may range from fully elastic to predominantly plastic. Ritter (3) identified three distinct types of behavior, shown schematically in Fig. 13.3, which are distinguished by differing deformation mechanisms of the film and substrate, though the debonding mechanism is the same for all three. Type I occurs when the deformation of the film remains elastic up until debonding occurs; in Type II the film deformation is predominantly plastic or irreversible at debonding but the film is not penetrated and Type III occurs when the indenter has penetrated the film before debonding occurs so that some indentation load is supported directly by the substrate. The failure type for a given system will depend on the film thickness and adhesion and the indenter sharpness. Blunt indenters and thick, poorly adhering films favor Types I and II while sharp indenters or thin, well-adhering films favor Types II and III. Between these types are regions of mixed behavior, particularly between Types I and II, where the film deformation is elastoplastic. Analysis of these regions is difficult so they should be avoided by suitable choice of indenter profile or sharpness. Pointed indenters, such as the Vickers pyramid indenter, are infinitely sharp and produce plastic deformation and can be used to produce Type II or Type III debonding. Elastic deformation, and hence Type I debonding, can be produced by employing sufficiently large radius spherical indenters. However, a small amount of plastic deformation may significantly perturb the elastic stress field under the indenter but may conveniently be detected by measuring the contact radius as a function of indenter load on both the loading and unloading cycle. Any discrep-

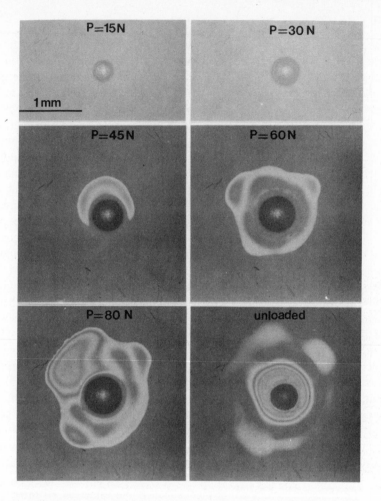

Figure 13.2:   Sequence of optical micrographs showing debonding of 98-μm epoxy film on a glass substrate with increasing indenter load. (a) 15 N, (b) 30 N, (c) 45 N, (d) 60 N, (e) 80 N and (f) after unloading. From Ritter et al. (3).

ancy between the two sets of measurements, particularly at low load, is indicative of plastic deformation that may not be readily apparent by visual inspection for residual deformation. Some preliminary experimentation is usually necessary to ensure a pure type. The three types will now be described in more detail and the corresponding equations will be given which enable the interfacial shear strength to be determined from the critical debonding load.

13.2.1 Type I - Elastic Film Deformation

Ritter et al. (3) have considered debonding under conditions of elastic deformation of the film when indented by a sphere. Figure 13.1 is a schematic showing the geometry of the contact. In principal any indenter shape that results in elastic deformation

THREE FAILURE TYPES

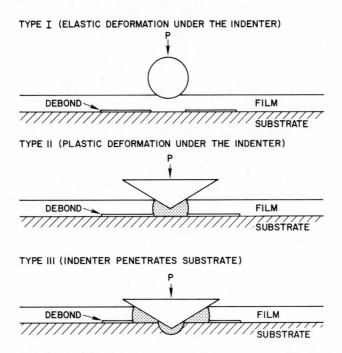

TYPE I (ELASTIC DEFORMATION UNDER THE INDENTER)

TYPE II (PLASTIC DEFORMATION UNDER THE INDENTER)

TYPE III (INDENTER PENETRATES SUBSTRATE)

Figure 13.3:   Schematic diagram of the three types of debonding.

could be used but spheres are readily available and their contact mechanics have been widely studied. The interfacial shear stress is a maximum inside the contact zone, typically at $r \sim 0.8a$. The large compressive normal stress across the interface inhibits interfacial failure (e.g., Ref. 4) so that failure initiates near the edge of the contact where the normal stress is zero. This has been observed experimentally by Ritter et al. (3). There are many analyses available for the elastic contact of a sphere on a thin compliant film attached to a rigid substrate (q.v., Ref. 5) the majority being numerical in nature and hence difficult to use. However, Ritter et al. (2) used an analysis due to Matthewson (6) which, though approximate since it is asymptotically correct for thin films, is unique in yielding results that are analytic and expressible in closed form. This makes the analysis of experimental data particularly simple compared with using more accurate treatments requiring complex numerical calculations. The error in Matthewson's analysis is typically a few percent which is not significant compared to the experimental error in the indentation technique. There exists a discontinuity in the interfacial shear stress at the contact edge, $r = a$, due to the approximate nature of the analysis, but a value for the shear stress, τ_i, averaged across the discontinuity, is given by

$$\tau_i = G\left[\frac{\alpha}{2}K_1\left(\frac{\phi a}{h}\right) + \frac{\nu a}{(1-2\nu)R} - \frac{\beta}{2}I_1\left(\frac{\gamma a}{h}\right)\right], \quad \nu < \frac{1}{2} \quad (1a)$$

$$\tau_i = G\left[\frac{17a^3}{24h^2R}\frac{K_1\left(\frac{\phi a}{h}\right)}{K_1\left(\frac{\phi a}{h}\right) - \frac{\phi a}{h}K'_1\left(\frac{\phi a}{h}\right)} + \frac{a}{4R}\right], \quad \nu = \frac{1}{2} \quad (1b)$$

where G and ν are the shear modulus and Poisson's ratio for the film, h is the film thickness, R is the indenter radius, and

$$\phi = \sqrt{\frac{6(1-\nu)}{4+\nu}}$$

$$\gamma = \sqrt{\frac{3(1-2\nu)}{2(1-\nu)}}$$

$$\beta = \frac{\left[\frac{\phi a}{h}K'_1\left(\frac{\phi a}{h}\right) - K_1\left(\frac{\phi a}{h}\right)\right]\left(\frac{h}{2R}\right)(1-6\nu)}{\left[\gamma K_1\left(\frac{\phi a}{h}\right)I'_1\left(\frac{\gamma a}{h}\right) - \phi I_1\left(\frac{\gamma a}{h}\right)K'_1\left(\frac{\phi a}{h}\right)\right](1-2\nu)}$$

$$\alpha = \frac{-4}{(4+\nu)\phi K'_1\left(\frac{\phi a}{h}\right)}\left[\frac{(1-6\nu)h}{2R(1-2\nu)} + \beta\gamma I'_1\left(\frac{\gamma a}{h}\right)\right]$$

with $I_i(x)$ and $K_i(x)$ being the ith order (i = 0, 1, . . .) modified Bessel functions of the first and second kind and $I'_i(x)$ and $K'_i(x)$ are their derivatives:

$$I'_1(x) = \frac{d}{dx}I_1(x) = \frac{1}{2}[I_0(x) + I_2(x)]$$

$$K'_1(x) = \frac{d}{dx}K_1(x) = -\frac{1}{2}[K_0(x) + K_2(x)]$$

The indenter load, P, and the contact radius, a , are related by

$$P = \pi a^2 G\left[\frac{2\nu h(1-6\nu)}{3R(1-2\nu)^2} + \frac{4\nu\beta h}{3a(1-2\nu)}I_1\left(\frac{\gamma a}{h}\right) + \frac{2(1-\nu)}{1-2\nu}\left(\frac{a^2}{4Rh} - \delta\right)\right], \quad \nu < \frac{1}{2} \quad (2a)$$

$$P = \pi a^2 G\left[\frac{a^4}{32h^3R}\left(\frac{6K_1\left(\frac{\phi a}{h}\right)}{K_1\left(\frac{\phi a}{h}\right) - \frac{\phi a}{h}K'_1\left(\frac{\phi a}{h}\right)} + 1\right) + \frac{9a^2}{8hR}\right], \quad \nu = \frac{1}{2} \quad (2b)$$

where

$$\delta = -\frac{\nu(4 + \nu)h\alpha}{12a(1 - \nu)}\left[\frac{\phi a}{h}K'_1\left(\frac{\phi a}{h}\right) + K_1\left(\frac{\phi a}{h}\right)\right]$$

Equations (1) and (2) may be combined to eliminate G to give the critical interfacial shear stress, τ_c, in terms of the critical debonding load, P_c, the contact radius at that load a_c, and the other variables. τ_c is of the form:

$$\tau_c = \frac{P_c}{\pi a_c^2}f\left(\frac{a_c}{h}, \nu\right) \tag{3}$$

Figure 13.4 shows the dimensionless quantity $\tau_c \pi a_c^2/P_c$ as a function of a_c/h for various values of ν. Each line is a universal curve for that value of ν. (Note that both τ_c and P_c are inversely proportional to R so that R dependence cancels.) h and R are readily measured; ν can be measured or estimated but it should be noted that Eqs. (1) and (2) are extremely sensitive to ν as ν approaches ½. This is not apparent from Fig. 13.4 because of the normalization to P_c which is also sensitive to ν. For example, Fig. 13.5 shows the variation of interfacial stress at the contact edge [Eq. (1)] as a function of the film Poisson's ratio, ν, calculated for $a/h = 5$. (The shear stress is normalized by the factor R/Ga in order to remove G and R dependence.) Therefore, ν needs to be known accurately for elastomers and other high Poisson's ratio materials. a_c can either be measured during the indentation experiment at the same time as P_c or it can be deduced from Eq. (2) in which case the shear modulus of the film, G, also requires measurement. The shear modulus of the film may be measured in various ways; Matthewson (6) detached strips of film and directly measured extension of the film as a function of applied load. Alternatively, if ν is known the shear modulus can be determined in situ by measurements of the contact radius as a function of indenter load using Eq. (2). For example, Fig. 13.6 shows load and contact radius data from Matthewson (6) for a 47-mm-radius sphere contacting a 2.15-mm thick rubber ($\nu = 0.5$) film. The theory line, calculated for a film shear modulus of 89 kPa, shows excellent agreement with the data.

The indenter radius should be chosen judiciously for either adhesion or modulus measurements. Equations (1) and (2) are only accurate for asymptotically thin films and so for better than 10% accuracy $a/h > 2$. Also, the maximum strain in the film, which is in the normal direction (z) and at the contact center, should not be large, that is, $\varepsilon_z < 20$ where

$$\varepsilon_z = -\frac{a^2}{2Rh} + \delta$$

These constraints define a region in the $(a/h, r/h)$ plane bounded by $a/h = 2$ and $\varepsilon_z = 0.2$ for which the stress analysis is accurate. This region is shown in Fig. 13.7 for two representative values of the film Poisson's ratio (a) $\nu = 0.5$ and (b) $\nu = 0.35$. For modulus determinations an indenter as large as is practicable should be used since then Eq. (2) is valid over the largest range of contact radius, a. Also shown dashed in Fig. 13.7 are loci for various adhesive strengths, $\tau_c/G = 0.01$, 0.02, and 0.05. For a particular film system an indenter radius for adhesion determination should be chosen that corresponds to a section of the τ_c/G locus which lies within the

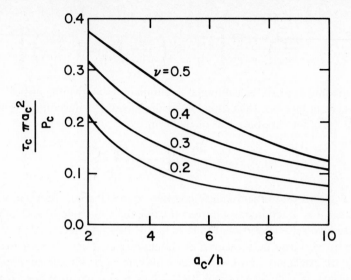

Figure 13.4:   Universal curves of the dimensionless interfacial shear strength for various values of ν.

region of accuracy. For higher Poisson's ratio films (e.g., Fig. 13.7a) a large indenter radius should be chosen, while for lower Poisson's ratio films with good adhesion (e.g., Fig. 13.7b) a smaller radius might be necessary since the τ_c/G locus crosses both boundaries of the region of accuracy leaving only a limited range of indenter radii for which the stress analysis is accurate.

To summarize, the interfacial shear strength, τ_c, may be estimated from straightforward measurements of P_c and a_c using Eqs. (1) and (2). This represents a major advantage of the indentation technique over other methods since all the important mechanical parameters of the film can be measured during the adhesion test. Not only is this very convenient but in a typical experiment the film system may be subjected to an environment which degrades the adhesion. Such an environment is also likely to alter the mechanical properties of the film but any such change is automatically taken into account by the indentation technique; thus, the test can be used to determine the effects of environment on adhesion.

In practice it is usually difficult to apply a thin film that is stress-free so that the effect of residual deposition stresses on the indentation experiment need to be understood. Ritter et al. (3) point out that, by the principal of superposition, a uniform residual stress has no effect on the interfacial shear stress throughout elastic deformation so that the measured adhesive shear strength is predicted to be independent of residual stress in the film for Type I debonding.

13.2.2 Type II - Plastic Deformation without Penetration

Matthewson (2) studied the case of predominantly plastic deformation of the coating beneath the indenter and made measurements of the critical load for debonding and the corresponding contact radius, for various film thicknesses and spherical indenter

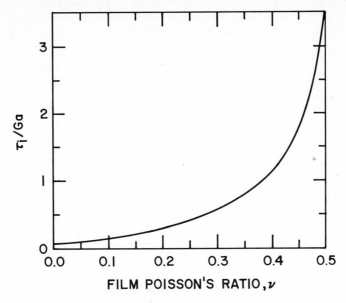

Figure 13.5: Variation of the normalized interfacial shear stress at the contact edge, $\tau_i R/Ga$, with film Poisson's ratio, ν, calculated from Eq. (1) using $a/h = 5$.

radii. He proposed a model for this system (Fig. 13.8) in which the plastic zone, assumed to be a cylinder of radius a, is replaced by a pressurized hole applying a radial stress, σ_r, to the surrounding elastically deforming film. Assuming no work hardening and the Tresca yield criterion, σ_r is given by

$$\sigma_r = Y - H \qquad (4)$$

where Y is the compressive yield strength of the film material and H is the mean contact pressure, $P/\pi a^2$, which can be taken to be the hardness of the film. Matthewson assumed that $H \sim 3Y$ but Ritter et al. (3) pointed out that $H \sim 2.25Y$ is more appropriate for polymeric film materials (7), the former expression being more appropriate for metal films. The latter expression gives $\sigma_r = -0.56H$ which Matthewson then uses as a boundary condition at $r = a$ for the elastic solution for stresses outside the contact region (6). The interfacial shear stress is a maximum at the boundary $r = a$, and, denoting values of parameters at the critical point for debonding by the subscript c, the interfacial shear strength is given by

$$\tau_c = \frac{-0.56H_c}{\dfrac{K'_1\left(\dfrac{\phi a_c}{h}\right)}{\phi K_1\left(\dfrac{\phi a_c}{h}\right)} + \dfrac{\nu h}{a_c \phi^2}} \qquad (5)$$

where

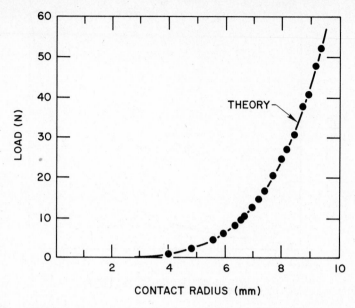

Figure 13.6: Measurements of load and contact radius for indentation of a 2.15-mm-film of rubber ($\nu = 0.5$) by a 47-mm-radius sphere. The theory line is calculated from Eq. (2) using $G = 89$ kPa (after Ref. 6).

$$H_c = \frac{P_c}{\pi a_c^2} \tag{6a}$$

or for a Vickers pyramid indenter

$$H_c = \frac{P_c}{2b_c^2} \tag{6b}$$

where b_c is half the diagonal length of the film/indenter contact area. Equations (5) and (6a) give that τ_c for a ball indenter is of the form

$$\tau_c = H_c \cdot f\left(\frac{a_c}{h}, \nu\right) \tag{7}$$

The hardness, H_c, is determined from the indenter load and contact radius (or diagonal) at debonding. In general, H_c can depend on the film thickness and in such cases will be higher than the hardness of bulk film material due to the confinement by the rigid substrate. However, provided a_c is not too much greater than h, then H_c can be taken to be approximately independent of h. Given that τ_c and ν are expected to be independent of h, then Eq. (7) predicts that the aspect ratio of the plastic zone, a_c/h, is a constant at debonding so that Eq. (6a) predicts that P_c/h^2 is a constant. Matthewson's adhesion test data for polyester films on abraded glass substrates do indeed show a parabolic relationship between critical stress for debonding and film

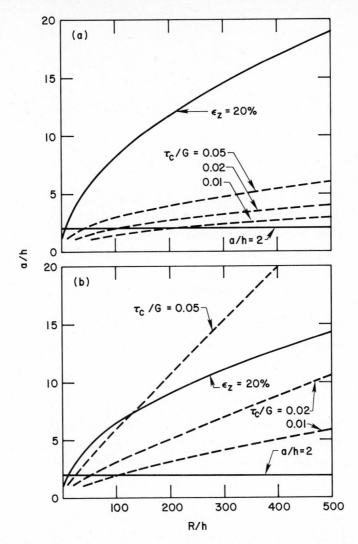

Figure 13.7:   Region of accuracy for the contact stress solutions for an elastic film on a rigid substrate indented by a sphere; calculated for film Poisson's ratio (a) $\nu = 0.5$ and (b) $\nu = 0.35$.

thickness. For example, Fig. 13.9 shows data for indentation by a 4-mm-diameter sphere on polyester films of various thicknesses in the 100- to 1000-μm range on abraded glass substrates. Again, like Type I debonding, all important mechanical properties of the film (its hardness) are measured during the indentation experiment; necessarily providing values that are appropriate for the indentation conditions and film thickness.

It should be noted that Eq. (5) does not explicitly contain the indenter radius, R. In fact the indenter profile is not relevant to the analysis since it is effectively masked

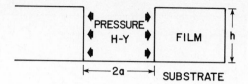

Figure 13.8:   Schematic diagram of the model for Type II debonding (after Ref. 2).

by the plastic zone in the film. A nonaxisymmetric indenter, such as the Vickers pyramid, could even be used without loss of accuracy.

In contrast to Type I debonding, a uniform residual stress in the film, σ_{res}, does affect the interfacial shear stress for Type II debonding; in this case, the numerator of Eq. (5) is offset by the residual stress and should be replaced by $-(0.56H_c + \sigma_{res})$ (2). A uniform tensile residual stress in the film reduces the critical load for debonding since the residual stress also contributes to the net interfacial shear stress. Ritter et al. (3) suggest this difference between Types I and II may be used to measure the residual stress since any difference in interfacial shear strength measured under Type II conditions, compared to Type I conditions, must be due to the residual stress which can then be estimated by the modified form of Eq. (5).

13.2.3 Type III - Plastic Deformation with Penetration

When the film is penetrated at debonding, the substrate supports a portion of the applied indenter load as shown in Fig. 13.10 for a Vickers indenter. The interfacial shear strength can still be determined using Eq. (5) but H_c should be replaced by H_c^f, the film hardness, and a_c replaced by b_c, the half-length of the indenter/film contact area diagonal. However, because of the penetration of the film the hardness is not simply related to the load by Eq. (6b). A model proposed by Howes and Ryan (8), shown schematically in Fig. 13.10, assumes the portion of the load supported by the film is given by the film hardness, H^f, multiplied by the projected contact area on the film, $2(b^2 - c^2)$, where b and c are the half-lengths of the diagonals in the film and substrate contact zones. Similarly the portion of the load supported by the substrate is $2H^sc^2$ where H^s is the substrate hardness so that the total load is

$$P = 2H^f(b^2 - c^2) + 2H^sc^2 \qquad (8)$$

Note that b and c are related via the indenter geometry and the film thickness. Measurements of P_c and b_c or c_c (the values of P, b, and c at debonding) then provide H_c^f and b_c for substitution in Eq. (5). However, if $H^f \ll H^s$, Eq. (8) involves determining the small difference of two large numbers and may give unreliable results. In this circumstance, it is preferable to determine H^f by direct measurement of P and c prior to penetration of the film.

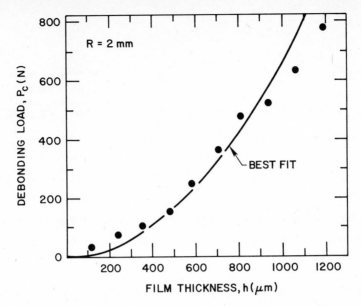

Figure 13.9:   Critical load to debonding as a function of film thickness for Type II debonding of polyester films on abraded glass for a 2 mm radius spherical indenter. The best fit line is for a parabolic relationship. After Ref. 2.

13.3 THE PROPAGATION STAGE OF FAILURE

After initiation of debonding, an interfacial crack grows into a roughly circular shape until some equilibrium size is reached. Further increase in the indenter load increases the size of the crack and, in principle, the interfacial fracture resistance can be calculated from the load/crack size relationship. The situation is complex to analyze but some progress has been made. Marshall and Evans (9) have considered indentation by a Vickers pyramid which results in plastic deformation of the film (Type II initiation).   They model the debonded region of the film as a clamped circular plate which has buckled away from the substrate (Fig. 13.11a). By considering changes in the elastic energy stored in the buckled plate, they determine the strain energy release rate, ζ.   By assuming that the criterion for propagation of the interfacial crack is that the strain energy release rate should exceed some critical value, ζ (the usual criterion for crack growth in a homogeneous material) it may be shown that the indentation load P and crack radius c are related by

$$\frac{P^{3/4}}{c} = \text{const} \tag{9}$$

Rossington et al. (10) confirmed this relationship for ZnO films on silicon. Marshall and Evans also are able to account for residual deposition stresses in the film.

   Marshall and Evans (9) draw an analogy between the debond crack and the lateral cracks that form in a homogeneous solid on **unloading** due to residual stresses around the plastically deformed zone.   In other words, they assume the debond crack

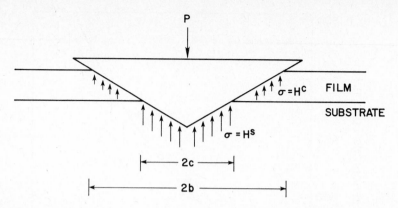

Figure 13.10:   Schematic diagram of the load sharing model for Type III debonding (after Ref. 8).

forms on unloading and the buckled region of the film is not in contact with the substrate at the center of the indentation (Fig. 13.11a). In fact this analysis is not appropriate for most polymeric film systems since it is clear from the experiments of Matthewson (2) and Ritter et al. (3) that the debond crack grows while the indenter is loaded and does not grow further upon unloading. Figure 13.11b shows a more reasonable model for this situation where the buckled film is in contact with the substrate at the center of the contact. However, the buckling equations describing the film deformation for the models of Fig. 13.11 will be similar but with different boundary conditions applied at the center. Therefore, the buckling analysis of Evans and Hutchinson (11) and as used by Marshall and Evans (9) could provide a reasonable qualitative description of the debond crack in polymer films, as well as a basis for development of a more appropriate analysis.

13.4 SUMMARY

The indentation technique provides a method of determining the adhesion of polymeric films which is both simple and convenient, only requires small specimens, and which determines all important film mechanical properties in situ during the experiment. Most importantly, the technique is unique in being able to provide the interfacial shear strength and fracture resistance; both of which are required for a complete fracture mechanics description of the adhesion. However, the technique is relatively new and requires further investigation, especially in the analysis of the propagation stage.

The technique is self-consistent in that it provides adhesion estimates that do not depend on the debonding type or film thickness as exemplified by Fig. 13.12 which shows adhesion data from Ritter et al. (3) for both Type I debonding by a spherical indenter and Type II debonding by a Vickers pyramid for a range of film thicknesses. While the adhesive strength is not expected to depend on film thickness and deformation type, estimates of apparent adhesion using other test methods frequently do show

(a)

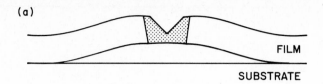

(b)

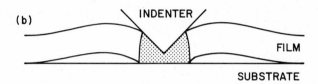

Figure 13.11:   Schematic of the buckled film due to indentation induced fracture: (a) the model of Evans and Hutchinson (11) and (b) a more appropriate model for debonding of polymeric films.

dependencies on these parameters, usually due to the poor definition of the details of the debonding process.

13.5 REFERENCES

1.   K. L. Mittal, "Selected Bibliography on Adhesion Measurement of Films and Coatings," *J. Adhesion Sci. Tech.*, 1, 247 (1987).

2.   M. J. Matthewson, "Adhesion Measurement of Thin Films by Indentation," *Appl. Phys. Lett.*, 49, 1426 (1986).

3.   J. E. Ritter, T. J. Lardner, L. Rosenfeld, and M. R. Lin, "Measurement of Adhesion of Thin Polymer Coatings by Indentation," *J. Appl. Phys.*, in press (1989).

4.   M. W. Vratsanos, E. L. Thomas, and R. J. Farris, "The Adhesive Behavior of Poly(p-phenylene benzobisthiazole) (PBT)/epoxy Composites," *J. Mat. Sci.*, 22, 419 (1987).

5.   K. L. Johnson, **Contact Mechanics,**, 136, Cambridge University Press, Cambridge, U.K., (1985).

6.   M. J. Matthewson, "Axi-symmetric Contact on Thin Compliant Coatings," *J. Mech. Phys. Solids*, 29, 89 (1981).

7.   D. M. Marsh, "Plastic Flow in Glass," *Proc. Roy. Soc.*, A279, 420 (1964).

8.   V. R. Howes and M. A. Ryan, "The Effect of Protective Coatings on Indentation Phenomena for Glass," *J. Aust. Ceram. Soc.*, 22, 13 (1986).

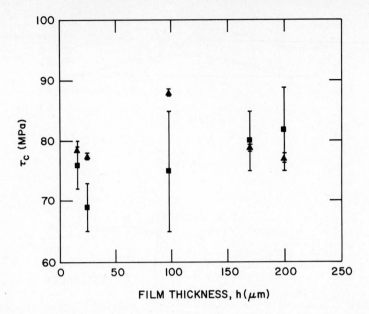

Figure 13.12: Adhesive shear strength measurements on epoxy films of various thicknesses using (■) a 2-mm-radius ball indenter (Type I debonding) and (▲) a Vickers indenter (Type II debonding). After Ref. 3.

9.   D. B. Marshall, and A. G. Evans, "Measurement of Adherence of Residually Stressed Thin Films by Indentation. I. Mechanics of Interface Delamination," *J. Appl. Phys.,* 56, 2632 (1984).

10.   C. Rossington, A. G. Evans, D. B. Marshall, and B. T. Khuri-Yakib, "Measurements of Adherence of Residually Stressed Thin Films by Indentation. II. Experiments with ZnO/Si," *J. Appl. Phys.,* 56, 2639 (1984).

11.   A. G. Evans and J. W. Hutchinson, "On the Mechanics of Delamination and Spalling in Compressed Films," *J. Solids Structures,* 20, 455 (1984).

SUBJECT INDEX

A

Ablation, 249
Absorption, light, 233
   coefficient of, 233
Accelerating particle, 151
Accelerator, 140
AC conductance method, 215
Acoustic analysis, 241
Activation energy, 268, 270, 277
Adhesion, 109, 289, 316
Adhesion measurement, indentation test, 347
   pull test, 348
   peel test, 348
Admittance, 126-128
   loss effects, 128
   plot, 127
Amide/imide polymer, 46, 49
Amino acids, 336, 339
Amplifier, lock-in, 242
Amplitude reduction factor, 103-105
   effects on reflectance, 105
Analyzers, electron energy, 291, 294
   mass, 306, 307
   double-focusing mass, 307
   quadrupole mass, 307
   cylindrical mirror, 294, 295
   concentric hemispherical, 294, 295
Angle of incidence, 98, 100, 106
Annealing, 249, 316
Arrhenius rate equation, 268
Atomic force microscopy (AFM), 327
Auger electron spectroscopy (AES), 260, 313,
  315, 316

B

Bending-beam materials, 31
Bending-beam setups, 32
Biological materials, 335
Boltzmann constant, 268, 269
Boundary layer, see transition layer
Box-car integrators, 242
Breakdown voltage, 272
Broadening, extraneous, 296
   instrumental, 292
   line, 292

C

Calorimeter, pyroelectric, 238
Capacitance, 271, 272
   importance of minimizing, 129
   interplay with losses, 129
   shunting effects, 127, 128
Charge, static, 296
Charge carrier, 268, 272, 273
   components, 267
   crowding, 267
   domains, 283
   release, 267
   space, 270, 276, 278, 279
   storage, 267
   trapping, 278
   transfer, 267, 268, 273
   transport, 268
Charge distributions, 232
Compound resonator, 133
   Lu and Lewis description, 133
   Mechanical model, 134

JDR